Dutch Deltas

Brill's Studies in Maritime History

VOLUME 1

The titles published in this series are listed at *brill.com/bsmh*

Dutch Deltas

Emergence, Functions and Structure of the Low Countries' Maritime Transport System, ca. 1300–1850

By

Werner Scheltjens

BRILL

LEIDEN | BOSTON

Netherlands Organisation for Scientific Research

This book is the result of a research project financed by NWO, the Netherlands Organisation for Scientific Research.

Cover illustration: "Boot" (Holzschnitt 1972) by Jacob Pins; with kind permission of the Jacob Pins Gesellschaft Kunstverein Höxter e.V.

Library of Congress Cataloging-in-Publication Data

Scheltjens, Werner.
Dutch deltas : emergence, functions and structure of the Low Countries' maritime transport system, ca. 1300–1850 / by Werner Scheltjens.
pages cm. — (Brill's studies in maritime history, ISSN 2405-4917 ; volume 1)
Includes bibliographical references and index.
ISBN 978-90-04-27306-1 (hardback : acid-free paper) — ISBN 978-90-04-27307-8 (e-book) 1. Shipping—Netherlands—History. 2. Shipowners—Netherlands—History. 3. Deltas—Economic aspects—Netherlands—Commerce—History. 4. Economic geography—History. 5. Benelux countries—Economic integration—History. 6. Economic development—Benelux countries—History. 7. Netherlands—Commerce—History. 8. Benelux countries—Commerce—History. 9. Amsterdam (Netherlands)—Commerce—History. 10. Rotterdam (Netherlands)—Commerce—History. I. Title.

HE845.S34 2015
387.509492—dc23

2015017089

This publication has been typeset in the multilingual 'Brill' typeface. With over 5,100 characters covering Latin, IPA, Greek, and Cyrillic, this typeface is especially suitable for use in the humanities.
For more information, please see brill.com/brill-typeface.

ISSN 2405-4917
ISBN 978-90-04-27306-1 (hardback)
ISBN 978-90-04-27307-8 (e-book)

This book is printed on acid-free paper.

Contents

Preface

This book deals with maritime transport in the Age of Sail. More precisely, the most famous part—the shipmasters that constituted the Dutch mercantile fleet—are the subject of this book. Readers expecting to find new insights about Dutch maritime trade will look in vain. Our perspective is different: not trade itself, but the carriers that operationalised it are the topic of this book. The emergence of the Dutch maritime transport system, its functions and its structure are examined by asking the following questions: When and why did a maritime transport sector emerge in the Low Countries? What was its structure? How did it evolve? And, finally, how can the emergence, functions and structure of the Low Countries' maritime transport system be explained?

The thesis of the book is divided into five chapters. Chapters 1 and 2 provide a general overview of the geography of the Dutch maritime transport sector and of ports and their functions between ca. 1300 and 1850. Chapters 3, 4 and 5 go into detail about the composition of community-based populations of maritime transporters, their production of services and their local business culture.

Each chapter contains the following elements: (1) an introduction, in which its relation to the rest of the book is explained; (2) a methods section, in which the methodological tools employed in the chapter are introduced; (3) a main body, in which results of empirical research are presented, interpreted and explained and (4) a conclusion, in which the results are summarized and put in the context of the rest of the book.

The book is the result of two and a half years of research as a post-doctoral fellow at the University of Groningen and one additional year as lecturer at the University of Leipzig. I am grateful for the hospitable environment created by all staff members of the Department of Early Modern History, the Department of Economic and Social History and the Groningen Research Institute for the Study of Culture (ICOG) as well as by the staff of the Chair of Social and Economic History at the University of Leipzig. Last but not least, I am much obliged to Charlie Bush, who carefully proofread the manuscript, and to Hedwig Scheltjens for translating part of Chapter 1.

Parts of this monograph have been presented and published on various occasions. Parts of the introduction and of Chapter 1 have been presented at the International Congress of the Historical Sciences (Amsterdam 2010) and a FRESH Meeting on Economic Growth and Development before 1900 (Antwerpen 2011) before being published in *Scripta Mercaturae* (2013

[2009]) and the *Revue belge de la philologie et d'histoire* (2014).[1] Earlier versions of Chapter 2 have been presented at the 2nd Sound Toll Registers Online Workshop (Leeuwarden 2010) and an international workshop on "Gateways and Hinterlands" (Venice 2011). A presentation at the Business History Conference Annual Meeting (St. Louis 2011) and an article on the changing geography of the demand for Dutch maritime transport in the eighteenth century, published in *Histoire et Mesure* in 2012, constituted the basis for Chapter 3.[2] The methodological insights presented in Chapter 4 have been published more extensively in a book chapter, co-authored by Prof. Dr. Kurt Dopfer, and in a journal article, both published in 2012.[3] The empirical research results of Chapter 4 have been presented at the 3rd Sound Toll Registers Online Conference (Groningen 2012) and a sample study has been published in the *International Journal of Maritime History* in 2012.[4] Elements of Chapter 5 have been presented at the 1st Thematic Conference of the IASC on the "Knowledge Commons" (Louvain-la-Neuve 2012) and at the Conference "Design and Dynamics of Institutions for Collective Action. A Tribute to Prof. Dr. Elinor Ostrom" (Utrecht 2012).

Many of the insights compiled in this book have taken shape in close cooperation with Dr. Jan Willem Veluwenkamp, project leader of the Sound Toll Registers Online project and research group leader of the project *The Ascent of the Frisians. The Dutch Commercial System and the Market for Maritime Transport, 1550–1800*, funded by the Netherlands Organisation for Scientific Research (NWO). Without his stimulating and ever-supportive guidance, it is unlikely that the ideas about the emergence, functions and structure of the Low Countries' maritime transport sector would have developed the way they have. It is up to the reader to decide whether or not the present summary of these ideas has resulted in a convincing argument.

The contribution of this volume to historiography is threefold. It contains the first multi-faceted analysis of the Dutch maritime transport sector between ca. 1300 and 1850 that aims to explain its geography, operations and institutions. First and foremost, by explaining the rise, decline and functional development of communities of maritime transporters from an economic-geographical perspective, our understanding of the Dutch economy before 1850 is advanced considerably. Second, a number of the methodological tools

1 Scheltjens, "The volume," 83–109; Scheltjens, "Het ontstaan," 293–364.

2 Scheltjens, "The changing geography," 5–47.

3 Scheltjens and Dopfer, "Unified rule approach," 229–247; Scheltjens, "The operational structure," 357–389.

4 Scheltjens and Veluwenkamp, "Sound Toll Registers Online," 301–330.

described and applied in the text may serve as a basis for similar (comparative) analyses of the maritime transport sector in different parts of Europe. Finally, this volume contributes to a growing body of maritime-historical studies in which contemporary database management systems are used to explore and analyse previously inaccessible primary sources, leading to novel insights regarding the emergence, functions and structure of the Dutch maritime transport system before 1850.

List of Illustrations

List of Tables

List of Graphs

List of Figures

List of Maps

List of Abbreviations

BEH	Bremen, Emden and Hamburg
HDF	Home-delta Frequency
HMS	Home-market share
HPF	Homeport Frequency
LC	Location Coefficient
LDE	Lauwers-Dollard-Ems Delta
PDR	Pattern Density Rate
PF	Port Frequency
SMR	Scheldt-Meuse-Rhine Delta
STRO	Sound toll registers online
Tonkm	tonkilometer
VBL	Voorbijlandvaart
ZDZ	Zuiderzee Delta

Introduction

The vital contribution of the Dutch maritime transport sector to the emergence of what De Vries and Van der Woude have depicted as 'the first modern economy' seems to be undisputed.[1] However, knowledge of the Dutch maritime transport sector proper is rather scarce when compared to the vast literature on the Dutch Republic's international trade and maritime expansion. Very little is known about the 'carriers of trade' who made the first round of modern economic growth in the Dutch Republic physically possible. Who were they? How did they work? Did they operate alone or in groups, regularly or at random, to any destination or on fixed routes? And how can this be explained? A thorough analysis of the emergence, structure and functions of the Dutch maritime transport sector in the early modern period has not yet been executed. Consequently, there is an enduring lacuna in our knowledge of the Low Countries' economy before 1850.

Historiography

Often, maritime transport often only gets limited attention in historiographical analyses of *local, regional and international trade networks*. The rich historiography of city networks in the Middle Ages reveals little about maritime transport, which may be explained by the limited relative importance of overseas trade and transport and by the combination of economic functions, which was still quite common before 1400.[2] Generally speaking, it is accepted that the maritime transport sector gradually became a branch of the emerging services economy, that followed rules and developed structures that were different from those of international trade. Without explicitly dating this process, De Vries and Van der Woude place it in the period 1650–1800.[3] Before that

1 De Vries and Van der Woude, *The first modern economy*.

2 See: Ward, *The World*, 48: "In the twelfth century, most shipmasters owned their own ships, either participating in joint commercial ventures with a group of merchants assembled for each voyage, or arranging at their own expense the purchase and sale of cargo, the fitting-out of the ship and the hiring of a crew. As a member of a cooperative, the owner/master commanded the ship, acting as *primus inter pares* with the participating merchants working as crew, and would have received a charter fee or an enhanced share of the profits of the voyage. When working alone, all the risks and any profit (or loss) on the voyage would be his".

3 See: De Vries and Van der Woude, *Nederland 1500–1815*, 567: "Wij weten nog heel weinig over de prijzen van nieuwe schepen, de bedrijfskosten en de winstgevendheid van de scheepvaart,

 | DOI 10.1163/9789004273078_002

time, trade and transport could not clearly be separated from each other.[4] In historiographical analyses of local, regional and international trade networks, the presence of a transport sector is interpreted rather as the logical result of the development of such networks than as a *conditio sine qua non* for their emergence. As a result, the question of who provided transport services to these international trade networks remains unanswered. Nevertheless, it is clear that transporters physically realized the connections between the nodes of a trade network. As transporters did not necessarily come from the same place or region as the merchants they provided services to, but may have come from the surrounding countryside, there is a risk that the impact of changes in the structure of local, regional and international trade networks on the emergence, functions and structure of the Dutch maritime transport system remains under-exposed.

Shipping companies play a modest role in the *entrepreneurial history* of the Low Countries, even though their emergence in the fifteenth century is seen as one of the elements that distinguished the Dutch trade system from its competitors.[5] In historiography, the primary focus is on the shipping company as a financial instrument, which has led to the publication of a number of case studies about capital accumulation in shipping, the role of parts in ships (Nl. *Scheepsparten*) and of ship ownership.[6] The role of the shipmaster, and influence of changing ship ownership structures on it, has been studied much less intensively. Only a few historians have devoted considerable attention to the function of the shipmaster in international trade.[7] In the 1950s, Posthumus has sketched an emphatically negative image of the changing role of the shipmaster in shipping companies. From 'ship-owner of distinction' he turned into a 'hired man' (Nl. *zetman*) of the shipping company, i.e. a wage-earning shipmaster, who had lost a lot of the social status he had held previously.[8] Already in the 1970s, Hart expressed some doubt regarding Posthumus' assertion, but ultimately he did not do more than establish that Posthumus had overstated

---

maar zo lang de Republiek bestond bleef zij een van de hoofdsectoren van haar economie. Toen handel en scheepvaart meer en meer ontkoppeld raakten, werd de scheepvaart bovendien een meer in het oog vallende economische activiteit".

4 Lesger, *Handel in Amsterdam*, 50–64.

5 Hart, "Rederij," 106–110.

6 Asaert, *De Antwerpse Scheepvaart*; Brulez, "De scheepvaartwinst in nieuwe tijden," 1–19; Wegener Sleeswyk, "Omvang en financiering," 52–72; Gaastra, "Zeilende kustvaart," 154–165; Boon, "Schippers en hun schepen," 3–16; De Wit, *Leven, werken en geloven*, 73–82.

7 Posthumus, *De Oosterse handel*; Asaert, *De Antwerpse scheepvaart*; Lesger, *Handel in Amsterdam*, 50–64; Ward, *Medieval Shipmaster*.

8 Posthumus, *De Oosterse handel*, 109–123.

the role of the merchant-ship-owner, while underestimating the role of the shipmaster.[9] Later, Van Zanden and Boon have clearly opposed to Posthumus' negative portrayal, perceiving a positive effect of the shared shipping company (Nl. *partenrederij*) on the social status of the shipmaster.[10] A similar perception is found in a number of case studies about eighteenth-century shipping companies, which suggest that the shipmaster had a serving role with relatively large final responsibility.[11] However, the small number of available cases makes it impossible to generalize these findings. Moreover, predominant attention in these case studies is given to the shareholders, a cost-benefit analysis of the shipping company or the shipping routes of the ship; much less to the connection between shipping routes and the relations between shipmasters, factors and merchants, which Lesger has identified in Amsterdam's mid-sixteenth century gateway-system.[12] The emergence, functions and structure of the Dutch maritime transport system cannot be established on the basis of the analysis of shipping companies.

In *Urban and regional histories*, attention to the maritime transport sector is limited to general descriptions of its contribution to the local or regional economy. In most cases, a paragraph devoted to 'shipping' is added to paragraphs about trade, industry and crafts. The complex relation between trade networks and transport systems often figures in the background in urban and regional histories, for the benefit of broader attention to social aspects of the shipping business, and notably to issues of employment.[13] In their turn, *transport historians* primarily devote attention to infrastructural developments and their contribution to economic growth.[14]

Most fundamental knowledge about the structure of the Dutch maritime transport sector in the early modern period originates from the historiography of maritime labour in the Dutch Republic.[15] This area of research received an

---

9 Hart, "Rederij," 361.

10 van Zanden, *Arbeid*, 50–51; Boon, *Bouwers van de* zee, 107–123.

11 See: Bruijn, "Productivity," 174–194; 't Hart and Van Royen, "Het smakschip," 150–168; Reinders Folmer-van Prooijen, "Met de rapen in de pot," 222–232.

12 Lesger, *Handel in Amsterdam*, 50–64.

13 See, for example: Willemsen, *Enkhuizen*, 91–95.

14 See, for example: Blondé, "At the cradle," 89–111; Verhoeven, "Een divertissant somertogje," 78–97. Jan de Vries' classical study of the development of a regular domestic transport services system in the Dutch Republic deserves to be mentioned here. See: De Vries, *Barges and Capitalism*.

15 Lucassen, "Zeevarenden," 126–158; Lucassen, "A Multinational," 12–39; Van Zanden, *Arbeid*, 55–74; Van Royen, "The 'National' Maritime Labour Market," 1–9; Davids, "Maritime labour," 41–71; Knotter, "De Amsterdamse scheepvaart," 281–290; Van Royen, *Zeevarenden*;

enormous impulse from the works of Van der Woude and Faber, who sketched out the existence of a 'reservoir of shipmasters' that accounted for the carrying-trade between Amsterdam and northern/north-eastern destinations.[16] The geographical concentration of this 'reservoir' changed in the course of the early modern period from the Noorderkwartier—the rural area north of Amsterdam—and West Friesland, to Friesland around 1650 and the Province of Groningen around 1750.[17] Diminishing average ship size was seen as typical for this spatial dynamic.[18] Although the idea of a 'reservoir of shipmasters' has found broad resonance in historiography (even if with some nuances in its interpretation),[19] it should not be ignored that the image of the shifting 'reservoir of shipmasters' is framed in socio-geographical studies on seafarers and their labour markets in the Dutch Republic. In these studies, 'reservoir of shipmasters' has a much broader implicit meaning than the term itself suggests.

Building upon the work of, among others, Van der Woude and Van Royen, Van Zanden has pointed out the dominant role of the maritime transport sector in the modernization of the Holland economy during the late Middle Ages and the early modern period. Van Zanden's argument is structured around the concept of division of labour, which denotes a development in the organisation of labour. In this concept the combination of economic activities as a means of securing livelihood is replaced with an increasing and,—in the end,—total focus on one central activity that had to generate sufficient income for survival. The division of labour, especially in function of emerging merchant capitalism, is most apparent in the development of agrarian societies into 'modern' capitalist societies, in which one produces for the market and profits are used to pay for the services of others. Gradually, this led to the disappearance of the functional combination of agrarian and non-agrarian activities.[20] Wage labour became the norm. Merchant capitalism opened the door to the development of a services sector. For inhabitants of the coastal areas in the Low Countries, who have been characterised by their intense relation with the water, (maritime) transport services were a logical choice. Hence, the sectors in which

---

Boon, *Bouwers van de Zee*, 81–140; Van Lottum, *Accross the North Sea*, 126–159; Gorski, *Maritime labour*.

16 Van Der Woude, *Het Noorderkwartier*, 362–398; Faber, *Drie Eeuwen Friesland*, 264–292.

17 Hart, "Rederij," 113–114; Van der Woude, *Noorderkwartier*; Faber, *Drie eeuwen Friesland*.

18 Van der Woude, *Noorderkwartier*, 384; Van Royen, "Varend volk," 86–87.

19 The line of reasoning expressed by Van der Woude and Faber can be found in, among others: Van Royen, *Zeevarenden*; Van Zanden, *Arbeid*; Boon, *Bouwers van de zee*; Van Tielhof, *The 'Mother of all Trades'*.

20 Van Zanden, *Arbeid*, 52; De Vries, *Rural Economy*.

the division of labour caught on, were not only the agrarian sector and the textile industry, as De Vries seems to suggest,[21] but also—and maybe even pre-eminently so—peat-digging, fisheries and (domestic) carrying-trade.[22] During the late Middle Ages, in several combinations these activities were part of the proto-capitalist work cycle. In the peat-digging region located south of the river IJ, such work-cycles existed,[23] but North Sea-fishing and carrying-trade into the Baltic also seem to have been combined.[24]

For two reasons, one has to be cautious when research results based on labour market analysis are interpreted. First of all, it remains undiscussed when, why and how the so-called 'reservoir of shipmasters' emerged. The prevailing opinions are based on quantified shipping data concerning one geographical area (the Baltic). Without exception, the authors mentioned above have made use of Nina Ellinger Bang and Knud Korst's famous reference work.[25] The movement in time and space of the 'reservoir of shipmasters' therefore has a limited epistemological foundation, mostly because during quantification, valuable connections between the relevant data items in the original registers upon which Bang and Korst's *Tabeller over Skibsfart* were based, have been lost.[26] Secondly, seafarers and shipmasters should not be considered equal. Although a connection between the origins of seafarers and shipmasters on the merchant fleet can reasonably be supposed and substantiated, at least as far as the European carrying-trade is concerned,[27] it should by no means be concluded that a substantial part of the crew had the same domicile as the shipmaster, as Van der Woude claims.[28] The crew of the European merchant fleet was recruited in the environment of the shipmasters, which may include his immediate geographical surroundings as well as distant places with which

---

21 De Vries, *Rural Economy*, 119–174.

22 Van Zanden, *Arbeid*, 44–46.

23 Cornelisse, *Energiemarkten en energiehandel*, 226.

24 This is established by Willemsen for shipmasters from Enkhuizen. He bases his conclusions on Kranenburg's statement that, at that time, the *buis* (a type of ship) was still used for fishing as well as for the carrying-trade. See: Willemsen, *Enkhuizen tijdens de Republiek*, 69 and 94–95; Boelmans Kranenburg, De *Zeevisscherij*.

25 See: Bang and Korst, *Tabeller over skibsfart*.

26 For extensive source criticism regarding the *Tabeller over Skibsfart*, see: Brakel, "Schiffsheimat und Schifferheimat," 211–228; Brünner, "De waarde der Skibsfarttabellen," 269–280; Christensen, "Der handelsgeschichtliche Wert," 28–142; Heckscher, "Öresundsräkenskaperna," 1942; Unger, "De Publikatie," 147–205; Jeannin, "Les Comptes du Sund," 55–102.

27 Van Royen, *Zeevarenden*.

28 Van der Woude, *Noorderkwartier*, 375.

the shipmaster or his representatives had a connection.[29] Therefore, conclusions regarding the emergence, functions and structure of the Dutch maritime transport sector cannot be based solely on maritime labour market analysis. Nevertheless, research into the labour market for seafarers does shed light on an important aspect of the maritime transport sector, namely that of the recruitment of sailors. The spatial dynamics of the maritime labour market, which are discussed in the works of Van Royen and—more recently—Van Lottum, find their complement in this book.[30]

To sum up, there is no general image of the emergence, structure and functions of the Dutch maritime transport sector in the early modern period, despite its widely accepted importance for the Dutch economy and the significant attention paid to it in a large variety of historical disciplines. At least partly, this can be explained by the preferences expressed by historians of early modern Dutch trade and transport in their choice of research topics, source materials and methods. Economic historians prefer to find explanations of long-term developments in structure of production factors (cost of services and cost of capital). These methods do not convince, when an explanation is sought for the emergence, functions and structure of the Dutch maritime transport sector during the early modern period.

According to the cost of services-model, the spatial development of the 'reservoir of shipmasters' from the so-called Noorderkwartier (the area north of Amsterdam) in the sixteenth century to the Province of Groningen in the second half of the eighteenth century might be explained as a cost function. Originally, shipmasters in the Noorderkwartier were the cheapest. Later, their services became more expensive and 'cheaper' shipmasters from Friesland became more attractive. When their freight rates rose, shipmasters in the Province of Groningen appeared to be the most attractive executors of the transport services required at the ports of the Dutch Republic, above all: Amsterdam. If this was really the case, and shipmasters from Groningen actually did offer the most attractive conditions after 1750, then why do we not find them on trans-Atlantic routes? And if the choice for shipmasters from Groningen was in fact determined by the cost of their services, then why is their presence in the ports of Rotterdam, Middelburg or Dordrecht so limited? If these ports could rely on even cheaper transport services providers, why do we not find them in Amsterdam's shipping instead of shipmasters from Groningen?

29 Van Royen, *Zeevarenden*, 90–105.

30 Van Royen, *Zeevarenden*; Van Lottum, *Across the North Sea*.

Similarly, the cost of capital-model does not provide a solution. If we suppose that capital accumulation was the largest in Amsterdam and that it was much less in peripheral areas such as Friesland and Groningen, it could be expected that investors in Amsterdam, because of their greater financial means, would prefer to invest in larger ships than their colleagues in Friesland and the Province of Groningen, who lacked the necessary means. For this reason, then, ships in Friesland and Groningen would be smaller than ships in the immediate surroundings of Amsterdam. But, apparently, these small ships were sufficiently attractive to be used intensively. Why did the high demand for transport services by shipmasters from Friesland and Groningen not lead to an increase in the size of their ships? And why would investors from Amsterdam not invest in smaller ships themselves in order to keep the services sector in their own hands?

In short, 'mainstream' economic models of the types mentioned above lack a good understanding of the space in which the carrying-trade developed. Moreover, accepted divisions of the Low Countries in economic-geographical zones seem to emphasize the centrality of Holland in the economy of the late Middle Ages and early modern period. Postan, for example, distinguishes between four geographical regions in the Netherlands during the late Middle Ages: Zeeland, Friesland, South Holland and North Holland. Van Zanden distinguishes between a number of rudimentary defined economic zones in the Netherlands in around 1500: the south, which includes Zeeland, North Brabant and to a lesser degree Limburg; the eastern Netherlands, that mark the transition between the relatively densely populated south and the agrarian north of the Netherlands; and finally Holland, where already a dense population and many cities could be found.[31] By stressing the rise of Amsterdam and Holland at the cost of the southern Netherlands and the Rhine Delta, the economic development of other regions in the Low Countries is marginalized and the economic development of Holland is isolated from its direct spatial context. In this way, it can only be taken for a fact that the Low Countries in the late Middle Ages did not yet constitute one economic space, as they did in 1800.[32] How this process unfolded, remains unclear. Therefore, the goal of this book is to explain why a modern—professional and integrated—transport sector could emerge in the Low Countries between ca. 1300 and 1850, how, where and when this development occurred and how the resulting transport system functioned.

---

31 See: Postan, *Medieval Trade and Finance*, 219–220; Van Zanden, *Arbeid*, 30–32.

32 Van Zanden, *Arbeid*, 30–32.

Deltas

In order to answer these questions, overcoming the limitations of traditional services- and capital-cost-oriented approaches, an economic-geographical perspective is adopted. This perspective combines a survey of the regional economic origins of communities of maritime transporters, with analyses of their functions, operations, composition and institutions. Together, these analyses result in a general appreciation of the structure of the Dutch maritime transport system. Rephrased in spatial network analysis terminology, this book aims to advance our understanding of the 'lines' that connect the 'nodes' in international maritime trade networks. Many people were involved in actualizing these lines, setting sail to destinations all over and even beyond the European coastline, but their contribution to the emergence of the 'first modern economy' has been assumed *a priori* rather than having been thoroughly investigated.

The theoretical framework elaborated in this monograph could perhaps be seen as complementary to James Vance's well-known theory of the emergence of trade networks.[33] Vance focuses on the emergence of distribution networks, which are marked by the gradual rise of central places within the network as well as by increasing complexity as the networks develop. He identifies changes in the role of the nodes in the network and substantiated how central places, or gateways, could emerge from them. Some terminological overlap notwithstanding (specifically the term *gateway*), the theoretical model of Dutch Deltas addresses the lines that connect the nodes in the network and aims at explaining who 'operated' these lines, where these 'operators' were located and how they were related to the nodes. The locations that were involved in operationalising Vance's distribution networks overlapped partly with the nodes of these distribution networks, but for the major part, they did not, as will be explained in Chapter 1.

The goal of trade is to generate added value by exchanging goods; the goal of transportation is to generate added value by facilitating the exchange of goods. Throughout this monograph, trade networks are defined as sets of locations interconnected through the exchange of people, goods and information by communities populating these locations. Transport systems are complexes of physical attributes (rivers, roads, canals, seas, etc.) and communities populating them, thus allowing for the exchange of people, goods and information between the locations of a trade network. The delta is deemed to be an appropriate geographical unit for a comprehensive economic-geographical

33 Vance, *The Merchant's World.*

analysis of the Dutch maritime transport sector before 1850. Three deltas are determined (see Map 1; detailed maps of each delta are added in appendix): the delta constituted by the estuaries of the Scheldt, Meuse and Rhine (hereafter SMR) with Middelburg, Dordrecht, Rotterdam and perhaps also Antwerpen as its main *gateways*;[34] the Zuiderzee delta (hereafter ZDZ), around which Amsterdam, Hoorn, Enkhuizen, the ports of the mouth of the River Ijssel (Kampen, Zwolle), the Frisian ports and the Wadden Islands Texel, Vlieland, Terschelling and Ameland are located; and the Lauwers-Dollard-Ems delta (hereafter LDE), with Groningen and Emden as its central places and Dokkum as a modest third.[35] The Wadden Island of Schiermonnikoog is considered as part of the LDE delta. In the rest of the book, SMR, ZDZ and LDE will be referred to as the Dutch Deltas. All three Dutch Deltas underwent significant physical-geographical changes in the late Middle Ages: the amount of water in the estuaries rapidly increased; the deltas grew and became better suited for shipping. The composition of the water changed under the influence of rising sea levels. Salinization occurred in all three deltas and influenced their fish populations. Nevertheless, silting continued to be a problem in many ports in the Dutch Deltas.[36]

Characteristic of the Dutch Deltas are the presence of islands, a long, often broad coastline that constituted the crossing between water mass and dry land and the presence of the mouths of one or more rivers. Insofar as towns and villages were connected directly to the delta, they are considered part of this geographical unit. Groningen, for example, was part of LDE, even though it was never located directly on the seashore.[37]

The Dutch Deltas were surrounded by regions that built different political entities and that often defended opposite political and economic interests. SMR functioned as a 'traffic square' for Zeeland, Holland, Flanders and Brabant. Along ZDZ, Holland, Utrecht, Gelre, Overijssel and Friesland were the relevant political entities. LDE was surrounded by Stad en Lande and Ostfriesland. Although all Dutch Deltas were the continuous subject of political debates

34 In a recent study about the future of the Dutch and Flemish ports along the Scheldt, Meuse and Rhine, the estuaries of these three rivers are described analogously as the Dutch-Flemish Delta. See: Vanelslander et al., *Ruimtelijk-economische en logistieke analyse.*

35 For a geographical description of the regions surrounding the Dollard-Ems estuary, see: Knottnerus, "Räume und Raumbeziehungen," 11–42. The demarcation of LDE that is employed in this monograph is largely based on the work of Knottnerus, albeit that the small estuary of the Lauwerszee with its adjacent city of Dokkum is considered to be part of the same geographical unit.

36 Ypma, *Geschiedenis*; De Lange, *Eb en Vloed*; Schroor, *Geschiedenis.*

37 De Lange, *Eb en Vloed*, 24; Kerkmeijer, "Groningen," 138–139.

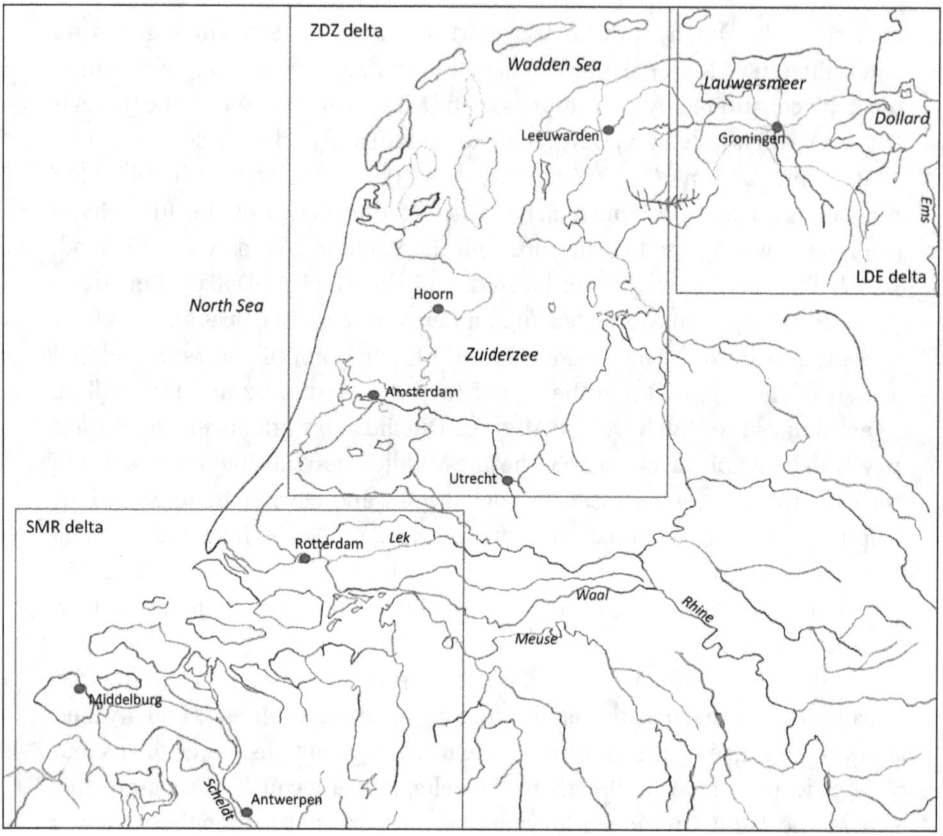

MAP 1 *Map of the Dutch Deltas. Based on the map of C. and J.C. Sepp "De VII Vereenigde Nederlandsche Provinciën", 1773.*

and economic competition, they were also areas of cooperation between individual places and regions.

Several types of direct connections were realised in a delta.[38] A distinction can be made between four types: (1) connections that stayed within the boundaries of the delta (domestic connections), (2) connections to locations along the coastline outside the delta (coastal connections), (3) connections that entered or left the delta (maritime connections) and (4) connections that crossed the delta, without having a location in the delta as final destination. Since the open sea could be reached through the delta by different types of

38 Direct transport connections do not require changing transport mode in order to deliver the cargo to its destination.

transporters, deltas were an ideal surface for the development of maritime transport services and thus for the emergence of spatial integration processes.

In this monograph, transport services obtain the predicate 'maritime' when they are not-incidental and when they cross the boundaries of coastal and domestic shipping. The boundaries of coastal shipping by shipmasters from the Dutch Deltas were defined by the Somme in northern France and the Elbe in northern Germany. Shipping is not-incidental when the registered community of the shipmaster appears at least twice in the relevant primary sources. A maritime transport community is a group of individuals that, according to registrations of domiciles in the sources,[39] belongs to the same place and supplies maritime transport services. All maritime transport communities active at a certain point in time in the Dutch Deltas taken together constitute the Dutch maritime transport sector at that particular time. A focus on shipmasters will be maintained throughout the entire book. Their isolation from a complex network of social and economic interactions, which includes relations with shipowners, merchants and other shipmasters, serves purely analytical purposes. Hence, the purpose of the book is not to analyse the economic activity of trading, but to gain better insight into the often neglected but equally important activity of transporting goods. As will be shown in the following chapters, maritime transport developed its own rules, regularities and usances.

Sources

The sources used in this volume comprise primary sources and secondary literature. The primary source-base consists of unedited and edited serial data sources (available in print and online) and of qualitative primary sources. The secondary literature used in this volume is primarily concerned with the regional economies and international trade networks of the Dutch Deltas' main and secondary gateways.

The first chapter on locations is based on published primary sources documenting shipping activity in the Dutch Deltas between ca. 1300 and 1800. Estimates of the size of the Dutch merchant fleet,[40] statistical observations of

39 The methodological issues related to domicile registrations will be treated extensively in Chapter 4.

40 Bruijn, "De vaart in Europa," 200–201; Van Royen, *Zeevarenden*, 13–20; De Vries and Van der Woude, *Nederland*, 469–475; Van Lottum and Lucassen, "Six cross-sections," 13–42; Unger, "The Tonnage," 247–261; Vogel, "Zur Größe," 268–333.

Dutch shipping[41] and analyses of the productivity of Dutch shipping[42] are of little help in this respect. The sources used in Chapter 1 are Bang and Korst's *Tabeller over Skibsfart*, a dataset of shipping to Archangel, based on the so-called *Galjootsgeldregisters*, which were used in Scheltjens' *De invloed van ruimtelijke verandering*, Lesger's dataset of the *convooigeldregisters*, De Buck's dataset on Archangel, which was published recently by Huygens ING, the appendices to Röhlk's *Schifffahrt und Handel zwischen Hamburg und den Niederlanden*, shipping lists in *Bronnen tot de Geschiedenis van den Levantschen Handel (1590–1826)*, shipping lists in *Bronnen tot de Geschiedenis van den Handel met Engeland, Schotland en Ierland 1150–1585*, Meilink's *Rekening van het lastgeld in Amsterdam*, Asaert's *Documenten voor de geschiedenis van de Antwerpse scheepvaart*, Oldewelt's *De scheepvaartstatistiek van Amsterdam* and Wätjen's *Die Niederländer im Mittelmeergebiet*.[43] A number of sources did not respond to the formal demands for further processing, mostly because the domiciles of shipmasters are not mentioned in them. Examples include the datasets of Postma on Suriname, Den Heijer about the West India Company, Klooster and Welling on the *paalgeldregisters*, and also Stuart Jenks' source publication.[44] Datasets that overlap with the Sound toll registers online (see below), such as Tangelder (1972) and Lindblad (1995) were not consulted.[45] In historiography, several lists and tables can be found that are based on the work of Bang and Korst; they were not consulted either. Finally, most printed aggregated overviews based on non-published sources could not be used because of a lack of information about shipmasters' domiciles. The same is true for summarizing descriptions without underlying quantification. The data regarding the areas

---

41 See for example: Unger, "De publikatie," 147–205; Snapper, "De zeescheepvaart," 15–30; Snapper, "Een voorlopige reconstructie," 118–129; Knoppers and Snapper, "De Nederlandse scheepvaart," 115–153; Oldewelt, "De scheepvaartstatistiek," 114–151; Knoppers, *Dutch trade*.

42 De Vries and Van der Woude, *Nederland*, 567–581; Van Zanden and Van Tielhof, "Roots of Growth," 389–403.

43 Bang and Korst, *Tabeller over skibsfart*; Sound Toll Registers Online: http://www.soundtoll.nl; Scheltjens, *De invloed*; Röhlk, *Schifffahrt und Handel*; Meilink, "Rekening van het Lastgeld," 187–230; Asaert, *Documenten*; Oldewelt, "De scheepvaartstatistiek"; Wätjen, *Die Niederländer*; Heeringa and Nanninga, *Bronnen*; Smit, *Bronnen*; Lesger, *Dataset van de Convooigeldregisters*; De Buck (†), Kerkvliet and Van Tielhof, *Amsterdamse notariële akten*.

44 The data sets of Postma, Den Heijer and Klooster are accessible via dans.knaw.nl. The data set of the *paalgeldregisters* of Amsterdam, compiled by Welling, can be found online as appendix to: Welling, *The Prize of Neutrality*. Furthermore see: Jenks, *Das Danziger Pfundzollbuch*.

45 Tangelder, *Nederlandse Rekeningen*; Lindblad, *Dutch Entries*.

of shipping activity, which was taken from the sources mentioned above and is summarized in Appendix 1, was categorized as follows. These categories will be used throughout Chapter 1.

A predominant source in all chapters are the Sound Toll Registers Online (hereafter STRO). The Sound Toll Registers are the records of the toll levied by the king of Denmark on the passage of ships through the Sound, the strait between Denmark and Sweden connecting the North and Baltic Seas.[46] They are stored by the Danish National Archives. The more than 700 volumes of the Sound toll registers that have been preserved include a practically uninterrupted series from 1574 to 1857 and some scattered records for the period between 1497 and 1574. They hold information on about 1.8 million passages. For each individual passage, both westward and eastward, the Sound toll registers contain the passage date, the name of the shipmaster, his domicile, his port of departure and—from the mid-1660s—his port of destination, the composition of the cargo and the toll paid.

TABLE 1 *Categorization of the main areas of shipping activity*

Region	**Area of shipping activity**
Africa	Morocco and beyond
Arctic	Archangel, Murmansk
Atlantic—south	France (below the Mouth of the Seine), Spain, Portugal
Atlantic—north	Norway
Atlantic	Greenland, Iceland, Faroe Islands
Atlantic—west	America, Caribbean Islands
Baltic	Baltic Sea
North Sea—east	Between Emden and Kiel; "Kleine Oost"
North Sea—central	northern Netherlands
North Sea—south	southern Netherlands / northern France (including the Mouth of the Seine)
North Sea—west	British Isles
Mediterranean	Mediterranean Sea
Indian	VOC-Area

46 The Sound Toll Registers Online project has been introduced at several conferences and in a number of publications. The following description of the source is based on: Scheltjens and Veluwenkamp, "The Sound Toll Registers Online," 301–302.

STRO should neither be confused with the well-known printed tabular summaries of part of the original registers,[47] nor with the so-called "Dutch Sound Toll Registers" (also known as: Nederlandse Sonttabellen). The latter are nothing more than a much abridged version of the original Danish toll registers, compiled by the Dutch consul at Helsingør between 1714 and 1781.[48] Typically in the Dutch Sound toll registers each passage is summarized on one line of text and only the main cargo on board is mentioned, without providing information about its volume or the taxes paid.

A pioneering effort in the creation of historical databases of maritime shipping was executed by a research group headed by Pierre Chaunu in the early 1950s. His group of French historians made an attempt to compute the period from 1784 to 1793 of the Danish Sound toll registers. Two journal articles notwithstanding, the project has not resulted in the publication of an electronic database.[49] In the 1970s, the Danish historian Hans Christian Johansen undertook a similar effort, which eventually led to the release of an electronic database of the Danish Sound toll registers for the years 1784 to 1795, also published on microfiches as an appendix to Johansen's dissertation.[50] This database has recently been converted to contemporary standards for historical databases by the Dutch historian George Welling.[51] STRO is a relational database set up to make the Sound toll registers' data instantaneously available to all via www.soundtoll.nl. In many ways, STRO is the delayed continuation of Johansen's work.

To some extent, STRO may not seem to be the ideal source for studying SMR and LDE, since the orientation of LDE's and SMR's trade was different to that of ZDZ, with its evident orientation towards the Baltic. As a consequence, critics might say that the image of the Dutch maritime sector in SMR and LDE is biased and the size of the maritime transport sector may be underestimated. While this is true, it should be noticed that not maritime trade, but the maritime transport sector is the topic of this monograph. So, whereas a part of SMR's and LDE's trade relations with the outside world may not be captured by STRO, it can reasonably be expected that STRO contains at least a representative part of the locations of SMR's and LDE's maritime transport sector, as well

47 Bang and Korst.

48 For a more detailed description of the Dutch Sound toll registers, see: Scheltjens 2009: 82–88.

49 Desfeuilles, "La Navigation," 492–520; Desfeuilles, "La Navigation," 56–74.

50 Johansen, *Shipping and Trade.*

51 http://www.let.rug.nl/welling/sont/johansen.htm.

as vital information about their emergence and their functions (in the form of the transport movements executed by masters domiciled in SMR or LDE).

The methodological tool employed in Chapter 1 is the location coefficient (LC), a measurement tool used primarily in transport geography. The LC establishes the share of a certain good in total freight transport. Additionally, LC may also be employed "to appreciate the importance of an economic activity for a community compared with the importance of the same activity within a defined larger area".[52] The advantage of LC is that it reduces the bias in absolute numbers, that is created by the predominance of STRO as a source for historical maritime shipping. In Chapter 1, LC is used in order to estimate the relative importance of various areas of activity for the different cities, 'vlekken' and villages in the three Dutch Deltas. The formula is as follows: LC = (Mti/ΣtMti)/(ΣtMt/ΣM), in which Mti represents the total number of ship movements of a maritime transport community t in area of activity I, as found in the consulted sources listed above. Mt is the sum of the registered ship movements of a transport community in all areas of activity and M is the sum of all registered ship movements of all transport communities in all areas of activity. The result for LC is a number that allows a distinction between less than proportionate share (LC < 1), proportionate share (LC = 1) and predominant share (LC > 1). Both the sums for each area of activity and for each individual community are taken into account in the calculation of LC, which makes the absolute numbers relative to each other. The values of LC are summarized for all individual maritime transport communities and for each time-period in Appendix 1. Clearly, LC is nothing but a rough indicator of the predominant type of transport services supplied by a certain community during a certain period of time. For example, 2.42 is the value for LC_Arctic for the maritime transport community of Ameland in 1700–1800. This figure resulted from the following calculation: ΣtMt=14571, ΣtMti=1919 and ΣM=271883 (see Appendix 1 for details). Since the values for M and Mt have been added to the appendix, the value of Mti can be reconstructed through a simple calculation. In this example, the value of Mti is 249. The precise calculation looks as follows: LC_Arctic = (249/1919)/(14571/271883)=2.42.

In chapters 2 and 3, quantitative techniques are applied to analyse long-term developments in the functions of ports (chapter 2) and structural changes in the production of maritime transport services (chapter 3). The methodological approach of this monograph has perhaps one important limitation: the description of long-term structural development in the Dutch maritime transport sector *as a whole*, which is central to Chapters 1 to 3, does not leave

52 See: Rodrigue et al., *The Geography of Transport Systems*, 157–158.

room for examples from the lives and actions of the shipmasters that are analysed. In fact, adding information about individual shipmasters would generate considerable bias to these chapters: neither do single cases hardly ever fit perfectly within the observed structural developments, nor do they have much added value for the scientific argument. A more serious issue perhaps is that, by including such case studies in descriptions of long-term structural developments, two different methodologies are mixed up, one of a quantitative, the other of a qualitative nature, which distorts the quality of the scientific argument. As a consequence, Chapters 1 to 3 may make a rather abstract and theoretical impression on the reader, which I hope to be able to lighten somewhat, by including some cases in Chapters 4 and 5.

In Chapter 4, a different use of STRO is pursued, using the names of shipmasters and their registered domiciles to execute a case study analysis of the maritime transport community of Hindeloopen and its composition (in Chapter 5), as well as to tackle the issues of ambiguous domicile registrations (in STRO) and of 'virtual' migration in times of trouble. While it would be a mistake to claim to have fully identified these masters, the standardisation of their names (and of their movement patterns) has brought identification much closer and—in fact—allowed production of a reasonable estimate of the number of individual shipmasters belonging to 32 domiciles in the Dutch Deltas between 1670 and 1850 (see Chapter 4). That their names are not mentioned in the statistics of Chapter 4 is merely a matter of space. In the next section on metrology and data conversion, I briefly discuss how the raw data from STRO has been converted to tonkm estimates in order to facilitate the quantitative analyses of Chapter 3. In the last section of this introduction, I describe the name standardisation procedures that lay at the basis of the population analysis in Chapter 4 and 5.

Conversion

Several steps were taken to homogenize the raw data from STRO and prepare them for further use. This section contains an overview of these steps and explains how the raw data from STRO was converted into volumetric estimates measured in tonkm. Tonkm is a commonly used measure in transport geography and transport economics, particularly in its maritime branches.[53] The

53 Rodrigue et al. specify that "The ton-km (or ton-mile) is a common measure expressing the realized freight transport demand. Although both the passenger-km and ton-km are most commonly used to measure realized demand, the measure can equally apply for

measure is applied widely in studies estimating the degree of specialisation of individual ports or clusters of ports as well as the level of integration between ports. The measure consists of two parts: tonnage, which indicates the quantity of goods shipped, and kilometer, which measures the distances over which the goods are shipped. Tonkm estimates of shipping activity thus comprise both a volumetric and a geographical perspective.

Estimating shipping activity in historical time-periods preceding the use of standard measures and nomenclatures of goods is not an easy task and most historians refrain from its execution, arguing that the vast variety of weights and measures employed in pre-modern times makes a conversion to metric values impossible. Indeed, each weight or measure represented different values depending on the place where the measurement took place and what product was measured. Moreover, the enormous variety of products is equally problematic. Some products are unknown now and many had functions that cannot easily be 'translated' into present-day nomenclature. The biggest challenge, however, is that of the translation of the names of products in one, common language in order to facilitate comparative analysis, aggregation, or the reconstruction of commodity chains. Generally speaking, weights and measures and product denominations in the pre-statistical era share one fundamental characteristic: their meaning was regionally—and often even locally—defined. Any attempt to convert pre-modern data on shipping activity must therefore deal with metrology and product nomenclature first.

Not many historians have undertaken such an endeavour, and those that have either made far-reaching simplifications, which severely limited the potential use of their standardised dataset,[54] or they restricted their conversion to case-studies with no ambition to generalize the results. In an environment where historical research is still predominantly an individual effort, it is clear that time remains an important barrier to the issues involved with weights and measures and product nomenclature. It is therefore not surprising that in many historical studies estimated values of pre-modern weights and measures are merely given in footnotes or appendices, often as a simple

---

transport supply." See: https://people.hofstra.edu/geotrans/eng/ch7en/conc7en/ch7c4en.html. It is in the latter meaning that tonkm is used in this monograph. The first tonkm estimates of the volume of Dutch shipping between 1500 and 1800 have been published in 2009 by Van Zanden and Van Tielhof. See: Van Zanden and Van Tielhof, "Roots of Growth," 389–403. For a critique of their method, see: Scheltjens, "The Volume," 73–101.

54 For example, see: Van Zanden and Van Tielhof, "Roots of Growth," 389–403. Critique in Scheltjens, "The Volume," 83–109.

dictionary reference, leaving the actual conversion, with its local specificity, unresolved.

However, the potential gains from converting pre-modern weights and measures into standardised metric values can hardly be overestimated, especially when trade and transport are under consideration. Contrary to rough indicators of shipping activity, such as declared tax income, total ship size or number of ships calling at a port, tonkm estimates, which assume conversion of pre-modern weights and measures from the bottom up, offer the advantage of comparability at various analytical levels, while having a great potential for further statistical analysis. This point is quintessential to this study. Without tonkm estimates it would be impossible to make substantiated statements about changes in the composition of cargoes, changes in the relative importance of maritime transport communities on certain routes or about changes in shipping efficiency. Moreover, it would be impossible to compare the regional development of shipping activity across time and space. Describing the emergence, functions and structure of the Dutch maritime transport system in the early modern period without the use of tonkm estimates would be extremely difficult.

Despite its important advantages, tonkm estimates also have their limitations. First of all, not every dataset contains the elements necessary to make conversion possible. Therefore, as said, the location coefficient has been used in Chapter 1 in a rough and simplified manner, serving merely as a basic indicator for some large-scale developments that are treated in detail in the following chapters. Second, tonkm estimates do not provide any information about the value of the distributed goods. Finally, the reliability of tonkm estimates depends entirely on the reliability of the source on which they are based. Inaccuracies, mistakes and false declarations influence the reliability of the estimates. Likewise, when electronic databases of historical records are used as the basis for conversion, one should be aware of additional errors due to misreading, omission or repetition.

In this monograph, tonkm estimates of the shipping activity between the North Sea and the Baltic play a central role in Chapter 3. The data upon which these estimates rely are taken from STRO, which are fully available in digital form for the years 1633 until 1857. The focus of the analysis will be on the period 1670–1850, to which the following method for conversion has was applied.[55]

---

55 This methodology has been described in detail in Scheltjens, "The Volume" and in Scheltjens, "Konvertierungsmöglichkeiten" (forthcoming). The former describes a first attempt to convert STRO data for the years 1784–1795. As more and more data from the

The preparations that were necessary to convert 'raw' STRO data to tonkm estimates comprise: (1) calculation of distances between ports, (2) standardisation, homogenisation and categorization of textual data items, like product names, weights and measures or names of ports, and (3) conversion of numerical data to the decimal system.

Calculation of distances between ports was probably the easiest part of the conversion. Standardized geographical data was provided by George Welling, who has created a geo-referenced data set of all places mentioned in STRO, which includes both the original spelling variations, a standardized place name and additional geographical categories.[56] Some of these categories refer to historical regions; others refer to present-day provinces and countries. The larger geographical categories in Welling's thesaurus of place names correspond to the 'small regions' and 'big regions' elaborated earlier by Hans Christian Johansen. Distances between ports were estimated using data published by the National Imagery and Mapping Agency of the U.S. Government as the main source.[57] The data provided in *Pub 151, eleventh edition* of 2001 are distances between ports and junction points or other ports, expressed in international nautical miles of 6.076 feet.[58] Since the ports mentioned in this source are limited to ports of international significance at the present day, not all distances could be calculated directly from it. The exact location of a number of pre-modern ports is unknown and many others—especially small ports that have lost their maritime functions a long time ago—are not mentioned in *Pub 151*. A dual strategy has therefore been developed in order to estimate distances to and from these ports. First, the distances that could be calculated directly using the information provided in *Pub 151* were processed. Then, the distances were extrapolated to the regional level using the historical regions defined in Johansen and Welling.[59] Ports that are not listed in *Pub 151* but belong to a region of which at least one port is listed, obtain the distances related to the listed port in the same region. When many ports of one region are listed in *Pub 151*, the predominant port in terms of shipping frequency is preferred. For

---

Sound toll registers became available online, this preliminary effort was expanded to cover the period 1670–1850.

56 The thesaurus of place names is incorporated in the online search engine of STRO. See: www.soundtoll.nl.

57 I am indebted to Dr. George Welling (University of Groningen) for bringing this source to my attention.

58 Junction points are "positions where many routes converge and through which ships pass when sailing from one major area into another" (*Distances between ports*, 1).

59 Johansen, *Shipping and Trade*; Welling, *Version history*.

example, the distance between Zoutkamp and Pillau is calculated using the data available for Emden and Königsberg.

Due to the large number of cargo registrations in STRO (almost 4 million), homogenization and categorization of product data was a complex undertaking. Two major issues had to be dealt with: reduction of the amount of variation in the "raw" data and simplification of the product description. Reduction was achieved through homogenization of source-related variations, such as spelling variations in the product denominations, and of database-related errors, such as typing mistakes. Simplification was achieved through a process in which cargo descriptions were divided into their constituent parts, with the isolation of the main product denominator as its final aim.

The homogenization of weights and measures and of quantities in the 'raw' STRO data was dealt with in an analogical way. In the case of quantities, descriptions written in full text, roman numbers, fractions or a combination of these, were converted to their equivalent in the decimal system. Reduction of variation in weights and measures was achieved through a dual process. First, homogenization of variant spellings was pursued; then, denominations of equivalent weights and measures in different languages were linked with the help of the information provided by Horace Doursther in his *Dictionnaire universel des poids et mesures anciens et modernes*.[60] Horace Doursther's dictionary, published in 1840, comprises the works on historical weights and measures of well-known predecessors like Kelly, Nelkenbecher and others,[61] while adding a vast amount of information that results from a study of commercial reports, tariff lists and the like.[62] Though recognized by Bob Allen and Tommy Murphy as an extremely exhaustive source providing metric equivalents for all the measures,[63] Doursther's dictionary is still largely unknown in the scholarly community. It has been used solely as a reference work from which short explanations of a specific measure and its value at one or a few specific places were drawn. The general scope of Doursther's dictionary remains largely unexploited.

Next to being extremely exhaustive and global in scope, there is another reason for selecting Doursther's dictionary over other dictionaries and compilations of pre-metric weights and measures, and that is language. While reference works for specific geographic regions, micro-regions and even places may

60 Doursther, *Dictionnaire*.

61 Nelkenbrecher, *Taschenbuch eines Banquiers*; *Le cambiste universel.*

62 Doursther, *Dictionnaire*, ii.

63 Allen and Murphy, *Before the Metre*.

certainly be more accurate and detailed than Doursther's dictionary, they are in most cases detached from the international context of weights and measures, providing only their local names, without referencing to their 'international' equivalents. In a study like ours, this would be a major disadvantage. Doursther's dictionary regroups weights and measures under one heading, provides (cross-)references to translations of these measures into various languages and adds information about the local names of measures in the conversion details. Doursther's dictionary was put to use as an 'engine' for the standardisation of weights and measures in STRO.

Registration of cargo items in the Sound was recorded on the basis of freight letters.[64] Cargo items and their measures were translated into Danish, registered quantities were copied and the sum of customs due was calculated separately for each cargo item. While customs duties were calculated according to a number of rules described in various customs treaties, there is no mention of any conversion of weights and measures used in the freight letters to local (Danish) equivalents. On the contrary, the 1645 Treaty of Christianapolis / Kristianopel and its confirmation in 1701 stated that the weight of measures of goods upon which custom payments are due is that of the place where the good had been loaded.[65] This allows us to *localize* the weights and measures found in the Sound traffic database, stating that the registered point of

64 Scherer, *Der Sundzoll*, 63–64.

65 The original Dutch text of the 1645 and 1701 treaties is as follows: "Den Tol in den Orisont sal van de Ingezeetenen der vereenigde Nederlanden [...] geheeven ende betaald werden, welverstaande dat de Lasten, Maten en Gewigten van Waaren en Koopmanschappen, welke volgens de voorschreeve Tollrollen bij het Last, Mate of Gewigte moeten betaalen, sullen werden geconsidereert en genoomen op een en dezelfde voet als die werd gereekent ter plaatse daar die waaren, die vertollen zijn geladen [...]." The rest of article 2 is dedicated to the rules of calculation mentioned above. It is stated that "[...] omme dat geen verschil sal vallen van de uytrekeninge van wijnen, soo zal een vat Bourdeause wijn in de Zond geestimeerd worden op twee en vijftig Rijxdaelders, en een vat Nantoise op Vijff en veertig Rijxdaelders. Spaense wijnen aengaende een Pijpe Mallagaese off andere Spaense wijnen op vijff en veertig Rijksdaelders ende sal volgens dese tauxatie den tol betaelt, werden nae de Christianopelse Tol Lijste, dat is den dertigsten penning". See: *Tractaat noopende de tollen in den Orisont en Noorwegen, tussen sijn Koninklijke Majesteit van Denemarken en Noorwegen, ter eenre, en de Hoog Mog. Heeren Staaten Generaal der vereenigde Nederlanden ter andere zijde, geslooten tot Koppenhage den vijftienden Juny 1701. Mitsgaders de Tolrolle van de jare 1645, in gevolge van het tweede Articul van dit jegenwoordig Tractaat.* In 's Gravenhage, Bij Paulus Scheltus, ordinaris drukker van de Hoog Mog Heeren Staten Generaal der vereenighde Nederlanden. Anno 1701. Met privilegie. 31 pp.

departure of the ship is the point of reference for establishing the metric equivalents of the weights and measures declared at the Sound customs office.[66]

The actual conversion process was a bottom-up process that started with the individual Sound passages and their corresponding cargo registrations. Registrations in STRO were matched with the information on weights and measures found in Doursther's dictionary and with the information on distances between ports found in *Pub 151*. Tonkm estimates were calculated for each individual cargo item passing through the Sound in order to preserve their full potential at different levels of aggregation. The matching process itself consisted of seven steps and involved all standardized variables as well as the categorization of goods into a number of types and the categorization of places into a set of different regions. The matching procedures were the following:[67]

1. Boolean matching
2. matching based on identical measure and location, with product similarity (using type of good as matching category)
3. matching based on identical measure and product, with geographical similarity (using region as matching category)
4. matching based on identical measure, with location selected according to a pre-defined set of rules and product specifications either missing, similar or considered irrelevant
5. matching based on measurement similarity
6. standardization of non-weights and non-measures
   - Boxes and cases
   - Pieces (*stykker, tylt, dusin, skok*, etc.)
7. exclusion of *ad valorem* taxations

66 Obviously, a certain degree of uncertainty remains. It is possible that a ship departed from a different place than the one given at the Sound as point of departure. Also, it may be possible that the ship loaded additional cargo items at ports located between the registered point of departure and the Sound. In both cases, the Sound toll registers usually do not provide these data. It must be stressed, however, that both cases mentioned would involve providing false or incomplete documentation at the customs office in Elsinore. Numerous scholars have tackled this issue, providing very different indications of the percentage of fraud. Calculation of matching scores of registrations in the Sound traffic database and similar sources outside of the Sound might possibly narrow down the range, but this lies well outside the scope of the present paper.

67 Described more extensively in Scheltjens, "The Volume"; Scheltjens, "Konvertierungsmöglichkeiten".

The converted data set that has been used for the empirical analysis in Chapter 3 contains conversion to metric tonnes of about 85% of all cargo registrations in STRO between 1670 and 1850.

Name Matching

Changes in the composition of shipmasters' populations are the central topic of Chapters 4 and 5. The starting point for this chapter was the standardisation of personal names of shipmasters registered in the Sound. The name standardisation process that was executed to prepare the analysis in Chapters 4 and 5, was conceptualized as a two-stage process.[68] In the first step, all names, first names and patronymics of relevant shipmasters were selected[69] and an existing thesaurus of Frisian name variations was linked to it.[70] This process was executed for 32 maritime transport communities in the Dutch Deltas, of which members were registered in STRO between 1670 and 1850 (more details about the selection of these communities in Chapters 3 and 4). As a result, a large part of the first names, patronymics and surnames of the selected maritime transport communities was matched with a corresponding standard form. In the next step, the non-matched names, first names and patronymics were added to the existing thesaurus as new variations. Like before, the frequency of each variation was taken as a guide for limiting the number of variations that had to be dealt with, while at the same time achieving the highest possible efficiency during the process. In other words, the variations with the highest frequency were dealt with first and the process was cut off at a minimum frequency of 3. Whenever necessary, new standard names were added to the list. During this time-consuming, partly manual process, the decisions that were made were primarily based on similarity of the 'unmatched' variant with other 'matched' variants. For 'new' standard names, the repository of the Meertens Institute was used as a primary source of reference.[71] Upon completion of this

---

68 Based on methodology elaborated for research assistants and doctoral students at the University of Groningen.

69 In the thesaurus, last names and patronymics are gathered together in one table, since, typically, the distinction between the two cannot easily be made in pre-modern times.

70 Thesaurus of Frisian personal names. Version used in this monograph is version 7, dated July 2010.

71 Attributing new standard names was preferably done by relying on existing sources. In other case, elimination of spelling variations led to a standard form that was then added to the thesaurus. For Dutch sources on first and last names, see: Meertens Instituut.

matching step, the majority of variant first names, patronymics and surnames of the selected maritime transport communities were matched with their corresponding standard forms.

However, the process was not complete after these two steps. All the typical errors that occur at the data input phase—be it in the original registers or in the electronic database based upon them—inevitably lead to residual errors, non-recognised standard names or ambiguous standardisation results. The standardisation that was completed in the first two steps must consequently be regarded primarily as an effort to reduce variation, just like in the case of weights and measures, products and names of places in the Sound toll registers.

Of course, it would be possible to pursue more than just the elimination of variation in the personal names registered in STRO, for example by taking into account more contextual data, probably even from other sources such as notarial deeds, ships passports (Nl. *zeebrieven*) or other toll registers like the so-called *Paalgeldregisters*.[72] Such further processing was not executed here, since the value it would have added to achieving the research goals of the present volume is rather limited. Charter parties (Nl. *Chertepartijen*), for example, are qualitative sources that offer a detailed insight into the planning of individual voyages or in the settling of accounts upon completion of (problematic) finished voyages.[73] In such sources, only a limited number of probably rather exceptional cases are documented. In fact, the collection of charter parties of the early seventeenth century, published by Winkelman in the 1970s and 1980s, corresponds to only about 6% of all Sound traffic during that same period. Setting up matching procedures in order to combine data from charter parties with STRO would perhaps lead to a better understanding of changes that might have occurred during the shipmaster's voyage, but such findings would have to be validated with the utmost care, avoiding generalizations based on very small and perhaps even unrepresentative samples.

72 Relevant sources include, but are by no means limited to the *paalgeldregisters*, the ships' passports (Nl. *zeebrieven*) in e.g. Amsterdam, Stavoren or Workum. Additionally, it would also be possible to link the data of STRO with the muster rolls kept in Amsterdam and Groningen. The latter are particularly interesting, since they have been made available in an electronic database, hosted by the Noordelijk Scheepvaartmuseum in Groningen. A prerequisite for matching such different sources is that names are standardised in a compatible way for each of these sources. For details about the location of the archival sources mentioned here, see Bibliography.

73 Schöffer, "De Vonnissen", Winkelman, *Bronnen*, Christensen, *Dutch trade*.

The situation is rather different when an attempt is made to match STRO with other toll registers and a few scholars have tried to execute such a record linking process.[74] There are two recurring problems, however, related to such matching procedures. The first and foremost problem is time. Most sources that may be matched with STRO cover much shorter time-periods, which essentially limits the potential of matching procedures of this kind to provide some sort of reliability test for STRO.[75] Additionally, the names may be registered differently in different sources. In the case of the comparison between the Dutch Sound toll tables—a summary of the original Danish registers compiled by representatives of the States General in Helsingör between 1709 and 1780[76]—and the shipping lists of the St. Petersburg Times (Ru. *Sankt-Peterburgskie Vedomosti*)—the Russian Empire first newspaper, established in 1728[77]—these differences were quite clear: in the latter, the names were written in Cyrillic. In the case of the *paalgeldregisters* of Amsterdam for the years 1771–1787, which were entered into an electronic database by George Welling, the differences in name registration are more subtle. In the *paalgeldregisters*, many first names and patronymics were abbreviated. Only the initials remain. Of all registrations in the *paalgeldregisters* between 1771 and 1787, about 60% give only the initials of the shipmaster. The corresponding registrations in STRO mostly have the names of shipmasters written in full, which makes nominal record linkage rather problematic, especially for automated matching procedures, such as used by Welling in 1988.[78] It seems that his rejection of automated matching procedures based on the unsatisfactory results (only 3% successful matches) must be criticised for two reasons: the adopted method was suboptimal and the data sets were not sufficiently compatible. Still, we can agree with

74 Knoppers, *Dutch trade*; Johansen, "Scandinavian shipping"; Welling, "Price-supply relations"; Welling, *The Prize of Neutrality*; Scheltjens, *De invloed*.

75 Reliability tests for the Sound toll registers and its electronic equivalent STRO have been undertaken by several scholars. For examples, see: Welling, "Price-supply relations"; Christensen, *Dutch trade*; Alanen, *Der Außenhandel*. Different methods, such as the comparison of aggregated volumes or passage frequencies, have been applied by Ojala, Moreira and Karvonen, "Assessing the Reliability" (forthcoming).

76 Dutch Sound toll tables. For details about this source, see: Scheltjens and Van Koningsbrugge, *Van onze reporter*; Scheltjens, *De invloed*. The earlier part of the original tables resides in the Dutch National Archives in The Hague (until 1766); the later part of the tables can be found in the Archives of the Directory Board of Muscovy Trade in Amsterdam (until 1781).

77 Morozova, "Organizacija i obrabotka," 24–37.

78 Welling, "Price-supply relations".

Welling, that automated name matching procedures should be dismissed, but not primarily for the reasons, that he described. Rather, they should be dismissed, because none of the existing name matching algorithms are capable of dealing successfully with pre-modern (Dutch) first names, patronymics and surnames.[79] In this monograph, further matching of the selected data set with other sources has not been pursued.

---

79 For critical surveys of relevant nominal record linkage procedures, see: Floud, *An introduction*; Wrigley, *Identifying people*; Bloothooft, "Assessment of systems".

CHAPTER 1

Locations

Understanding the regional economic origins of communities of maritime transporters is necessary to gain insight into the locations and functions of the transport communities that—literally—gave body to the Dutch maritime transport sector. In this chapter a long-term descriptive analysis of the geography of the Dutch maritime transport sector between ca. 1300 and 1800 is presented. Communities of maritime transporters in the three deltas are identified on the basis of published and unpublished primary sources. The spatial dynamics of these communities of maritime transporters are analysed by delta and in 100-year time-periods with the aid of the location coefficient, a methodological tool that is widely used in transport geography (see introduction). This chapter describes the rise and decline and the main areas of activity of maritime transport communities in each of the three deltas, the professionalization of the maritime transport sector and the process of centralization of the Dutch maritime transport market and of integration between the three deltas. The goal of this chapter is to provide a general outline and frame of reference for the following, more analytically detailed chapters.

Overview

This section gives a succinct overview of the geography of the maritime transport sector in the Low Countries between ca. 1300 and 1800. There is no doubt that this overview is incomplete. The sources available are fragmentary, and only to a limited extent accessible through (digital) source publication.[1] The overview of the geography of the maritime transport sector in the Low Countries can only be construed in a thoroughly aggregated form; loss of detail is unavoidable. The overview presented below is, thus, a first attempt to describe the geography of the maritime transport sector in the Low Countries from ca. 1300 to 1800 in order to reconstruct the beginning of an integrated maritime transport space in the Low Countries.

In total 328 places were identified where maritime transport services were part of the local economy between ca. 1300 and 1800. In some places the maritime transport sector was continuous; in many other cases it was a temporary

1 An overview was presented in the Introduction.

 | DOI 10.1163/9789004273078_003

phenomenon that came about at one point, only to disappear again a while later. This pattern of emerging and disappearing is partly known in the historiography (see the previous section), but the developments described are only valid for part of the Low Countries and part of the period described in this article.

Seventy-five of these 328 places can be labelled as cities. The other places are not cities. They are mostly small villages, in some cases it concerns villages that in time grew into 'vlekken'—large villages with a (limited) number of urban (central) functions of which trade and industry predominate.[2] The distinction between cities, villages and 'vlekken' is based on the work of Rutte and Van der Aa. Rutte identifies nine different 'urban landscapes' in North Holland between the eleventh and the fifteenth centuries (equalling ca. 120 cities in total). In addition, Stabel provides an overview of cities in Flanders in the late Middle Ages.[3] The cities in the coastal areas as described by Rutte correspond mostly to the sixty cities in Zeeland, Holland and West Friesland, that Cox researched with regard to the role of the granting of city rights.[4] According to Rutte, obtaining city rights was neither sufficient nor necessary to use the term 'city', arguing that a place being a city depended on a varying combination of legal (city rights), morphological (city walls), economical (market) and societal characteristics (trade).[5] The identification of villages and 'vlekken' was based on the famous geographical dictionary of Van der Aa.[6]

2 'Vlek' is a Dutch term that is hard to translate. It has two German equivalents: 'Flecken' and 'Marktflecken'. Based on Krünitz' famous encyclopedia, 'Flecken' may be described as a category between a village and a town; a large village with urban functions, but often without city rights. For 'Marktflecken' it is stated in particular "(…), daß man unter Marktflecken eigentlich nicht ein Dorf, das das Recht hat, einen oder mehrere Jahrmärkte zu halten, verstehen darf, sondern einen solchen kleinen Ort, der das Privilegium hat, verschiedene Arten Handwerker, wie in Städten, unter sich zu haben, ja auch einigen Handel treiben zu können". See: Johann Georg Krünitz, *Oekonomische Encyklopädie oder allgemeines System der Staats-Stadt-Haus-und Landwirthschaft*, 1773–1858, 242 vol. Consulted online at: http://www.kruenitz1.uni-trier.de/.

3 For an overview of cities in the Low Countries before 1400 and for a definition of the term 'city', see: Rutte, "Stadslandschappen," 143–169. An insightful discussion of the emergence of the urban network of the Dutch Republic can be found in: 't Hart, "De Nederlandse Republiek," 54–61; Stabel, *Dwarfs among Giants*. Of the cities in the Low Countries before 1400 particular attention was given to cities located in the coastal districts.

4 Cox, *Hebbende privilege van stede*.

5 Rutte, "Stadslandschappen," 146–147.

6 Van der Aa, *Aardrijkskundig woordenboek*.

The characteristics described by Rutte were sufficient to establish international, maritime trade relations and thus to create the need for maritime transport services within the cities.[7] As far as coastal and rural societies are concerned, regional economic development in the form of a growing local production in agriculture, fishery, forestry or the extraction of minerals (e.g. peat, chalk or clay) would appear to be necessary to establish maritime transport communities.

Categorising places as city, 'vlek' or village was not always as straightforward. The two main sources—Rutte on the one hand, and the geographical dictionary of Van der Aa on the other—provide contradictory information on a number of places. Rutte classifies Schagen as a city, while Aa considers it a 'vlek'. At the same time Aa's description of Woudsend suggests it is a 'vlek', but this assumption is never made explicit.[8] In other cases, Van der Aa describes a place as a 'rural town'. And there are several cases in which a place is not defined as a village, vlek or city, but as a fortress, neighbourhood, hamlet, polder or island. These cases were considered to be non-urban communities, that were categorised under the general term of 'villages'. However, in the case of Delfzijl, described as 'fortress', classification as village does not seem to do justice to its scale or economical activities in the eighteenth century.

Apart from the temporal development, described in the next sections, the geographical distribution of these 328 places is disproportionate over the three Dutch Deltas (see appendix 1 for an overview of all places). Ten toponyms could not be identified.[9] Eleven places did not belong to one of the three estuaries; they are situated at the coast of the open sea (in this case the North Sea). Out of these places there are six villages (Bergen (NH), Callantsoog, d'Hoef, Kamperduin, Lombartszijde, Yzermonde), four cities (Blankenberge, Duinkerken, Nieuwpoort, Oostende) and one vlek (Den Helder). The largest concentration of maritime transport communities was obviously concentrated around the Dutch Deltas, in some cases situated along the coastline, often more inland.

---

7 It is worth mentioning again in this context, that the distinction between shipmaster, merchant and freighter was hard to make before the end of the sixteenth century. Later developments in the ownership structure of ships and cargoes make this easier, but a fundamental treatment of this development would require much in-depth microlevel research and therefore lies outside the scope of this monograph.

8 Van der Aa, *Aardrijkskundig woordenboek*, XII, 641.

9 These toponyms are: Esens, Haukynswaard, Oldendam, Orno, Oudebom, Ralduch, Rinchoove, Tarndijk, Walls en Zuiderdam.

The majority of cities where maritime transport services were at some point part of the local economy are situated in SMR, mainly medieval cities in Flanders, Brabant, Zeeland and South Holland. Maritime transport services were more or less part of the economy of these cities during the researched period, except in those cases where physical-geographical circumstances (silting or—on the contrary—flooding) countered the existence of maritime shipping. Villages were also involved in delivering maritime transport services, but their number was rather limited, especially in comparison to the number of maritime transport communities in villages along ZDZ. The share of villages in the maritime transport services of SMR seems to be minimal—as far as the available sources allow for reliable assumptions to be made. It is striking that not one village seems to have had a long term maritime transport community. A few villages in SMR provided maritime transport services before 1400. A much smaller number did so in the fifteenth, sixteenth and seventeenth centuries. The eighteenth century is characterized by a growing number of maritime transport services in the villages around SMR. The sole vlekken in SMR, Roosendaal and Westenschouwen, only had maritime transport communities for a short period of time in the late fifteenth and sixteenth centuries.

In ZDZ, by far the largest of the Dutch Deltas, the participation of villages, vlekken and cities in providing maritime transport services was fairly universal. In total thirty-seven cities were identified, situated along the coastline of ZDZ. The nine 'vlekken' around ZDZ are all located in the north of ZDZ, except for Schagen, which Rutte characterizes as a city. Blokzijl is the most southern point; the other vlekken are spread out in Friesland, partly at the coast, partly inland. Most striking are the 130 maritime transport communities in the villages around ZDZ, notably the concentration of these villages on the south, west and north side of ZDZ. Only in Vollenhove at the east side of ZDZ a few villages with maritime transport communities can be found.

LDE only has two cities—Dokkum and Groningen—and two vlekken—Appingedam and Kollum—where maritime transport communities were part of the local economy. The other thirty-six maritime transport communities belonged to the local economies of villages in the Veenkoloniën (*Eng.* Fen colonies) and, to a limited extent, villages located in the surroundings of the Lauwersmeer.

Rise and Fall

In this section, the rise and decline of maritime transport communities in the Dutch Deltas is placed within the context of the development of regional economies and international trade. The section is split into periods of 100 years, so as to avoid using political or military events to mark the beginning or end of certain time periods. First, however, the period before 1400 is examined. Examples of places are mentioned in the text; an overview of all the places by period can be found in the appendices.

The development process of maritime transport communities in the Low Countries is complex. Not only does one have to pay attention to the distinction between trade networks and transport systems (see Introduction), the difference between city and countryside is also essential. Moreover, the availability of relevant sources is disproportionate, possibly suggesting that the maritime transport sector in some regions of the Low Countries is less explicitly present than should be the case. Especially the abundance of data from STRO makes it difficult to value the less prominent data from other regions.

To counter these shortcomings in the available source material we have chosen not to work with 'raw' data, which would result in an obvious positive bias for places registered in STRO, but to work with data converted into location coefficients (see Introduction). In this chapter, LC is used to determine the relative importance of several destination areas to various cities, 'vlekken' and villages in the Dutch Deltas.

Before 1400

Before 1400 maritime transport services were mainly concentrated in the 'old' cities of the Low Countries. They probably developed there under the influence of the early expansion of urban industries, such as breweries, potteries and other crafts, and of the international trade of (bulk) goods.[10] Insofar as this can be demonstrated by published primary sources, the cities of the Dutch Deltas that had maritime transport services before 1400 all possessed more or less the characteristics mentioned by Rutte—city rights, city walls, a market and a merchant guild. In SMR we have Aardenburg, Antwerpen, Arnemuiden, Axel, Bergen op Zoom, Brielle, Brouwershaven, Brugge, Damme, Delft, Dordrecht, Gent, Goedereede, Goes, Gouda, Hulst, Leiden, Middelburg, Reimerswaal, Rotterdam, Schiedam, Sluis, Steenbergen, Tholen, Veere, Vlaardingen, Vlissingen and Zierikzee. In ZDZ the cities with a maritime transport community before 1400 are Amsterdam, Deventer, Enkhuizen, Haarlem, Harderwijk, Harlingen, Hindeloopen, Hoorn, Kampen, Leeuwarden, Medemblik, Monnikendam, Stavoren, Utrecht, Workum and Zutphen. Friesland is also mentioned in 1294 as the home of shipmasters on

10 Berkenvelder, "Frieslands handel," 136–187; Unger, "Beer," 113–127; Hybel, "Introduction," xvii: "(…) strictly speaking, international bulk trade was not an innovation of the mid-twelfth century". Unger and Hybel describe how international trade in bulk goods developed between 1150 and 1400 from an incidental activity to regular trade.

the route between England and northern France, but without mentioning a specific place name.[11] Groningen was the only city in LDE that had a maritime transport community before 1400. In the toll register of Newcastle-upon-Tyne of 1294–1297 one Groningen captain is mentioned by name.[12] The ports of the cities of Blankenberge, Duinkerken, Nieuwpoort and Oostende were located outside the Dutch Deltas, and alongside storage and transhipment points, these also constituted communities of shipmasters.

Before 1400 the main destinations of shipmasters from the above-mentioned cities were located at the frontier between coastal trade and maritime shipping. Especially prominent is the route on the northern French coast (Calais, Rouen) and the northern German coast (Bremen and Hamburg).[13] The 'truly' maritime destinations were located along the English east coast, in south-west France (La Rochelle and Bordeaux), in Norway, on Schonen and in the southern part of the Baltic (Lübeck and Danzig).[14] The position of foreign (Hanze) shipmasters on the maritime transport market of the Low Countries was very strong in this period, but local captains dominated the inner and estuarine routes.

The participation of villages and cities around SMR, ZDZ and LDE in freight transport between the ports of their own deltas can easily be substantiated.[15] Shipmasters of the 'old' cities of SMR, ZDZ and LDE also took part in this distribution, thus providing evidence for the fact that inland and estuarine transport were more developed in the fourteenth century, while sea transport only had a limited scope. In their earliest development stages maritime transport connections in Friesland and Groningen were focused on the Baltic, rather than on England and France, but there is not enough data for a quantification.[16]

In the period before 1400 an exceptionally large number of villages (and so-called 'smalsteden' or small villages with limited city rights) in SMR also offered maritime transport services, next to estuarine transport services, while villages in ZDZ and LDE were not yet of any importance. The development stadium of

---

11 Sneller and Unger, *Bronnen*.

12 Smit, *Bronnen*, 71–77 there 73.

13 Merchant-shipmasters from Holland and Friesland were active in beer exports from Hamburg in the fourteenth century. See: Unger, "Beer," 116 and 122. Scattered information regarding the domiciles of shipmasters active in the distribution of beer from Hamburg in the second half of the fourteenth century can be found in: Smit, "De registers," 1–97.

14 Ketner, *Handel en scheepvaart*, 4–12.

15 For estuarine transport in ZDZ, see: Smit, "Het Kamper Pondtolregister," 209–296. For scattered information about estuarine transport in SMR, see: Unger, *De tol van Iersekeroord* and Smit, *Bronnen*.

16 Poelman, *Bronnen*.

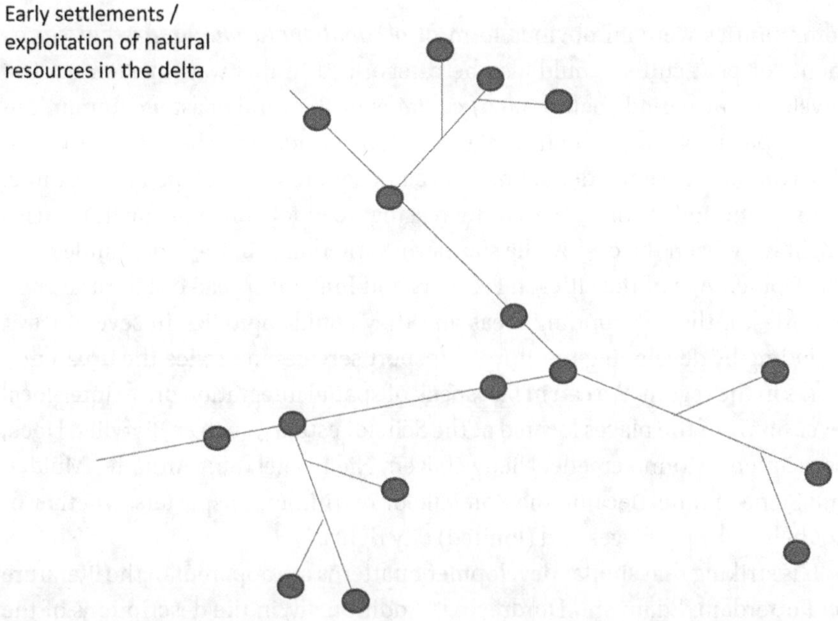

FIGURE 1 *Schematic overview of the location of the earliest transport communities in a delta.*

the economy in the Low Countries, of which the core before 1400 was located in the cities on the southern side of SMR (Flanders, Brabant), is reflected in the participation of rural coastal communities in the delivery of maritime transport services. Although relevant source material is only scarcely available, on the basis of the available literature on the geography and the local economies of these places we can formulate a careful hypothesis about the development of maritime transport in these rural coastal communities.

Initially, rural coastal communities were interested in transporting their local produce to (potential) buyers (see Figure 1). Whether they were located directly at the water front or more inland, rural coastal communities had a simple social structure, completely centred around their own, local production. In certain research studies on shipping communities some of these coastal and rural communities are denoted with the term *occupational community*, meaning that the whole local society was orientated on one single branch of industry.[17] Such communities—not more than settlements in their earliest form—developed in the Low Countries at places where sea or land offered the possibility to gain an income and thus to survive. Fishing

17 Based on: De Wit, *Leven, werken en geloven*, 15.

communities were an obvious form of *occupational communities* but settlements of peat-cutters could also be categorized in this way. In their earliest development period (before 1400) captains in the rural coastal communities hardly participated in the international goods trade. For the most part their ships did not leave the delta, unless urban economies developed in the direct vicinity and unless there was an increasing need for transport facilities. This might have been the case in the southern Netherlands in the late Middle Ages. The Golden Age of the cities in Flanders and Brabant spread out to rural communities in the surrounding areas and they could capitalize in several ways, including by developing maritime transport services to service the urban networks in the region. We can in fact speak of spatial integration on an inter-local level. Some of the places located at the Scheldt estuary, such as Biervliet, Hoek, Lapschuere, Monnikereede, Nieuwehaven, Nieuwvliet, Sint-Anna-ter-Muiden and Slepedamme, became the domicile of maritime transporters, whether or not helped by privileges and (limited) city rights.

It is striking that similar development patterns are apparent in the literature on Rotterdam, Edam and Dordrecht.[18] Additionally, in the descriptions of the locations in Van der Aa's *Aardrijkskundig Woordenboek* similar development patterns are clearly recognisable. These are processes differentiated in time, but the core of the development of maritime transport communities in these cities was the same as that of village communities. This indicates both a fundamental difference and similarity in the development of villages and cities. The cities, later forming the core of the Dutch trade system, were once only villages; their reason for existence was initially identical. That some settlements such as Amsterdam, Rotterdam, Dordrecht and others developed into cities, while others did not, thus not becoming part of urban networks, was due to a variety of factors of which location and the development of a local merchant class were probably the most important. In the first half of the fourteenth century Amsterdam's trade was mostly ran by farmers and fishermen, who had moved there because of the prime location of this settlement for the execution of their commercial activities. In the second half of the fourteenth century they developed their own merchant guild. This is also the time when the transport supply in Amsterdam started to grow enormously and Amsterdam shipmasters became active in maritime transport.[19] At the end of the fourteenth century Amsterdam shipmasters called at places such as Danzig, Great Yarmouth,

18 See: Van der Schoor, *Stad in* Aanwas, 27–41; 71–93; 117–126; 166–176; Boschma-Aarnoudse, *Tot verbeteringe*, 119–125; Van Herwaarden et al., *Geschiedenis van Dordrecht*, 179–216.

19 Ketner, *Handel en Scheepvaart*, 12.

Lynn, La Rochelle and Bordeaux.[20] From then on, the distribution function of Amsterdam obtains its international fame.

1400–1500

After 1400, an increase in transport services for estuarine and coastal shipping can clearly be determined in villages where the economy was driven by the local production of fish, dairy or peat. A similar rise also occurred in a growing number of cities where the economy was driven by the local production of goods and the development of regional distribution functions.[21] A process of labour division in the local community, facilitated by the regional economic development on the one hand, and the rise of international maritime trade on the other formed the basis of these developments.

Contact with the outside world comes to the fore as a necessary condition for the opening up of local coastal communities. This contact can be specified as economic interaction, that initiates inter-local exchange of production between different *occupational communities*. In economic history, such development is connected with the transition from *subsistence* agriculture to market-orientated production.[22] Society slowly acquired a proto-capitalist character, in which wage income became necessary. While the complexity of local *occupational communities* grew under the influence of economic exchange with their immediate surroundings, the necessary conditions for the development of processes of functional specialisation were created. These were the development of the labour cycle and labour specialisation, which are related.[23] The development of the labour cycle is the transition from a combination of seasonal labour activities to one main activity. Labour specialisation refers to the further development of the main activity, where different labour functions, which originally were part of one occupation, slowly emancipate

20 See: Poelman, *Bronnen*; Sneller, *Bronnen*; Smit, *Bronnen*.

21 Cf. Smit, "Kamper Pondtolregister," 211–212.

22 See: De Vries, *Rural economy* and Van Zanden, *Arbeid*.

23 Specialisation is a well-studied topic in Dutch historiography. De Vries, Van Zanden and Bieleman have written standard works about specialisation in the agricultural sector. Veluwenkamp has made an important contribution to the study of entrepreneurial behaviour and the development of international trade networks in the early modern period in which significant attention is devoted to the role of specialisation. The topic also figures prominently in research into the development of the workcycle. Finally, in several maritime-historical studies about Dutch shipping in the early modern period, the concept of specialisation is used, often referring implicitly to the specialisation index that is known from transport geography. See: De Vries, *Rural Economy*; Van Zanden, *Arbeid*; Lucassen, *Naar de kusten*; Veluwenkamp and Klein, "The role of the entrepreneur," 27–54.

and are taken up by separate occupational groups. The development of transport services for third parties was one possible outcome of these processes of labour division and, especially in the case of *occupational communities* with an intrinsic transport need, such as fishing communities and settlements in peat-digging areas, the emancipation of the transport function was a logical consequence. Paraphrasing De Vries, one could describe this process as the transition from *subsistence* transport, which merely existed for the benefit of the local community, to *commercial* transport, that was not only at the disposal of the local community, but could also be hired by 'third parties'.

Due to the emerging economic interaction with the surrounding areas, the rural coastal communities became less isolated and gradually became part of a regional society (see Figure 2). Then, their geographical location and the development of their own merchant class facilitated the emergence of new 'central places', whose regional dominance was often confirmed in the form of city rights. For two reasons, these new 'central places' were important for carriers coming from local production communities. First, because their development brought about population growth, which led to growth in trade and a rising demand for transport services. Second, because thanks to their geographical and economical-political position merchants in regional 'central places' had access to other regions and could establish an interregional goods trade. They mostly used carriers from their own communities to execute this trade. But carriers from surrounding rural communities could, geographical and technological circumstances allowing (i.e. if they had ships to do so), participate in the interregional goods trade. Rural communities could develop into full-fledged services suppliers; their carriers no longer needed to be merchants as well.[24]

The result of these emerging new regional 'central places' (such as Edam) and the opening up of their surrounding rural coastal communities is twofold: (1) the 'old' cities were no longer the only cities housing maritime transport communities and (2) some transport communities grown out of original *occupational communities* received the title 'maritime'. Their action radius expanded so strongly that they were no longer only active in their own delta, but also outside, in coastal shipping and sometimes even in maritime shipping, although their share in the latter remained limited.

As before, the available source material for the fifteenth century is fragmented and not quantifiable. It is, however, possible to distinguish a number of destination regions. Destinations in the North Sea to the west (England and Scotland) and to the east (Kleine Oost), along the Atlantic Coast and to the

---

24 Ketner, *Handel en Scheepvaart*, 97–100.

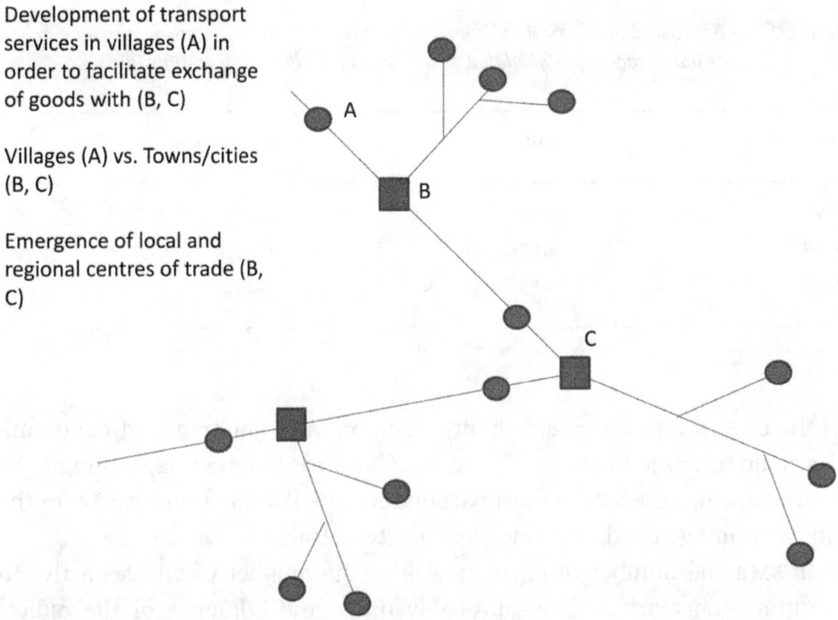

FIGURE 2 *Schematic overview of the development of local transport services and the development of local and regional trade centres in cities.*

Baltic are well documented.[25] With the help of LC a distinction can be made in the share of the different destination regions in the maritime transport services of the Dutch Deltas, though the sources force us to limit this distinction to a comparison between the North Sea and Baltic. Although the available data is limited, LC suggests that the maritime transport sector in SMR mainly concentrated on destinations along the Atlantic Coast and England (LC_Baltic = 0.15 and LC_North Sea – west = 1.53, see appendix 1.2. for details of individual places), while the relative importance of the Baltic for the maritime transport sector was largest in ZDZ (LC_Baltic = 2.13 and LC_North Sea – west = 0.29, see appendix 1.2. for details on individual places). In the case of LDE there are no data available to calculate the location coefficient. These results confirm the traditional split between the western focus of the southern Netherlands and the eastern focus of the northern Netherlands (see Table 2).

25 E.g. Asaert, *Documenten*; Poelman, *Bronnen*; Smit, *Bronnen*; Sneller, *Bronnen*; Studer, *The port books*.

TABLE 2 *Average LC values for the shipping patterns of the maritime transport sector per estuary, 1400–1500. Calculated on the basis of detailed data in Appendix 1.2*

Delta	Baltic	N.Sea – west
Outside	0,00	1,63
SMR	0,15	1,53
ZDZ	2,13	0,29

In the case of LDE no specific information on maritime transport communities could be found in published sources. The areas around LDE, including the Lauwersmeer, were for the most part undeveloped at that time, except for the city of Groningen and, to a much lesser extent, Dokkum.

In SMR the number of cities as well as the number of villages active in maritime transport declined, probably under the influence of the radical physical-geographical changes in the first half of the fifteenth century in the Scheldt estuary. A large number of cities, including Axel, Reimerswaal, Sluis, Gent, Steenbergen and Tholen, lost their prime locations for maritime shipping and no longer appear in the excerpted transport registers; their decline was final. The villages (and 'smalsteden') of Kortgene, Lapschuere, Maarland, Monnikereede, Nieuwvliet, Slepedamme and Sint-Anna-ter-Muiden also disappeared. Maritime shipping in the remaining villages seems to decline in the fifteenth century. Especially the decline of the villages of Nieuwehaven (located at the Bay of the Honte), Biervliet and Renesse is striking. Judging by their activities in the sixteenth century, the decline of the villages of Westkapelle, Katwijk, Noordwijk, Zoutelande and Zwartewaal seems to be due to limitations in the available source material. In the fifteenth century there seems to have been a growing concentration of maritime transport services in a limited number of cities in SMR, most importantly Antwerpen, Veere, Schiedam, Gouda, Zierikzee, Brouwershaven and Middelburg. Also in the fifteenth century a first 'vlek' emerged in SMR: Westenschouwen, located in the utmost west of the island Schouwen. Westenschouwen was "[...] early on the most famous and flourishing 'vlek' of Schouwen, of which the port did not only provide a good opportunity for inland shipping, but also for maritime shipping, so that plenty of merchant ships to the east and west, and especially a lot of herring fishers, set sail from there [...]".[26]

26 Van der Aa, *Aardrijkskundig woordenboek*, XII, 289–290.

New villages emerged mostly in ZDZ, while existing cities expanded and increased their maritime transport activities. The villages in ZDZ were mainly located in the regions of Waterland (e.g. Broek, Landsmeer, Ransdorp, Zuiderwoude and Zunderdorp), Dregterland (e.g. Grootebroek and Schellinkhout) and Kennemerland (e.g. Wormer). The Wadden Islands, Texel and Terschelling, also started to provide maritime transport services. The location of these maritime transport communities in the direct vicinity of cities, such as Enkhuizen (Dregterland) and Amsterdam, Edam, Hoorn and Monnikendam (Waterland), which were developing slowly into more than regional trade centres, indicate that access to trade centres and their respective trade networks was essential for the economic development of these villages. Since the amount of maritime transport services also grew in the cities itself, it seems that the surrounding rural areas acted as some sort of additional supply area for maritime transport services. The rural areas obtained their new economic function as an extension of those already in existence, which were primarily concerned with delivering local produce to regional markets. In first instance delivering maritime transport services must have been complementary to other activities in the labour cycle for rural coastal communities such as Broek in Waterland, Ransdorp or Wormer.

The economic geography of the Low Countries in the fifteenth century facilitated the process of commercialisation of transport: there were grain surpluses that could be imported and there was a demand for fish and cattle produce that could be exported.[27] A growing demand for transport services was the logical result of the expansion of cattle and dairy production in Holland north of the IJ, in Friesland and in Groningen, but also of the developing, export-orientated breweries in Delft and Gouda.[28] Different occupational groups answered to the growing demand for transport services. The process of division of labour, that is known to have started in the late Middle Ages in agriculture, was in fact a general phenomenon that appeared in different occupational groups such as the agrarian sector, fishery, mining and shipping.

For the further development of the Dutch maritime transport sector, it is important, that this new situation, where maritime transport services could be delivered in a growing number of places, created a broad basis for operational specialisation in transportation on specific routes and of specific goods. It is

27 Van Zanden, *Arbeid*, 43. In the fifteenth century, grain was imported to Holland from the Islands of Zeeland, the lands of Gelder, Gulik and Kleve, England, the Baltic and the Somme region. See: Sneller, "De Hollandsche korenhandel," 161–178; Van Tielhof, *De Hollandse Graanhandel*; Gras, *The evolution*.

28 Boekel, *De zuivelexport*; Sneller, "De Hollandsche korenhandel," 175.

likely that merchant-shipmasters from the ports located in the Dutch Deltas already specialised in shipping on certain routes during the fifteenth century, although there is no consensus on this subject in the literature. The romantic image of the shipmaster as an adventurer is still popular. Asaert, however, correctly emphasises the importance of competence, workmanship and the oral passing of knowledge in the activities of Antwerpen shipmasters in the fifteenth century, and as such he draws a picture of the shipmaster as an entrepreneur operating within the boundaries of his knowledge and ability and not as an adventurer bound for an undetermined destination over the sea.[29]

The ship movements of SMR, ZDZ and LDE functioned within two large axes of goods transport. A first axis connected the north of Germany via the Dutch Deltas with England and the French coast.[30] A second axis connected the Iberian peninsula and the French-Portuguese Atlantic coast via the Dutch Deltas with northern Europe (Norway and the Baltic Sea). Another axis could be identified, running from Friesland to the 'Kleine Oost' and a second one overland from the 'Kleine Oost' to Flanders and Brabant. Transport connections between the Dutch Deltas undoubtedly existed in the fifteenth century. The origin of products in the markets of the Ijssel and Brabant, for example, bore witness to a broad participation of cities north and south of the IJ in the interregional exchange of goods.[31]

While in the fifteenth century cattle, dairy and grain transports occurred between the northern Netherlands, on the one hand, and Flanders and Brabant, on the other hand, the Dutch Deltas, and the cities and villages belonging to them, mostly operated independently. The transportation of goods within the Dutch Deltas and to 'foreign' destinations was more intense than between the different deltas. The geographical space in which the shipmasters operated was relatively limited and only partially shared by the different deltas. The transport connections between north and south and west and east mainly ran via the route 'binnen-dunen' or overland, and only went by sea to a limited extent.[32] Not much is known about the intensity of such contacts between the Dutch Deltas in the fifteenth century; it can be assumed that the share of shipmasters domiciled in one delta in the shipping on routes between the other deltas (i.e. 'third' routes) was almost non-existent. While there were

29 Asaert, *De Antwerpse Scheepvaart*, 213.

30 For grain trade with the Somme region, the following places were particularly important: Rouen, Abbeville, St. Valéry and Amiens. See: Sneller, "De Hollandsche Korenhandel".

31 See also: Lesger, "Intraregional trade," 189–199; Smit, "Kamper Pondtolregisters".

32 Sneller, "Wijnvaart," 198; Smit, *Beneden-Maasgebied*, II, XVII; Ibelings, "De route "binnendunen," 221–254 and 323–339. See also: Unger, "Beer," 121.

contacts, there is no evidence of an integrated Dutch maritime transport system before 1500.

1500–1600

International, maritime trade developed in cities with a sufficient demographic base, but also in storage and transhipment centres of overseas products and in regional *gateways*, where the economy was based on the distribution of local and regional production. This development process went hand in hand with the increasing 'maritimization' of intra-European streams of goods, strengthening the locational advantage of some cities compared to others. The breakthrough of the 'maritimization' of intra-European streams of goods took place in the first half of the sixteenth century. The globalisation of the goods market in the sixteenth century has probably strengthened and accelerated this process of maritimization. International trade and maritime transport were closely connected to each other: growth in international trade equalled growth in maritime transport. This was apparent not only on an urban level, but also in the settlements that had developed into regional *gateways* (whether or not having acquired city rights), where the demand for foreign products originated in the regional economy.[33] The rise of international streams of goods not only added to the growth of the number of maritime transport services in cities and rural areas, but also caused a structural change in the division of maritime transport supply between cities and rural areas.

We have already seen that the spatial structure of the maritime transport sector between 1400 and 1500 had evolved from being mainly concentrated in 'old' and 'new' cities and focused on services for urban networks to a structure driven by economic growth in which maritime transport services by rural communities also had a place. The development of domestic and international shipping out of fisheries, cattle and peat trades played an important role in this. After 1500 no new cities developed (although there are some exceptions) until the beginning of the nineteenth century.[34] Instead, there was a striking intensity in the expansion or construction of ports between 1570 and 1650,[35] combining with the development of domestic and international systems of

33 For example, this was the case in Edam in the second half of the fifteenth century and in Middelburg during the same period, see: Boschma-Aarnoudse, *Tot verbeteringe*; Sneller, "Wijnvaart"; Unger, "Middelburg".

34 Rutte, "Stadswording," 154 note 21. Cox' research ends in 1484, when Purmerend received the so-called *keurrecht*. See: Cox, *Hebbende privilege van stede*.

35 't Hart, "De Nederlandse Republiek," 57.

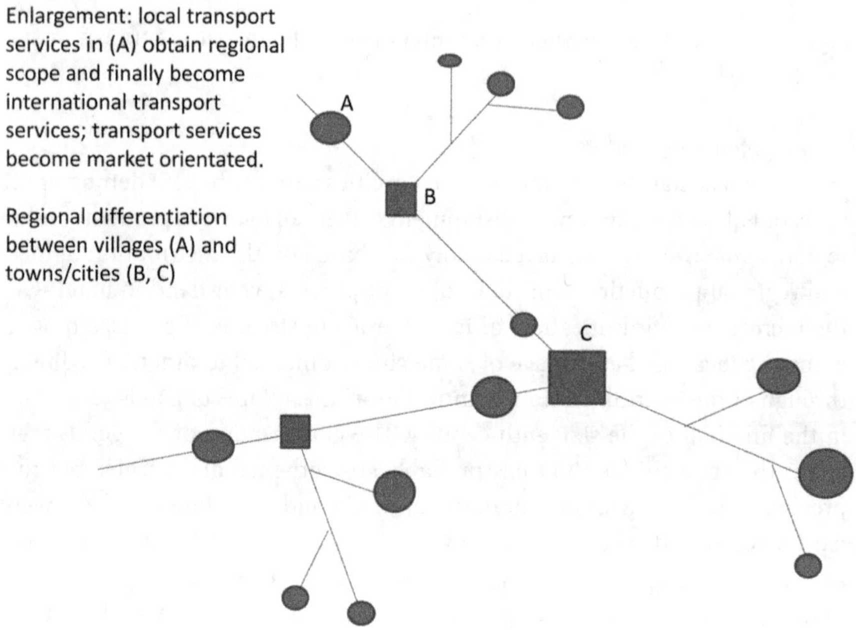

FIGURE 3 *Schematic overview of the further development of transport services in villages and cities and the effects of the emergence of market-orientated transport.*

barges, i.e. regular cargo and passenger transport services.[36] The period of urban development in the Low Countries in the late Middle Ages was followed by a period of infrastructural development.

Cities remained the catalysts for the infrastructural changes, and in the course of the sixteenth century they started to function more and more as *gateways* for information.[37] This facilitated sailing on 'third routes', which manifested itself in triangular transport patterns—and almost simultaneously—in transport patterns in which the domicile of the shipmaster was no longer part of the transport route (see Figure 3). While in theory all 'third route' shipping was part of a triangular pattern, it is useful to make the distinction. In a triangular pattern, a shipmaster leaves his domicile with (or without) cargo to a place functioning as port of departure from where he continues to his final destination. From this destination, he then returns to his domicile. In 'third route'-shipping the last step is different: the shipmaster does not return to his domicile, but goes back to the port of departure or to another destination, belonging to the urban network of the port of departure. This is

36 De Vries, *Barges and Capitalism*; Fuchs, *Beurt- en wagenveren*.
37 Lesger, *Handel in Amsterdam*, 209–249; Stabel, *Dwarfs among Giants*, 137–175.

TABLE 3 *Average LC values for the shipping patterns of the maritime transport sector per delta, 1500–1600. Calculation on the basis of Appendix 1.3*

Delta	Arctic	Atl. – N.	Atl. – S.	Atl. – W.	Baltic	Medit.	N.Sea – C.	N.Sea – E.	N.Sea – S.	N.Sea – W.
LDE	0,00	0,00	0,00	0,00	1,23	0,00	0,93	0,07	0,00	0,12
Outside	0,00	0,00	0,00	0,00	1,16	0,00	0,51	0,00	0,00	20,28
SMR	0,00	0,00	0,04	0,00	1,00	0,00	0,59	0,01	0,00	22,59
ZDZ	1,54	2,87	0,38	0,00	1,26	0,11	0,78	0,31	0,91	1,96

possible since every port developed more or less extensive *gateway* functions and thus played a part in the information provision of the shipmaster. The significance of complex triangular and 'third route'-patterns was still limited at the end of the sixteenth century and their development differed between the Dutch Deltas. The best-known transport patterns of the latter type are usually referred to in the literature with the term 'voorbijlandvaart', but this term only refers to a fraction of the total shipping on 'third routes', notably to those routes that went straight from the Baltic to southern Europe (or in the opposite direction) without calling at a Dutch port.

The so-called 'voorbijlandvaart' is especially apparent in the LC values for the Atlantic coast of France, Spain and Portugal (*Atl. – S*). Only the maritime transport sector in ZDZ has a limited, less than proportional share in this trade. The table above indicates clearly that the Baltic (*Baltic*) and the Dutch Republic (*N.Sea – C.*) belonged to the most common shipping areas of the Dutch maritime transport sector in the sixteenth century. The respective shares of the Dutch Deltas on these areas were more or less proportional; the maritime transport communities of all three Dutch Deltas participated in shipping to these areas and none of the Dutch Deltas distinguished itself with an extraordinary concentration of activities on these areas. The involvement of maritime transport communities located outside the Dutch Deltas in the shipping to and from the Dutch Republic (*N.Sea – C.*) was rather limited. Their share in the shipping to and from England and Scotland (*N.Sea – W.*) on the other hand was very large. This was also true for the maritime transport sector in SMR; with regard to the previous period this does not change. In the course of the sixteenth century new trade areas were added to the work environment of the maritime transport sector (*Arctic* and *Medit.*), while the first quantitative data became apparent for shipping to and from Norway (*Atl. – N.*). It is remarkable that the northern (*Arctic, Atl. – N.*) and southern shipping areas

(*Medit, Atl. – S.*) appeared to be serviced completely by shipmasters from ZDZ; they had a dominant share in shipping to and from Archangel and Norway, while in relation to the other destinations shipping to and from the south was underrepresented in their activities. The maritime transport sector of LDE was focused on the Baltic and the Dutch Republic, whereas in SMR it was mainly focused on England and Scotland. The maritime transport sector of ZDZ appeared to have gone through a dual development with the Baltic, the Dutch Republic and England as 'nearby' destinations and the Arctic, Norway, the Atlantic coast and the Mediterranean Sea as long distance destinations (see Table 3).

In LDE the number of maritime transport communities remained limited to the cities of Dokkum and Groningen. The 'vlek' of Appingedam and Ameland started to rise simultaneously with the other Wadden Islands in the second half of the sixteenth century and therefore—taking into consideration the difficulty in attributing this Wadden Island to either ZDZ or LDE—was finally added to ZDZ. Shipmasters from Delfzijl and Farmsum were also sporadically present in LDE. The extent of maritime transport activities was generally limited and appeared to have mirrored the status of the regional economic development of LDE: there was only a restricted need for maritime transport services and LDE was only limitedly integrated with the outside world in the sixteenth century. This explains why delivering transport services to international trade centres in neighbouring regions was not yet a possibility.

Just as in the previous period, maritime transport in SMR in the sixteenth century was mainly an urban affair. Only nine villages in SMR had their own maritime transport communities. While they had their golden age in the sixteenth century, their share in the maritime transport sector of SMR remained limited. The maritime transport communities of Biervliet, Domburg, Katwijk, Noordwijk, Scheveningen, Ter Heide, Westkapelle, Zoutelande and Zwartewaal seem to disappear (mostly) for good after 1600, at least insofar as the available sources reveal. Earlier, the number of villages with maritime transport communities in SMR had already decreased from twenty-four in the fourteenth century to twelve in the fifteenth century. The cities in SMR where the maritime transport sector developed most strongly in the course of the sixteenth century were Rotterdam, Schiedam, Delftshaven, Vlissingen, Dordrecht, Veere, Gouda, Middelburg, Zierikzee, Arnemuiden and the 'smalstad' Brouwershaven. Some cities outside the Dutch Deltas, such as Duinkerken, Nieuwpoort and Oostende, could temporarily profit from the economic upswing during the second half of the sixteenth century.

The golden age of maritime transport communities in cities of SMR can be attributed partly to developments in the energy supply. As ships transporting peat in Holland south of the IJ got bigger, shipmasters turned more and more often into salesmen who had their own ships and traded in peat and timber.[38] In the same period in Delfland and Schieland peat extraction became more and more orientated to export to Zeeland, Brabant and Flanders.[39] Rotterdam mostly exported to Zeeland; Gouda to Brabant and Flanders.[40] Gouda and Haarlem developed in this period into a regional peat distribution centre. Where the peat they exported came from cannot be determined exactly, but it is most likely that it came from north-west Overijssel and to a lesser extent from Friesland.[41]

In Dordrecht shipmasters from Gouda and Haarlem took over the peat supply in 1530–1540, because the traditional peat areas in South Holland were already exhausted by that time.[42] In the same period an important expansion of the international carrying trade took place in Gouda and Haarlem. In this case the registrations in the Sound of shipmasters from Gouda and Haarlem are illustrative.[43] Since the Sound toll registers for this period do not specify destinations westwards, it cannot be determined where the return cargo transported by Gouda and Haarlem shipmasters was going. One can assume that they mainly focused on transport for their own city. There was also a modest trade on 'third routes' in Haarlem, as evidenced in the convoy contributions from Amsterdam in 1580 and 1584.[44]

---

38 Cornelisse, *Energiemarkten*, 156–157. Note that this development seems to contradict Posthumus' insights into changes in the ownership structure of ships. Cornelisse, however, speaks of relatively small but seaworthy ships that were not taken into account by Posthumus.

39 Diepeveen, *De Vervening*, 137.

40 Diepeveen, *De Vervening*, 129–143. Diepeveen mentions the origin of shipmasters from Zeeland, Flanders and Brabant in the register of Gouda in 1568/69. Shipmasters from Flanders came from Bergen op Zoom, Antwerpen, Breda, Leur, Lier, Mechelen, Brussel and Roosendaal. From Brabant: Moerbeke, Gent, Biervliet, Brugge and Sluis; from Zeeland: Middelburg, Arnemuiden, Goes, Vlissingen, Ierseke, St. Maartensdijk, Zoutelande and Bommenede.

41 Gerding, *Vier Eeuwen Turfwinning*, chapter 5 and chapter 6. See on page 170: "In Noord-West Overijssel was de turfwinning midden zestiende eeuw van die orde dat er in Blokzijl en Zwartsluis vanaf 1560 tol werd geheven op alle turf die daar passeerde. Het is aannemelijk dat een groot deel van deze turf over de Zuiderzee werd geëxporteerd".

42 Cornelisse, *Energiemarkten*, 153.

43 Quite differently, Baasch notes that shipmasters from Haarlem, Gouda and other towns no longer (?) set sail for Hamburg as early as in the 1620s. See: Baasch, "Hamburgs Seeschifffahrt," 295–420.

44 See Lesger, *Dataset van de Convooigeldregisters*.

Cities and villages in SMR had a rather strong focus on transport activities to the West of the North Sea. The Baltic was to a lesser extent part of the shipping patterns of the transport communities in these cities. Furthermore, it is remarkable that shipmasters from Bergen op Zoom and Brugge participated in the goods exchange between the North Sea, Baltic and the Low Countries. The maritime transport sector in Antwerpen did not seem to develop further in the course of the sixteenth century.[45] The Antwerp trade network probably relied more and more on the rapidly expanding transport services supply in places closer to the sea in SMR. The rise of the 'vlek' Roosendaal in the sixteenth century deserves to be mentioned in this context. Roosendaal owed its economic upswing to its favourable geographical location with regard to Antwerpen. By building a port, and supported by the 1451 freedom to 'execute all sorts of trade', the village could develop into a 'vlek' where the inhabitants gained their income mainly from agriculture, trade and shipping.[46] Due to problems with the water supply (the lack of water), the rise of Roosendaal came to an early end.[47]

In ZDZ an explosive growth of the number of maritime transport communities in coastal and rural populations could be witnessed on the east coast of the Zuiderzee around 1540 and in West Friesland and Friesland around 1550. The number of maritime transport communities in villages around ZDZ rose from eleven in the fifteenth century to seventy-seven in the sixteenth century, while the number of cities around ZDZ rose from fifteen to thirty-five. The distribution of maritime transport communities expanded to such extent that it becomes difficult to distinguish clear areas of concentration. In Friesland, which just like Holland, had made the transition from farming to cattle and dairy production, the demand for transport services increased rapidly, making it an *incentive* for various fishing communities to commit to the carrying trade. The fast rise of shipmasters from the Wadden Islands around 1570 is documented in historiography. They were originally fishermen, but jumped to the chance and with their relatively small ships rapidly laid claim to shipping on the Baltic.[48]

45 Of course, the position of Antwerpen as a trading port was severely damaged by the closing of the River Scheldt. The repercussions of this event on the Antwerp maritime transport market, however, cannot easily be established, since—to my knowledge—no suitable shipping registers are available that would allow assessing the survival of this sector in the late sixteenth century.

46 Van der Aa, *Aardrijkskundig woordenboek*, IX, 633–638.

47 Van der Aa, *Aardrijkskundig woordenboek*, IX, 638.

48 Van Royen, *Zeevarenden*, 95–98; Van Zanden, *Arbeid*, 45–46.

In Overijssel, where the physical geography was less favourable to making a transition from farming to cattle breeding, the demand for transport services had a local/regional character and the specialisation process took off in the peat trade (especially in Blokzijl, Kuinre and Zwartsluis). Next to the striking growth in the number of maritime transport communities in villages and cities around ZDZ, the early rise of certain 'vlekken' is remarkable. In the case of the 'vlekken' of Schagen, Lemmer and Makkum, the rise is still very premature; Molkwerum, however, owed its economic development in the sixteenth century almost entirely to the supply of maritime transport services. "During the golden age of maritime shipping one could occasionally go from home to home without encountering any men, except for old ones," noted Van der Aa.[49]

After 1600 some of these communities would slowly grow into fully-fledged maritime transport communities that gradually moved away from the distribution of local produce and for whom transport services supply developed its own dynamic. Other maritime transport communities on the other hand would disappear completely in the early seventeenth century, e.g. in the cities of Deventer, Harderwijk, Kampen, Haarlem, Gouda and Alkmaar, on the island of Wieringen and in the villages of Ransdorp, Uitgeest, Uitdam, Huizum, Heem, Grootebroek and Graft.

1600–1700

In the fifteenth as well as in the sixteenth centuries, external *shocks* were often a *trigger* for the expansion of estuarine transport services into maritime transport services. Such *shocks* could have a political character, such as the Dutch Revolt, but also a climatological background, such as crop failures.[50] The geographical location of external *shocks* was less important; proximity to the delta was not necessary. While external *shocks* could lead to expansion of the action radius of transport communities, other mechanisms were necessary to give such expansion a permanent character. Without continuous interaction with the *gateways* of the delta to which a certain maritime transport community belonged, a lasting expansion of estuarine transport services into maritime transport services was not possible. Also, the expansion could not become permanent if the relationship with the *gateways* of the delta disappeared. Maritime transport communities in cities and villages were largely dependent on the efficiency of the information centres of the deltas to which they belonged.

---

49 Van der Aa, *Aardrijkskundig woordenboek*, VII, 1033.

50 A relevant study that examines the influence of political unrest and war on international trade and transport is Snapper, *Oorlogsinvloeden*.

This changed the role of cities. Thanks to their central position in trade networks and transport systems, it was not only information and knowledge about markets and prices (of goods and services) but also about the supply and quality of transport services that accumulated in cities. The informative function of cities influenced the further evolution of the relationship between cities and rural areas and of the Dutch maritime transport sector. The general growth of Dutch maritime transport services was crucial for the development of *gateways* as centres of trade and transport organisation. This development led to transport systems with complex structures in which the transport communities of cities and villages were active *full-time* in delivering transport services.

These communities no longer operated only between their 'own' central *gateway* and other places, but also between two other cities belonging to the network of their main information centre. "Despite the urban domination in the political structure there were no big contradictions between cities and rural areas",[51] but more of a city-rural continuum. This was the case in the coastal areas, where cities "were part of a dense network where every domination over the rural areas was disputed by the neighbouring centres".[52] Rural and coastal communities became part of larger societal structures via their maritime transport communities.

Around 1660 a new phase dawned in the spatial structure of the Dutch maritime transport sector. The period of infrastructural development had reached its peak around 1650. Amsterdam and Rotterdam started to dominate the urban relations in the Dutch Republic and "many functions previously executed by the whole urban area" now shifted to these centres.[53] The continuous development of Amsterdam and later Rotterdam into centres of European freight distribution was demonstrated by continuous population growth and in the growing accessibility of these cities.[54]

Because of the dominance of Amsterdam and Rotterdam, rural communities and other cities in the same delta gradually lost their autonomy in the organisation of international trade.[55] This led to the final division between transport systems and trade networks. Most cities lost their independence as information centres for trade and transport. Their place was taken by one central *gateway*, functioning as a centre of information about maritime labour

---

51 't Hart, "De Nederlandse Republiek," 59.

52 't Hart, "De Nederlandse Republiek," 60.

53 't Hart, "De Nederlandse Republiek," 71.

54 't Hart, "De Nederlandse Republiek," 57.

55 This shift is part of a broader process situated in the first half of the eighteenth century by 't Hart, of which the contours already became visible in the second half of the seventeenth century. See: 't Hart, "De Nederlandse Republiek," 71.

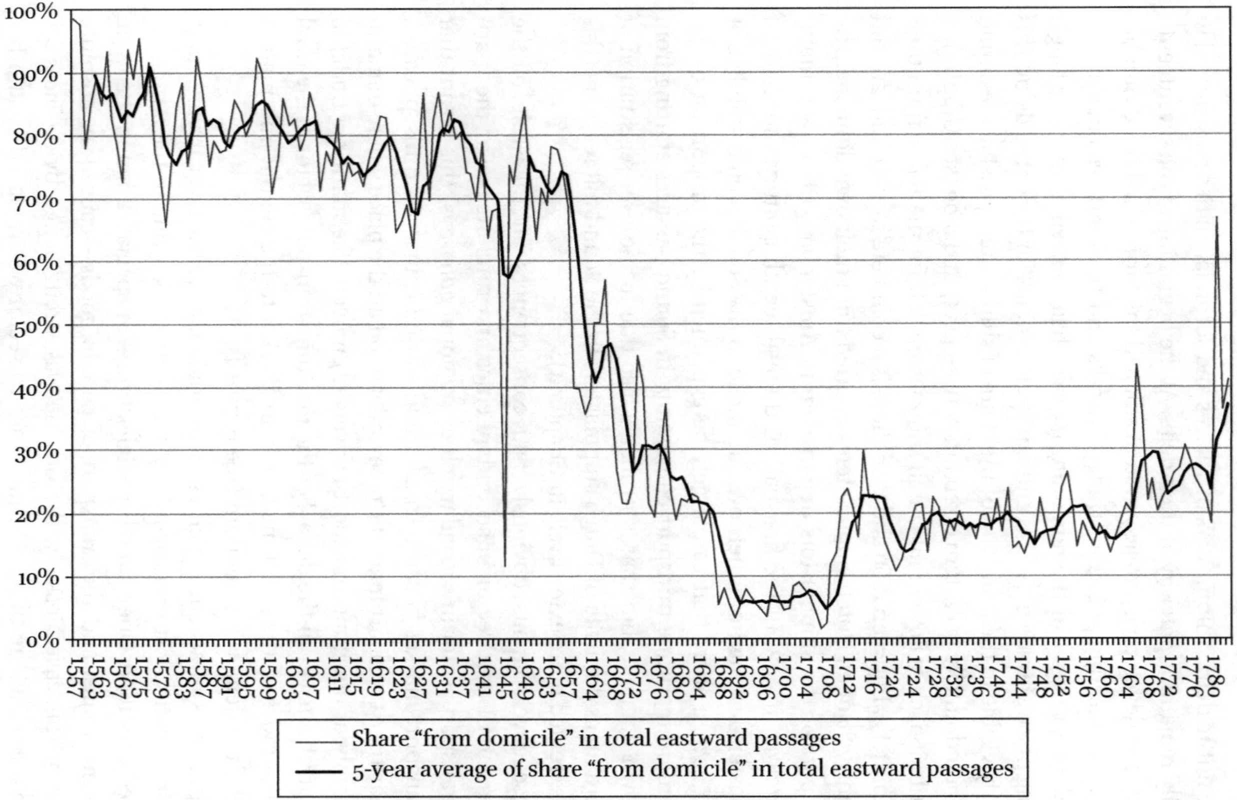

GRAPH 1 *Share of the number of sound passages by Dutch shipmasters in eastern direction, with departure from their own domicile versus the total number of sound passages in eastern direction.*

SOURCE: BANG & KORST, TABELLER AND STRO.

for the entire delta. Shipmasters of villages and secondary cities now turned to the only remaining *gateway* of their delta. At the beginning of the seventeenth century, the developing *gateway* system still had the form of a limited number of regional 'meeting points' of trade networks and transport systems, but by around 1700, the role of the remaining *gateway* had evolved to that of a single meeting place of trade networks and transport systems for the entire delta. This is clearly visible in the evolution of the share of shipmasters passing the Sound from their domicile in eastern direction between 1557 and 1800 (see Graph 1).

Until 1660, circa 80% of all Dutch shipmasters left for the Baltic from their own domicile. Between 1660 and 1700 there is a complete turnaround: around 1700, only 5% of all Dutch shipmasters departed from their own domicile. The larger part of the shipmasters departed from Amsterdam, the only remaining *gateway* in ZDZ. In the first half of the eighteenth century, the share of shipmasters leaving from their own domicile to the Sound would stabilize at 15% to 20%, a share that can largely be explained by the proportion of the Amsterdam maritime transport community in Sound passages starting from their own domicile, but that is probably also due to the over-registration of Amsterdam as domicile of Dutch shipmasters in the Sound toll registers (this topic is treated more extensively in Chapter 4).

These developments obviously had consequences for the scale of the maritime transport sector in secondary cities and rural areas. For the maritime transport sector in secondary cities (regional *gateways*) this meant that the transport market became smaller and in due course even disappeared. The maritime transport market (to be understood as: the place where demand and supply of maritime transport converged) moved, like maritime trade, to the central *gateway* of the delta. For the maritime transport sector in the rural areas the move of the maritime transport market to the central information centre of the delta was important (see Figure 4).

Shipmasters of secondary cities, who originally served the carrying trade of their own city, lost their own local transport market due to the structural changes after 1660. Operational specialisation—or specialisation in the trade on certain routes and the transport of certain goods—caused an absolute decline in this case. Possibly, the decline was strengthened by *business life cycle*-patterns, where the maritime transport sector was replaced by another business activity in bad times and thus could disappear within one generation. Boon and Willemsen consider such *life cycle*-patterns to be essential and ascribe them an important role in the decline of the West Frisian maritime cargo trade at the end of the seventeenth century.[56] This was less of a problem for the central *gateways* as they had a much more diversified services supply.

56 See: Boon, *Bouwers van de Zee*; Willemsen, *Enkhuizen*.

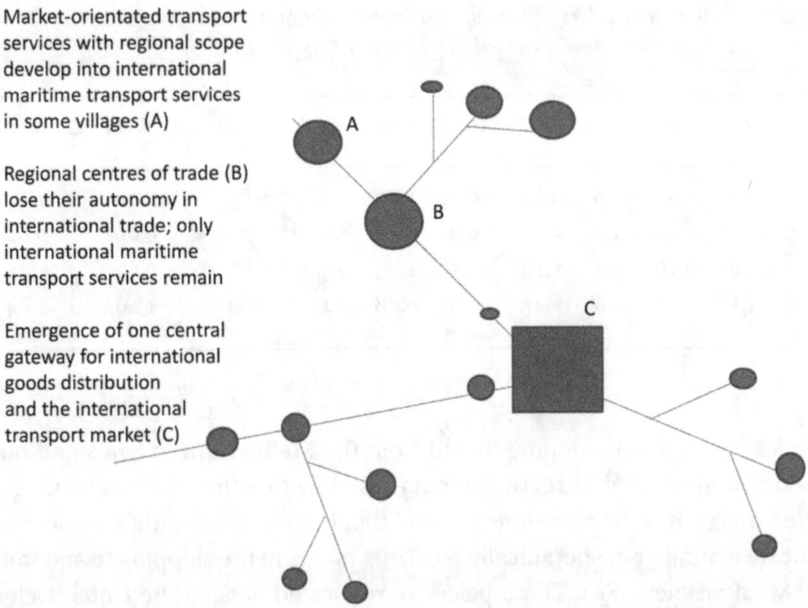

FIGURE 4 *Schematic overview of the location of maritime transport services in a delta with dominant gateway.*

Similarly, for shipmasters of villages, the change was not necessarily dramatic. The transport market moved, but their activities stayed the same, namely: supplying services to 'third parties' who (mostly) did not come from their own domicile. In this case, the operational specialisation could have a positive effect, leading to maritime transport communities being able to build a strong position on the international cargo trade on certain routes. As was the case with secondary cities, operational specialisation could also have a negative effect when changes in the market, caused by external *shocks*, quickly led to the loss of their market position.[57] The needs of the *gateway*, and the fulfilment of these needs, which was bounded by economic-political and physical-geographical circumstances, determined the routes and goods of the maritime transport communities serving this *gateway*.[58]

There are no surprises in the shipping patterns by delta (see Table 4). The pattern already visible in the earliest times continued—with a proportional participation in the carrying trade between the Baltic (*Baltic*) and the Dutch Republic (*N.Sea – C.*). The less than proportional activity in these areas by shipmasters from transport communities outside the Dutch Deltas (*outside*) is also

57 For examples, see: Scheltjens, "The Changing geography," 3–47.
58 See: Scheltjens, "The influence," 115–149.

TABLE 4 *Average LC values for the shipping patterns of the maritime transport sector per delta, 1600–1700. Calculated on the basis of Appendix 1.5*

Delta	Arctic	Atl.	Atl. N.	Atl. S.	Atl. W.	Baltic	Medit.	N.Sea – C.	N.Sea – E.	N.Sea – W.
LDE	0,03	1,31	0,16	0,42	0,00	1,02	0,00	0,83	7,04	1,15
Outside	20,71	0,00	0,00	0,66	0,00	0,52	61,64	0,53	0,00	0,00
SMR	0,08	0,16	5,23	1,15	0,00	1,03	15,61	0,87	0,52	1,91
ZDZ	4,18	0,30	0,46	1,45	10,36	0,88	12,58	0,82	1,26	2,64

visible. The data for shipping to and from the Mediterranean Sea stand out, but the total number of registered data is so low that they give a slightly distorted image. It is the case though—and that is remarkable—that the sources indicate a number of sporadically recurring places in the shipping to and from the Mediterranean Sea. These places were located outside the Dutch Deltas (*outside*), in SMR or in ZDZ. The carrying trade of LDE did not play a part in the Mediterranean.

It is worth the effort to look further into the maritime transport communities active in the shipping to and from the Mediterranean in the seventeenth century. Remarkably, a more than proportional activity of shipping to and from the Mediterranean was in the hands of maritime transport communities in cities of SMR and in the villages of ZDZ. The number of cities in ZDZ was rather limited, although Amsterdam and the most important cities in Waterland and West Friesland were obviously present. When we limit ourselves to the cities and villages in ZDZ it is remarkable that they did not exhibit a uniform pattern. A number of places had a large share in the shipping to and from the Mediterranean (e.g. Monnikendam, Oosthuizen, Krommeniedijk, Graft, Medemblik—details in appendix 1.5.); others did not have a share at all (e.g. Harlingen, Vlieland, Terschelling—details in appendix 1.5.).

Moreover, in a number of places a more than proportional participation in shipping to and from the Mediterranean went hand in hand with a similar situation with regard to shipping to and from Archangel. This was the case, for example, in the earlier mentioned places of Monnikendam, Oosthuizen and Krommeniedijk, but not for Graft. Finally, it is interesting that for the maritime transport communities of the 'central places' in ZDZ—Amsterdam, Hoorn and Enkhuizen—a difference in the proportional division of the transport activities per shipping area can also be observed. Amsterdam and Enkhuizen had a more than proportional share in the shipping to and from Archangel and the

Mediterranean; in the case of Hoorn this was only true for the Mediterranean. Such differences might be explained as the expression of some degree of specialisation on certain routes, which took shape at the local level.

For rural maritime transport communities such specialisation was a way to obtain and secure a place in the transport market. This finds expression in the extremely high or low LC values. For urban maritime transport communities the concentration on one or a few shipping areas was less urgent, at least for as long as their city managed to keep an autonomous position in the international trade networks. LC values are therefore less explicit. As far as Hoorn is concerned, one can assume that there was a shift of focus in the dominant shipping areas. In the late sixteenth century the focus was on the Arctic, the southern European Atlantic Coast and the Mediterranean, but in the seventeenth century the focus on the Arctic and the Mediterranean declined. It was replaced by a certain differentiation of maritime transport services on different areas along the North Sea and the Atlantic Coast, which was often less than proportional (details see appendices 1.4. and 1.5.). Maybe the maritime transport sector—in the light of the changing functionalities of cities as central places—was looking for a new élan?

In addition to the changes on long distances voyages, the strong concentration of maritime transport services on Bremen, Hamburg and the 'Kleine Oost' by shipmasters of LDE (*N.Sea – E.*), on England and Scotland by shipmasters of ZDZ (*N.Sea – W.*) and on the south and north Atlantic Coast by shipmasters of SMR (*Atl.S. and Atl.N.*) are worth mentioning. The data indicate that there were clear differences between the Dutch Deltas in destinations located closer to the Dutch Republic.

In LDE, Dokkum and Groningen remained the only cities with a supply of maritime transport services. In the villages there were no important changes. The only maritime transport community that really flourished in the seventeenth century was that of Ameland; the scale of its maritime transport community could easily match that of the other Wadden Islands. An explanation for the modest scale of the maritime transport sector in LDE in the seventeenth century can probably be found in the status of its regional economic development. Groningen remained the only central place in the whole region, and Dokkum lost a lot of its already diminished economic importance when the Admiralty left in around 1650. Because of this and because of the continuing silting of the Dokkumer Ee, Dokkum transformed into a land-orientated city, where maritime activities were no longer a necessity.[59]

---

59 Schroor, *Geschiedenis van de stad Dokkum*.

In SMR, where especially Zeeland suffered from difficult economic-political circumstances at the end of the sixteenth century, the number of villages with maritime transport communities drastically declined. Only Langendijk and Katwijk still had very modest maritime transport communities, but their activities were not more than sporadic. The cities in SMR provide a varied picture, formed by the concentration of maritime transport services in a limited number of cities, on the one hand, and the decline of maritime transport supply, on the other. The result is that Rotterdam and Schiedam clearly stood out in (registered) size of their maritime transport services compared to the other cities in SMR and they seemed to have developed further, while Delftshaven and Dordrecht experienced decline, probably at the cost of Maassluis, Middelburg, Vlissingen and Zierikzee. The maritime transport services disappeared almost completely from the urban economies of Brielle, Delft, Gouda, Gorinchem, Sluis and Veere. In absolute numbers this means that of the forty-one villages and cities in SMR with a maritime transport community in the sixteenth century, only seventeen remain in the seventeenth century.
The developments around ZDZ are similar, although because of the shift of maritime transport to new locations in the delta the absolute number of maritime transport communities remains almost the same (109 in 1500–1600 and 111 in 1600–1700). In the following description of the development of the maritime transport sector in ZDZ we will pay extra attention to the changing role of secondary cities, the geographical shift of the maritime transport sector and the 'de-urbanization' of the maritime transport sector. These three developments became visible in the second half of the seventeenth century.

In ZDZ, all cities, except for Amsterdam, are actually secondary cities. In total there are eighteen (versus thirty-two in 1500–1600): Alkmaar, Bolsward, Driel, Edam, Enkhuizen, Franeker, Harlingen, Hindeloopen, Hoorn, Leeuwarden, Medemblik, Monnikendam, Purmerend, Sneek, Stavoren, Workum, Ylst and Zaandam. The cities around ZDZ that no longer had a maritime transport community in the seventeenth century are: Amersfoort, Arnhem, Beverwijk, Deventer, Elburg, Haarlem, Harderwijk, Hasselt, Kampen, Muiden, Naarden, Nijmegen, Sloten, Weesp, Zutphen and Zwolle. Of the eighteen remaining cities only Driel and Ylst were new.

Graph 2 demonstrates that the percentage of shipmasters serving their own domicile on passages through the Sound (as port of departure eastwards and/or as port of destination westwards) transformed dramatically in the course of the seventeenth century. Until about 1660 shipmasters of secondary cities in ZDZ started between 20% and 40% of their Sound passages in their own domicile. After 1660 this share dropped to between 10% and 30% and after 1690 it was no more than 5%. In the same period shipmasters with Amsterdam

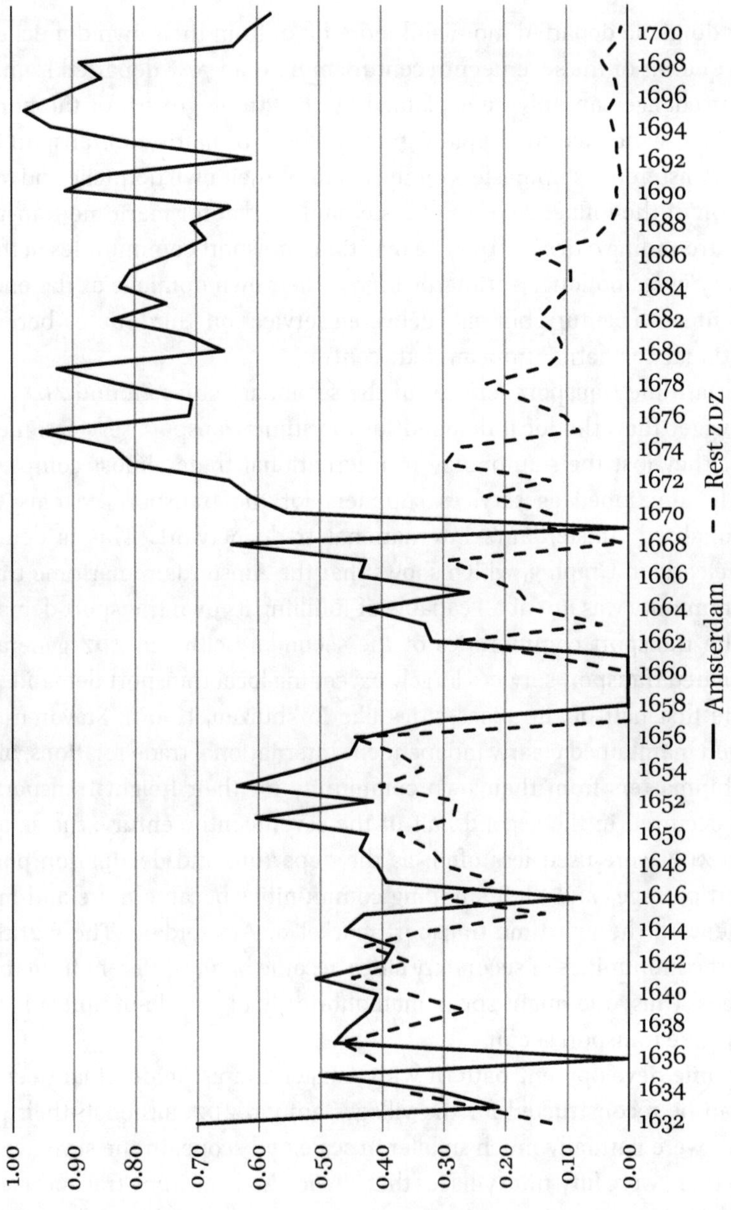

GRAPH 2 *Percentage of shipmasters who serve their own domicile on passages through the sound, 1632–1700.*
BASED ON STRO.

as their domicile departed more and more often from their own domicile; in the last quarter of the seventeenth century more than 70% departed from the city. This change can only be explained by the strong growth of the port of Amsterdam, which led to a rapidly growing focus of maritime transport services by Amsterdam shipmasters on the needs of their own domicile and to an expansion of the influence of the Amsterdam market for maritime transport to the surrounding cities in ZDZ. The maritime transport communities in these secondary cities no longer primarily served their own domicile at the end of the seventeenth century, but only delivered services on 'third routes' between international northern European trade centres.

The maritime transport services of the secondary cities around ZDZ grew much bigger than the local demand for maritime transport services, meaning that they lost their autonomy in international trade almost completely and only functioned as services suppliers for the transport systems that operationalised Amsterdam's international trade network. This is demonstrated clearly in Graph 3, which shows that the Amsterdam maritime transport community was not at all capable of fulfilling its own transport demand, while the transport communities of the secondary cities in ZDZ generated accumulated transport services largely exceeding local transport demand.

In the first half of the 1600s cities like Enkhuizen, Hoorn, Stavoren and Harlingen maintained clearly independent international trade relations, often using shipmasters from their own community for their freight transport. A change occurred in the second half of the seventeenth century: the smaller ports in ZDZ were used less often as the departure and destination points of Sound passages and their shipping communities became more and more dependent on the maritime transport market of Amsterdam. The maritime transport communities of secondary cities became barely different from those of villages. Thus, one might speak metaphorically of the 'de-urbanization' of the maritime transport sector.

The same development pattern with Amsterdam as predominant central place can be reconstructed for the villages around ZDZ, although their port activities were naturally much smaller in scale and scope. In the seventeenth century there were fifty-nine villages that already had maritime transport communities in the sixteenth century. Among these are the Wadden Islands Texel, Vlieland and Terschelling. Additionally, for almost two centuries a whole series of villages in the Noorderkwartier were part of the transport systems that Amsterdam counted on to develop its trade activities. In the seventeenth century twenty-six new maritime transport communities emerged, but their share in the maritime transport sector remained limited. Out of these twenty-six

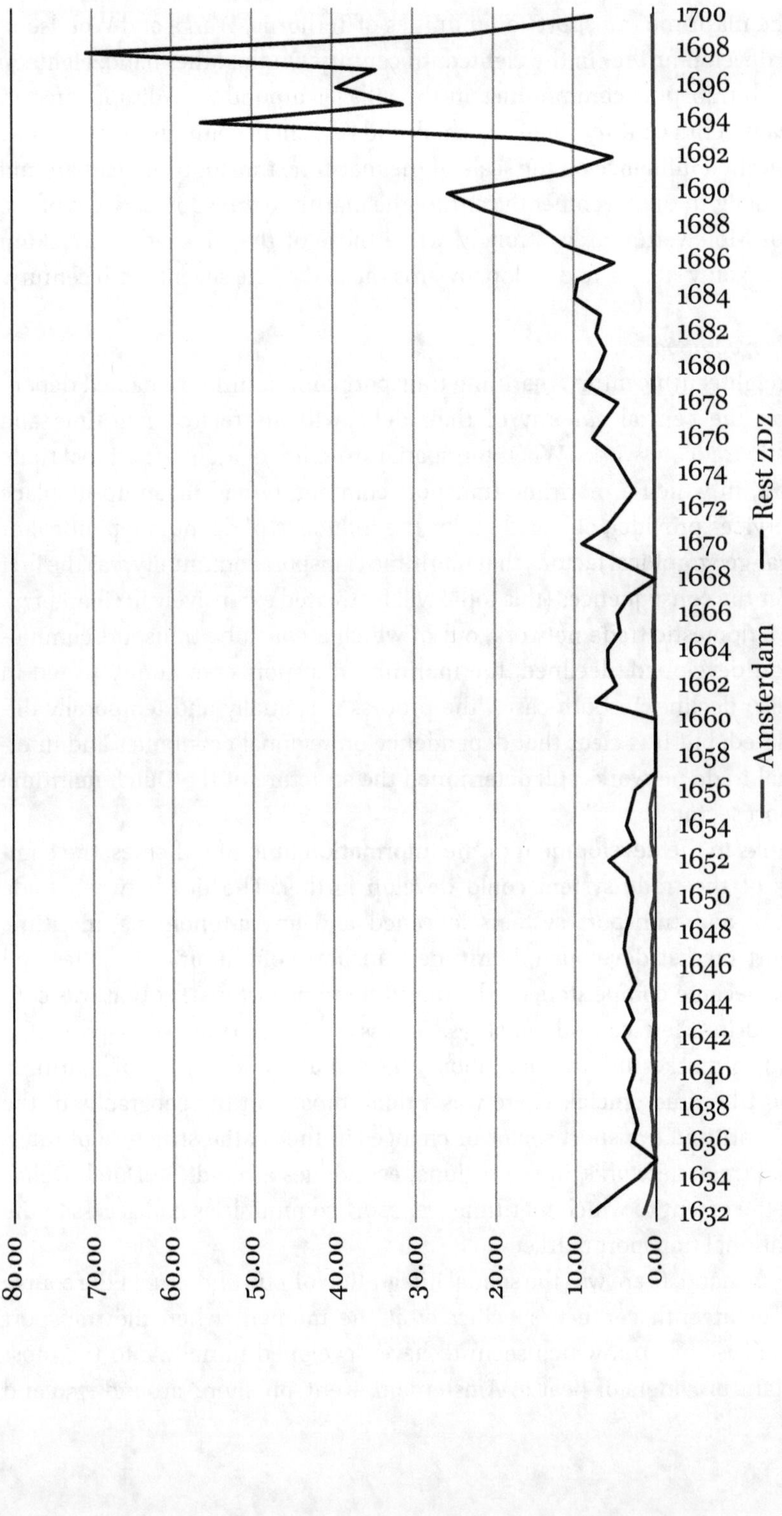

GRAPH 3 *Relation between the size of the maritime transport community and the transport demand of the domicile. Calculated on the basis of the number of sound passages, 1632–1700.*

SOURCE: STRO.

only the maritime transport communities of Terhorne, Warga and Woudsend would develop further in the eighteenth century. On the other hand, eighteen maritime transport communities in the villages around ZDZ disappeared in the seventeenth century. These are small and very small communities that only had a limited influence on the scale of the maritime transport services around ZDZ. Finally, in the seventeenth century the maritime transport services of the 'vlek' of Molkwerum grew strongly, while those of the 'vlekken' of Makkum and Joure only started to develop towards the end of the seventeenth century.

1700–1800

In the eighteenth century, maritime transport communities remained dependent on the central *gateway* of their delta with its related maritime and domestic trade networks. When the spatial structure of an international trade network, in which a maritime transport community had taken up its place as a services provider, changed under the influence of economic-political or physical-geographical factors, that maritime transport community was the first to suffer the consequences (this topic will be treated extensively in Chapter 3). When a domestic trade network, out of which a maritime transport community had developed, declined, the maritime transport community would in time also decline. In both cases, the process is spatially and temporally differentiated, but it is clear that dependency on regional economies and international trade networks still determined the structure of the Dutch maritime transport sector.

Thanks to the development of the information function of cities, the complexity of the trade system could develop further. The ties between trade networks and transport systems loosened and an 'autonomous' maritime transport market developed. Maritime transport communities in cities and villages became competitors on the maritime transport market that was concentrated in a few central *gateways*. Access to the maritime transport market and to market information about the demand and supply of maritime transport became crucial. There was a dual impact on the geography of the Dutch maritime transport sector of changes in the spatial structure of international trade networks, in the regional economies around the Dutch Deltas and in the extent to which maritime transport communities had access to the international transport market.

On one hand, there was the spatial integration of ZDZ and LDE in the course of the eighteenth century, specifically at the moment when the transport communities of LDE, which seem to have developed initially into the most important providers of peat to Amsterdam, went 'off shore' around 1750 and started to deliver services in the international carrying trade for the *gateway* Amsterdam. Similar spatial integration processes also took place around SMR. They led to a certain degree of horizontal integration between the Dutch Deltas (see Graph 4). The spatial structure of the maritime transport sector subsequently transformed into a fairly complex structure in which the *gateway* of ZDZ, Amsterdam, expanded its influence over the different Dutch Deltas. At the same time Rotterdam emerged as a second *gateway* in SMR, where maritime transport communities of ZDZ and LDE gained ground, in the first instance because they could answer the demand for peat (LDE) and dairy (ZDZ), later because the demand for services in the international carrying trade in Rotterdam grew strongly in the course of the eighteenth century. In this way, important advances were made towards the emergence of an integrated transport space in the Dutch Deltas.

On the other hand, the changes in the spatial structure of international trade networks caused the decline of a whole series of maritime transport communities, that had acquired a place in the maritime transport market in the course of the sixteenth and seventeenth centuries. This decline can best be explained as the negative consequence of path dependent processes, resulting in local shipping communities maintaining such a strong focus on certain routes that they were not able to react adequately when these routes declined because of external circumstances (see Chapter 3). There are other factors that also caused the decline of maritime transport communities though; migration to the large trade centres of Amsterdam and Rotterdam being only one.

The relative importance of different trade areas for each delta continued to change in the eighteenth century. Most remarkably, the Baltic (*Baltic*) and the Dutch Republic (*N.Sea – C.*) became less important for shipmasters from

TABLE 5 *Average LC values for the shipping patterns of the maritime transport sector per delta, 1700–1800. Calculated on the basis of Appendix 1.6*

Delta	Afr.	Arctic	Atl.	Atl. N.	Atl. S.	Atl. W.	Baltic	Indian	Medit	N.Sea – C.	N.Sea – E.	N.Sea – W.
LDE	0,00	0,18	0,00	0,34	0,52	0,00	0,97	0,00	0,15	0,98	2,75	0,91
Outside	0,00	15,29	0,00	0,00	0,54	0,00	0,36	0,00	21,96	0,34	8,25	0,00
SMR	0,36	1,80	1,67	0,49	0,65	0,20	0,70	2,87	5,96	0,63	7,47	0,56
ZDZ	163,42	4,91	0,10	2,52	0,97	0,02	0,97	0,00	1,16	0,91	1,06	5,73

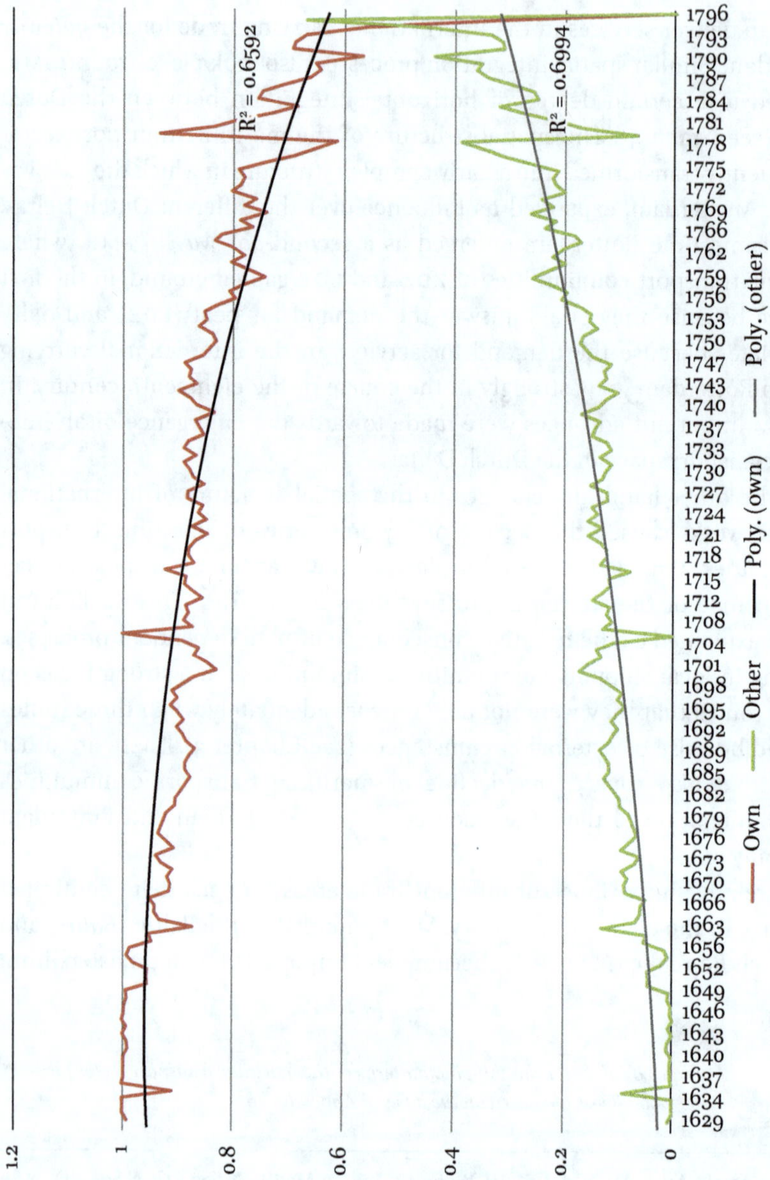

GRAPH 4 *Evolution of the relation between sound passages to and from the own delta (own) vs. sound passages to and from another Dutch Delta (other), 1632–1800.*

SOURCE: STRO.

SMR. That is also true for shipping to and from the northern and southern Atlantic Coast (*Atl.N. and Atl.S.*). Quite differently, the average LC values for the shipping patterns of the maritime transport sector in SMR suggest a more than proportional growth of the activities of maritime carriers from SMR on shipping to and from Greenland (*Atl.*), the Arctic (*Arctic*) and the Indian Ocean (*Indian*). The Mediterranean remained attractive (*Medit.*). Slightly in contrast with the rise of long distance destinations and maritime shipping is the growing importance of the 'Kleine Oost' as shipping area for maritime carriers of SMR (*N.Sea – E.*). This has probably to do with a differentiation within the maritime transport sector of SMR. Only the maritime transport communities of Gouda, De Lier, Rotterdam, Veere, Vlaardingen and Vlissingen have more than a proportional share of shipping to and from the 'Kleine Oost'. The maritime transport communities of Middelburg, Schiedam, Dordrecht and Zierikzee clearly demonstrated a different pattern (for details see appendix 1.6.). The decline in the importance of the Baltic for the maritime transport sector in SMR in fact stimulated the development of an integrated maritime transport space, even though this cannot be derived explicitly from the table above. When interpreted in the context of the previous graph, it may be assumed that freight transports from the Baltic by shipmasters from SMR declined in favour of growth of this transport by shipmasters from LDE.

The changes in the eighteenth century were most striking in LDE. In the cities of Dokkum and Groningen as well as in the villages around LDE maritime transport services increased strongly. In the 'vlek' of Appingedam and in the settlement of Delfzijl the supply of maritime transport services also increased considerably. Including Delfzijl, there were now twenty-nine villages in total around LDE with maritime transport communities. The community of Ameland remained the largest, although it declined strongly in the second half of the eighteenth century.[60] The decline of Ameland was compensated by the development of maritime transport communities around the Fen colonies (Nl. *Veenkoloniën*) around 1750, namely in Pekela and in Veendam and Sappemeer, although to a lesser extent.[61] The *Moorsiedlungen* on the other side of the Ems went through a similar development at the same time;[62] regional

60 See: Draper Van Duijl, "Shipmasters from the West Frisian Islands in the Sound Toll Registers, 1737–1800" (Soundtoll Working Paper 2012), in: http://www.soundtoll.nl/images/files/Draper%20and%20van%20Duijl.pdf (23 Mei 2013).

61 See also: Gaastra, "Zeilende kustvaart"; Kerkmeijer, "Groningen en de scheepvaart".

62 On this topic, see also: Scheltjens and Veluwenkamp, "Sound Toll Registers Online," 301–330; Kappelhoff, "Grundzüge der Wirtschaftsgeschichte," 319–475.

economic development around LDE was cross-border and had by no means a purely 'Dutch' character.

The development of maritime transport communities in LDE—in villages as well as in the cities of Dokkum and Groningen—is clearly related to the developments in ZDZ, where Amsterdam came to the fore as central *gateway* in the course of the seventeenth century and where the role of secondary cities changed dramatically under the influence of the economic expansion of Amsterdam, among other factors. It seems that the development of Dokkum and Groningen in the eighteenth century was similar to that of the secondary cities around ZDZ in the seventeenth century. None of these cities could rely on a broad autonomous international goods traffic for which it could use shipmasters from its own community; the maritime transport communities of Dokkum and Groningen served the international trade networks of Amsterdam and, to a lesser extent, Rotterdam. The same is true for the villages in LDE, which except for Ameland all gained importance in the eighteenth century. The Danish Sound toll registers indicate also in this case that Amsterdam and Rotterdam were the central access points to the international transport market. Shipmasters' domiciles only occasionally figured as port of departure or destination for passages through the Sound; the maritime transport services supply in the cities and villages around LDE largely exceeded local demand and were aimed at ZDZ, the Baltic and the east coast of the North Sea (Bremen, Hamburg).

There is no unambiguous explanation for the revitalisation of the maritime transport community of Dokkum in the course of the eighteenth century. Dokkum had transformed into a land-orientated city in the course of the seventeenth century, without much connection to the maritime world. Maybe this changed again under the influence of the reviving peat cultivation in the area around Drachten (Folgera-peat moors).[63] It is possible that, just like Haarlem, Gouda and Zwartsluis three centuries earlier, Dokkum developed a regional centre function in the peat distribution, which gave a new impulse to the transport industry of the city. Another possible explanation for the renewed interest in maritime transport could be the development of Leeuwarden, for which Dokkum acted as an outport. In this case the transport community of Dokkum could have had a development process similar to that of Arnemuiden in the sixteenth century: from storage and transhipment to regional distribution and maritime transport services. While there does not seem to be evidence

63 Gerding, *Vier eeuwen turfwinning*, 127.

supporting this *revival* of the maritime transport community of Dokkum, an explanation based on the regional economy seems most plausible.

In SMR the maritime transport sector went through a different development process, as far as it can be reconstructed on the basis of the available sources. Villages in SMR were only limitedly involved in maritime transport, although in comparison to the previous period there are more villages (twelve rather than four) supplying maritime transport services. The number of cities with their own maritime transport communities also rose slightly and, except for Brielle, Delftshaven, Schiedam and Vlissingen, all maritime transport communities grew in (registered) size. The largest concentrations of maritime carriers were located in Rotterdam, Dordrecht and Middelburg; smaller, upcoming maritime transport communities were located in the cities of Maassluis and Zierikzee. Outside the Dutch Deltas the maritime transport community of Oostende was characterized by a modest revival.

In ZDZ the polarisation of the maritime transport market in Amsterdam continued. Secondary cities had lost their autonomy in the international trade in the course of the seventeenth century and only functioned exceptionally as storage or transhipment points in international freight distribution. The maritime transport communities of these secondary cities were almost exclusively focused on Amsterdam. They were accompanied by a large number of villages and a few growing 'vlekken'; the composition of urban maritime transport services supply did not change in comparison to the previous period. In some cities maritime transport supply seems to have remained unchanged (e.g. Edam, Harlingen, Hindeloopen, Hoorn, Stavoren, Workum, Zaandam); in other cities there was a clear decline (Medemblik, Enkhuizen, Purmerend). A few smaller communities re-emerged (Bolsward, Leeuwarden, Sneek and Ylst). The most remarkable though is the steep rise in the size of the maritime transport community of Amsterdam in the eighteenth century. Between 1600–1700 it accounted for 8,201 registered ship movements in STRO; between 1700–1800 this number had increased to more than 83,000.

The phenomenon of ambiguity in the registration of shipmasters' domiciles in the Sound, where a non-Amsterdam shipmaster regularly received 'Amsterdam' as domicile (see more extensively the discussion of domicile ambiguity in Chapter 4), cannot entirely explain this spectacular growth. The growth of the maritime transport community of Amsterdam as well as the often recurring ambiguity in the domicile registrations perhaps point towards one and the same phenomenon: the monopolisation of the Dutch maritime transport market by Amsterdam. The composition of the cities in ZDZ did not change but the composition of villages did. Of the eighty-six villages in the period of 1600–1700, only fifty-nine remained in 1700–1800. Thirty-one new villages appeared. A large number of the fifty-nine known villages declined in the eighteenth century. Only the shipping communities of Woudsend, Warrega, Nieuwendam, Schellinkhout, Terhorne and Warder were characterized by strong growth in the eighteenth century. The 'new' shipping communities of Wartena, Oldeboorn and Heeg deserve to be mentioned; they were the only 'new' villages with a significant maritime transport community in the eighteenth century.

Conclusions

In this chapter I examined whether an integrated maritime transport space developed in the Low Countries during the early modern era. The research focused on three Dutch Deltas: SMR, ZDZ and LDE. During the earliest stages, the development of cities was the most important form of spatial development. Then, different aspects of spatial differentiation were discussed. Evidence was given for the fact that every delta had its own central places and that a clear distinction had to made according to the 'type' of transport community. It was concluded that a partly integrated Dutch maritime transport space existed at the end of the eighteenth century, shaped by two central *gateways* and with maritime transport communities spread over three different Dutch Deltas.

At this point, several important questions remain to be answered: what happened after 1800, when the Dutch Republic was placed under French rule? The Napoleonic Wars caused massive disturbances all over Europe and the continental blockade obstructed international trade for several years in a row during the first decade of the nineteenth century. And what about the industrial revolution that spread from England to the Continent, but did not gain ground in the Low Countries?[64] And what about Prussia and the Russian Empire, which both strived for dominance in Central Europe and which had already divided the Polish-Lithuanian Commonwealth in order to expand their economic power, especially with regard to the Baltic? How did this series of large external shocks affect the Dutch maritime transport sector?

In the next chapters, the survey of maritime transport communities is continued both thematically and in time. In Chapter 2, the functions of ports in the Dutch maritime transport sector are examined. In Chapter 3, an in-depth analysis of the production of maritime transport services by Dutch shipmasters

---

64 Mokyr, "The industrial revolution"; Pierenkemper, *Umstrittene Revolutionen*.

is pursued. In Chapter 4, populations of shipmasters are studied. All three chapters rely heavily on STRO. They provide more detailed insights about maritime transport communities in the Dutch Deltas and changes that occurred in their functions, operations and composition. Moreover, the detailed information in STRO, which was essential to carry out these analyses, is only available from 1670 until 1850. Thus, the following chapters are both a continuation and elaboration of the descriptive analysis of the geography of the Dutch maritime transport sector that was presented in this chapter.

CHAPTER 2

Functions

In this chapter, a method for the classification of ports participating in maritime transport in northern Europe is elaborated and applied to the foremost important primary data source regarding Dutch maritime transport communities in the early modern period: the Danish Sound toll registers. Accepting its inevitable limitations, we have chosen to elaborate our classification on the basis of function and frequency data for ports and homeports, although it may reasonably be argued that port frequency data is not a good indicator of the size of commodity flows since it ignores the possibility of changing ship size.[1] At the same time, however, Jackson warns against the use of distributed value as an indicator of the size of maritime transport, since it may be blurred by the transportation of exceptionally valuable goods.[2]

Method

In maritime history, several attempts have been made to get a grip on ports and their functions in trade networks and transport systems. A pioneering effort to classify ports according to their functions in large transport systems was executed by the late Frank Broeze, who made a useful, though predominantly descriptive distinction between gateways, entrepôts and hubs in the port system of the Asian Seas since 1750.[3] Jackson provides a rather rudimentary classification, making a distinction between major, secondary and insignificant (unimportant) ports.[4] Additionally, Jackson lists a number of elements by which ports can be described and classified, including: hinterland, purpose, people, shipping, financial capability, port-based industries, national support, port facilities and social life.[5] Furthermore, Jackson addresses two issues related to the classification of ports: should ports be assessed by the value of their trade? Or by the volume and commodity composition of trade?[6] While

1 Lesger, *Hoorn als stedelijk knooppunt*, 163.
2 Jackson, "Early-modern Seaport Studies," 25–27.
3 Broeze, "The Ports," 73–96.
4 Jackson, "The Significance of Unimportant Ports," 10–11.
5 Jackson, "Early Modern Seaport Studies," 23–24.
6 Jackson, "Early Modern Seaport Studies," 25–26.

 | DOI 10.1163/9789004273078_004

Jackson mentions many elements to theorize about ports and their great diversity, his viewpoints are never summarized in a clear and general classification. Polónia adopts a similar approach in 'Les petits ports dans le système portuaire Européen', arguing that every port has an internal dynamic that can be measured on the basis of a number of parameters.[7] These parameters are: tonnage, volume of cargoes, size of the merchant fleet, presence of maritime industries, labour force, capital forces and demographic growth. Unfortunately, Polónia refrains from explicating which of these indicators may be determinant for the definition of a port in a hierarchical ranking of ports.[8] Another classification of ports has been proposed by members of the French network of maritime historians (*Groupe d'Intérêt Scientifique d'Histoire Maritime*), who make a distinction between large (*grand*), secondary (*sécondaire*) and small (*petit*) ports.[9]

What unites all three proposed classifications is their attention towards 'small' ports, which are *not* seen as 'unimportant'. In this sense, these attempts to classify ports indicate a paradigmatic shift in contemporary (maritime) historical studies. However, while the research mentioned above most certainly contributes to the understanding of port systems in history, its potential outcome is limited to rather one-sided narration, in which quantitative data is not more than an illustration of qualitative evaluations. The classifications seem to miss a *formal starting point*, starting off directly with secondary descriptive parameters selected on the basis of the researcher's empirical insights. The current view also misses a *spatio-temporal context* in which changes in the frequency of ports as ports of departure or destination or as shipmasters' domiciles are related to changes in the supply and demand for maritime transport services at a port. Finally, the current view misses *functional diversity*: no clear distinction is made between trade networks, on one hand, and transport systems, on the other.

In general, the historiography of the spatial structure of maritime transport in the early modern period suffers from unwarranted quantification of historical ship movement data. Rather than maintaining a focus on the entire ship movement,[10] individual data items are quantified separately, at the cost of losing the connection between locations and their respective functions in

7 Polónia, "Les petits ports," 27–51.

8 Polónia, "Les petits ports," 33.

9 See for example: Le Bouëdec, "Small ports," 103–126; Le Bouëdec, "Intra-european coastal shipping," 89–107; Le Bouëdec, "Economie atlantique," 1–12.

10 A ship movement typically consists of a port of departure, port of observation and port of destination; the port of observation often being the same as either port of departure or destination. Many other data items may be included, like intermediate destinations,

a ship movement.[11] This is also the main reason why existing historical surveys have not been able to explain changes in the spatial structure of maritime transport in the early modern period.[12] One-sided quantification erases essential semantic and relational information. In the following sections, I will describe the evolution of ports and their functions in the Dutch Deltas, making use of two basic procedures in which both quantitative and qualitative elements are taken into account. Rather than 'counting' the frequency of a series of parameters (like Bang and Korst did in the *Tabeller over skibsfart*), the entire spatio-temporal trajectory is taken as a starting point for quantification. The relations between the various points of the trajectory are preserved. The links between points and their functions are not broken, as would be the case in *mainstream* quantification. Thus, the quantitative data resulting from this method maintain their significance for structural analysis. The reconstructed trajectory data that results from data gathering and database operations has the following general structure (see Table 6).

Registrations of ports and homeports (shipmasters' domiciles, *Dk.* hjemsted) are the starting points of identification, classification and visualisation of the spatial structure of northern European trade networks and maritime transport systems. The governing principle is that a port, its possible functions in a ship movement and their respective frequencies are key indicators of the supply and demand of maritime transport and its spatial and temporal characteristics. Homeport data are still often neglected in the study of trade networks and transport systems, but they can effectively be used to generate a novel classification of ports, since they add a functional perspective to port geography that cannot be derived solely from traditional port statistics. We are inclined to believe that the functional perspective obtained through the use of homeport data is more important than eventual ambiguity in homeport registrations (which is treated more extensively in Chapter 4). On top of that, there is the classic discussion about the homeport in the Danish Sound Toll Registers and the idea that the same shipmaster could have several homeports. While this is true, the uncertainty to which reference is made logically applies to the entire registration and not only to the homeport and, when the occurring variation is examined in the long-term, it becomes clear that these variations are mostly exceptions and that a dominant homeport can be identified in almost

cargo items, and the domicile of the shipmaster. To get an insight in the potential complexity of ship movement registrations, see: Dedieu et al., "Navigocorpus,".

11 Such has for long been the main critique on Bang and Knud Korst's *Tabeller over Skibsfart*, which are still popular among maritime-economic historians.

12 Van Royen, *Zeevarenden*; Van Royen, "Varend volk," 86–87; Van Der Woude, *Het Noorderkwartier*, 384.

TABLE 6 *The general structure of reconstructed trajectory data; legenda: Dep = point of departure, Obs = point of observation, Dest = point of destination; T1, T2, T3 = time intervals, calc = any statistical procedure (Count, Sum, etc.)*

<table>
<tr><th colspan="6">Reconstructed trajectory data</th></tr>
<tr><th colspan="3">Variables</th><th colspan="3">Attribute data</th></tr>
<tr><th>Entity data</th><th>Temporal data</th><th>Spatial data</th><th>T1</th><th>T2</th><th>T3</th></tr>
<tr><td>Shipmaster Homeport</td><td>Date</td><td>Dep Obs Dest</td><td>Calc</td><td>Calc</td><td>Calc</td></tr>
<tr><td colspan="3">Registration of movement</td><td colspan="3">Registration of frequency</td></tr>
</table>

every single case. In Chapter 4, this issue will be dealt with in more detail. For now, it suffices to say that the registrations of shipmasters from one domicile comprised a majority of shipmasters for which this domicile was their actual homeport and a minority for which it was an alternative domicile. Dominant (or: actual) domiciles had a median share in the total number of passages by its community members of 75.7%, with a standard deviation of 7.1%. As alternative domiciles, each had a share of 16.9% (standard deviation of 6.2%) in the total number of passages.

The value-added by the homeport as a core element in port classification lies in the fact that it recognizes the multiple functions of ports from the outset, providing a profoundly relational perspective on ports, in which both external relations (with hinterland and foreland) and internal relations (i.e. the port as homeport, port of departure or port of destination) are included. Moreover, by including homeport data in the classification of ports the importance of services functions is emphasized. Hence, the function of a port is also determined by the size and behaviour of the maritime transport community residing in the port. Indeed, not all homeports necessarily serve as ports of departure or destination, yet they all are communities, urban or rural, whose development is influenced fundamentally by their maritime function.

The subsequent analysis of ports will be based on five parameters in which frequency and functions are combined. These five parameters are port frequency, homeport frequency, home-market frequency, home-market share of supply of maritime transport and the transport-trade ratio of the port.

Port frequency is the total number of ship movement registrations in which a certain port is registered as port of departure or port of destination, during a pre-defined time period (usually one year), with distinction made between directions (incoming and outgoing). Port frequency data serves as an indicator of maritime traffic density at a port. The total number of ship movements starting at or going to a port during a pre-defined time period is the total *realized* demand for maritime transport at this port.

Homeport frequency is the total number of ship movement registrations in which a certain port is mentioned as homeport (shipmaster's domicile) in STRO during a pre-defined time period. Homeport frequency data serves as a proxy for the presence of maritime transport services in a port. It bears reference to a port's maritime transport community. It is not an estimate of ownership of a certain amount of ships by people residing in a certain location. Such estimate would require taking into account additional factors related to finance and investment. Homeport frequency is the total *realized* supply of maritime transport at a port during a pre-defined time-period. *Home-market frequency*, then, is the total number of ship movement registrations for which the homeport is also the declared starting point or given destination of the movement. Its calculation is the result of a combination of frequency and function data. If *home-market frequency* and *port frequency* are equal, then port activity is entirely facilitated by the port's own maritime transport community.

The *home-market share of supply of maritime transport* (X-axis on Figure 5) is the share that maritime transport services supplied to the home-market have in the total supply of services by the maritime transport community of a port. This home-market share of supply of maritime transport is the result of division of home-market frequency by homeport frequency. The home-market share of supply of maritime transport provides an insight into the dominant function of the maritime transport community of a port. A home-market share of supply that is equal to 1, indicates that all services supplied by the maritime transport community of a port, started or arrived at this port. Such maritime transport community is consequently categorized as *home-market oriented*. On the contrary, a home-market share of supply equal to 0 indicates that the port does not have its own demand and serves solely as the homeport of shipmasters providing services on the so-called 'third routes' introduced in Chapter 1. The maritime transport community of such ports is categorized as *service-oriented*. Arbitrarily, we might say that any calculation of the home-market share of supply of maritime transport that generates a result ranging from 0.00 to 0.49 indicates that the maritime transport community of the port is *service oriented*, while any result between 0.50 and 1 indicates that the maritime transport community of the port is *home-market oriented*.

The calculation of the frequency of a port as homeport in a ship movement registration as opposed to its frequency as a port of departure or destination lies at the basis of the *transport-trade ratio of the port* (Y-axis on Figure 5).

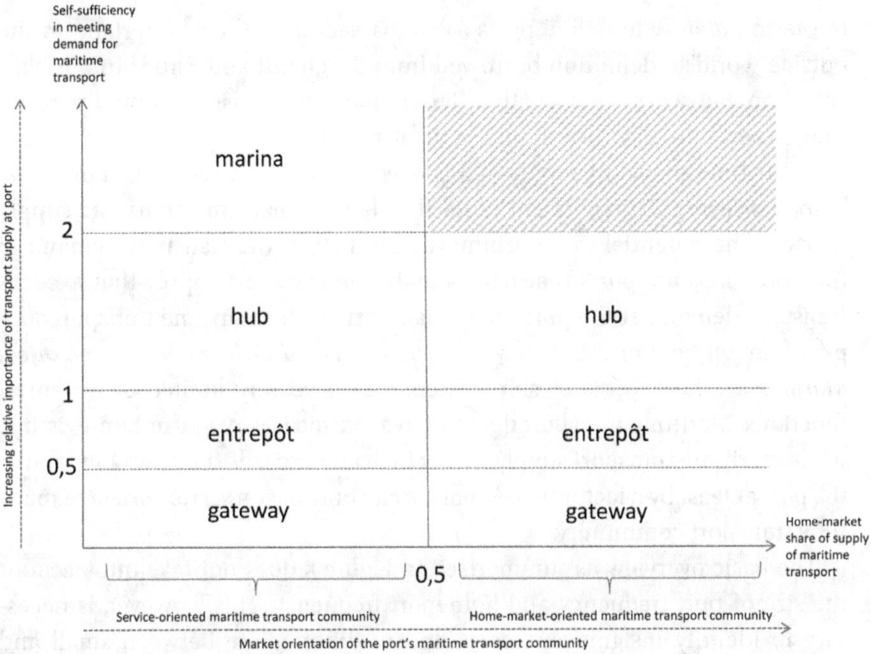

FIGURE 5 *Tool for the classification of ports.*

The transport-trade ratio is measured through division of port frequency and homeport frequency of a certain port. The result is a number equal to or higher than 0. A calculation result equal to 0 means that *realized* demand for maritime transport at the port was not met by shipmasters with this port as their domicile (in this case, the port does not have a maritime transport community). A number between 0 and 0.5 indicates that less than half of *realized* demand for maritime transport services is met by the maritime transport community of the port. A number between 0.5 and 1 indicates that more than half of *realized* demand for maritime transport services is met by the maritime transport community of the port. A number between 1 and 2 indicates that more than 100% and up to 200% of *realized* demand for maritime transport services could be met by the maritime transport community of the port. A number higher than 2 indicates that the maritime transport community of the port is as least twice as large as what is absorbed by *realized* demand for maritime transport at the port.

The home-market share and transport-trade ratio results allow the identification of four main classes of ports, which are to be understood as categories describing the role and function of the maritime transport community in a port and not solely as indicators of a port's trade. The four main classes resulting from the classification process are: gateway, entrepôt, hub and marina (see Figure 5). Throughout his research, the Dutch historian Clé Lesger has used the term *gateway* to define ports as 'ambassadors of their hinterland to the outside world'; a definition borrowed from Burghardt and Bird.[13] In our classification, *gateways* are ports that have a maritime transport supply that has the potential to absorb less than half of the transport demand at the port. The maritime transport supply of a *gateway* may be *service-oriented* or *home-market oriented. Entrepôts* are ports that have a maritime transport supply that has the potential to absorb more than half of the transport demand of the port. *Hubs* are ports that have a maritime transport supply that exceeds transport demand at the port up to a factor two. The maritime transport supply of an *entrepôt* or a *hub* may be *service-oriented* or *home-market oriented. Marinas* are homeports of shipmasters that have very limited or no portal functions. Maritime transport demand in a marina is very low or non-existing, while maritime transport supply is very high and exceeds transport demand at the port at least by a factor two. A marina can only have a *service-oriented* maritime transport community.

The basic overview as summarized in Figure 5 does not take into account the size of port frequency and homeport frequency. This, however, is necessary to identify insignificant ports and to differentiate between small and large ports. Such distinction can be made for port frequency and homeport frequency during pre-defined time periods, applying the rules specified in Table 7. The size of port activity in a gateway, entrepôt or hub is estimated using port frequency data, while the size of the maritime transport community in a gateway, entrepôt or hub is estimated using homeport frequency data. The size of a marina is estimated using homeport frequency data.

In addition to calculations for individual ports, the home-market share of the supply of maritime transport and the transport-trade ratio were also calculated for the Dutch Deltas and other European coastal regions. The following sections first describe structural differences and changes in the Dutch Deltas and then compare port functions in the Dutch Deltas with developments in the rest of Europe.

Transport Market Integration

Classification and estimation of the size and structure of port activity permits the description of the functions of ports in the Dutch Deltas in a formalized way. It has been made clear in Chapter 1 that the ports in the Dutch Deltas emerged from different regional economies, played different roles in regional

13 Lesger, *Hoorn als stedelijk knooppunt*, 15; Bird, *Centrality and Cities*, 115; Burghardt, "A hypothesis," 269.

TABLE 7 *Rules for adding size to the classification of ports*

	Size of port activity			
	Gateway	**Entrepôt**	**Hub**	**Marina**
Insignificant	PF < 1%	PF < 1%	PF < 1%	Not applicable
Small	1% < PF < 5%	1% < PF < 5%	1% < PF < 5%	
Large	PF > 5%	PF > 5%	PF > 5%	
	Size of maritime transport community			
	Gateway	**Entrepôt**	**Hub**	**Marina**
Insignificant	HPF < 0,5%	HPF < 0,5%	HPF < 0,5%	HPF < 0,5%
Small	0,5% < HPF < 2,5%	0,5% < HPF < 2,5%	0,5% < HPF < 2,5%	0,5% < HPF < 2,5%
Large	HPF > 2,5%	HPF > 2,5%	HPF > 2,5%	HPF > 2,5%

Legenda: PF = port frequency, HPF = homeport frequency, percentages are taken from annual total port frequency or homeport frequency respectively

trade and transport networks and logically also had different functions in the international transport market. It has already been pointed out that the emergence of an integrated Dutch maritime transport sector coincided with the establishment of central maritime transport markets in Amsterdam and Rotterdam. This development became particularly clear after 1650, several decades after Amsterdam's definitive rise as a centre of world trade. Generally speaking, three major developments can be discerned in the evolution of the maritime transport sector of ZDZ: (1) homogenization of the role of secondary ports and marinas, (2) the rise of Amsterdam's maritime transport community and (3) the absolute and relative decline of the maritime transport sector in ZDZ.

In Chapter 1, it had been pointed out that between 1650 and 1700 a structural change must have took place in the organisation of maritime transport in the Dutch Deltas. The secondary ports in ZDZ lost their relative autonomy in international trade, which finds expression in the changing structure of their voyages to and from the Baltic, as Graph 2 in Chapter 1 clearly indicates. Even if it is taken into account that pre-1669 westward voyages through the Sound did not specify the destination of the ship and that it is assumed that all ships went to Amsterdam and not to the shipmasters homeport, the changes that

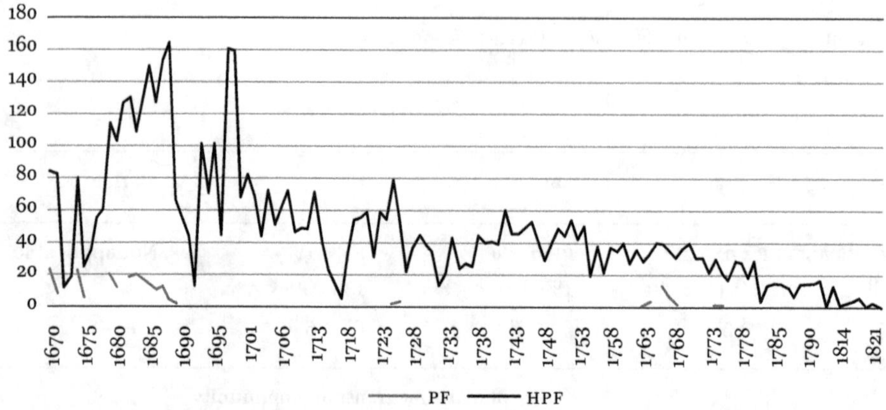

GRAPH 5 *Port frequency and homeport frequency of Hoorn, 1670–1821.*
BASED ON STRO.

occurred in the functions of ports in ZDZ are remarkable. Until about 1670, shipmasters from other ports in ZDZ used to start most of their voyages to the Baltic from their homeport, indicating the existence and perseverance of modest, but nevertheless autonomous international trade structures originating in ports such as Hoorn, Enkhuizen, Harlingen, Medemblik and Stavoren. In Medemblik and Stavoren, the loss of autonomous international trade links seems to have already occurred in the first half of the seventeenth century, while Hoorn and Enkhuizen retained some autonomy until about 1700.

The number of ports active in ZDZ, that were in shipping to and from the Baltic varied from thirty-two in 1692 and 1697 to fifty-one in 1689 and 1693. On average, more than forty ports annually provided maritime transport services. Remarkably, entrepôts are almost absent in the port structure of ZDZ, insofar as it can be reconstructed from STRO. All hubs in ZDZ were insignificant and their number never exceeded fifteen. The port structure of ZDZ was dominated by the continuous presence of one large gateway—Amsterdam—and several marinas, between eighteen and forty-five to be precise; on average about thirty. Of these marinas only a small fraction was large; some were small and the majority was insignificant on an international scale (which, however, does not imply that these ports were 'unimportant'). The port structure of ZDZ remained almost intact during the Great Northern War. The large gateway of Amsterdam continued to dominate the delta, while between thirty and forty marinas, of which maximum six were large, supplied maritime transport services between the North Sea and the Baltic. This does not mean that the Great Northern War had no impact on the port structure of ZDZ. Clearly, shipping was disturbed for several consecutive years; it has been substantiated elsewhere that merchants and shipmasters sought ways to avoid the Baltic.[14] The long-term effects, however, become clear only after 1720, when the number of large marinas declined from 5–6 in 1721 to 1–3 between 1722 and 1740 and 0–1 between 1740 and 1780. At the same time, the total number of marina's remained largely constant during the entire period up to 1780. Roughly between forty and fifty small and insignificant marinas were part of the port system of ZDZ between 1720 and 1780 (details in appendix 2.1.). None of the other types of ports—hub or entrepôt—was significant throughout this period. The Fourth Anglo-Dutch War of 1781–1783 and the Napoleonic era of 1795–1815 led to a reconfiguration of the port structure of ZDZ. Already in the years 1784–1794, it can be observed that almost all small marinas disappeared and that the number of insignificant marinas also started to decline. Transport services supply went down. At the same time, Amsterdam still was a large gateway, and in comparison with earlier periods, the total number of ship movements to and from Amsterdam did not drastically change yet. However, total shipping between the North Sea and the Baltic did, and shipmasters from ZDZ, registered in Amsterdam or in one of the delta's many marinas, clearly had a hard time finding ways to participate and profit from the general growth of Baltic shipping at the end of the eighteenth century. Again, the disturbances of war had severe long-term effects: after the end of the continental blockade, Amsterdam had lost its status of large gateway and only between five and ten insignificant marinas were left in ZDZ.

The process of homogenization was finalized when the ports of Hoorn and Enkhuizen, which still had some port frequency in 1670, joined the group of marinas with no port frequency at all around 1700. Between 1670 and 1700 the port frequencies of marinas, i.e. the number of ship movements starting at or going to a marina, declined from about 8% of all ship movements by shipmasters from the marina to less than 2%. The separation of transport systems populated by shipmasters living predominantly in marinas, on one hand, and trade networks marked by ships calling at and leaving ports, on the other, was completed before 1700. A very brief episode in the mid-1760s notwithstanding, none of the marinas would ever regain its previous (modest) autonomy in international trade. For the ports of Enkhuizen and Hoorn, and to a lesser degree for the much smaller port of Venhuizen, the port frequency and home-market frequency were almost equal; those few shipmasters that called at or left these ports, were often domiciled there. In the long run, it can be observed that, following the loss of autonomous international trade links,

14 Scheltjens, "The impact"; Veluwenkamp, *Archangel*; Kotilaine, "Competing claims"; Kotilaine, *Russia's foreign trade*.

the ports of Hoorn and Enkhuizen went into overall decline.[15] Their communities of maritime transporters diminished. In 1722, Hoorn was classified as large marina for the last time and between 1723 and 1774 it was listed mostly as a small marina. After 1775, Hoorn was an insignificant marina and shipmasters were only domiciled in Hoorn sporadically. The port of Enkhuizen lost its maritime significance even earlier. By 1670, Enkhuizen was no more than a small marina and after 1720 its role in international maritime transport services was definitively over. Shipmasters from Enkhuizen realised only a few passages through the Sound each year. The difference with the port of Harlingen, which was also a secondary port in ZDZ is remarkable. Even though the frequencies are small, there was a gradual but undeniable shift in the function of the port of Harlingen in the Dutch maritime transport system. Before 1700, Harlingen had a large maritime transport community which served its home-market as well as providing maritime transport services to third parties. Sometimes, this maritime transport community was classified as small, sometimes as large. Harlingen's port and homeport frequencies varied from year to year and seem to have been subject to fluctuations that were probably due to external circumstances. Nevertheless, a declining trend in homeport frequency, i.e. the frequency of Harlingen's maritime transport services, can be observed throughout the eighteenth century. After 1740, Harlingen became a small marina. From then on, a series of changes occur in the classification of the port of Harlingen, indicating that its function in the Dutch maritime transport system was also changing. To start with, it can be observed that, even though sporadic and with very modest frequencies at first, Harlingen's port frequency started to rise and experienced a breakthrough after 1800. Second, it is notable that the home-market frequency followed a different path. The home-market was only served sporadically, and during several years shipmasters from Harlingen did not provide services to their own homeport. In line with the general growth of maritime shipping in Harlingen in the first half of the nineteenth century, a steady, though modest, rise in home-market shipping can be observed. From 1815 onwards, Harlingen functioned mostly as an insignificant entrepôt.

The second major development in the port system of ZDZ is that of the growth of the maritime transport community of ZDZ's only gateway: Amsterdam. The port frequency of Amsterdam was subject to fluctuation in the course of the period represented in the Graph 7, but in general its port frequency never exceeded 2,000 ship movements destined to or departing from

15 The decline of Hoorn and Enkhuizen in the eighteenth century has been described by Lesger and Willemsen. Our focus is merely on the maritime transport communities of these ports, a topic they largely left aside.

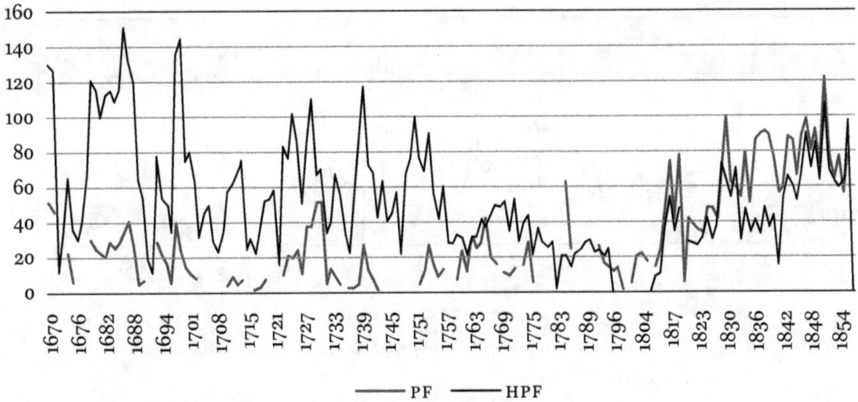

GRAPH 6 *Port frequency and homeport frequency of Harlingen, 1670–1857.*
BASED ON STRO.

Amsterdam for the Baltic. This lack of growth, but also the absence of clear decline, may be related to what De Vries has identified as the relative decline of the economy of the Dutch Republic.[16] The development of Amsterdam's maritime transport community in the eighteenth century contrasts sharply with its stagnating port development. Limited in size before 1700, Amsterdam's maritime transport community started to grow rapidly during the Great Northern War and it continued this almost uninterrupted growth until 1780, having reached a peak in 1777 when 738 ship movements of masters claiming domicile at Amsterdam were registered in STRO. Quite differently from what happened in Harlingen, home-market frequency kept pace with the growth of homeport frequency and for most years Amsterdam's maritime transport community functioned as a home-market oriented community (see Graph 7).

The rapid growth of Amsterdam's maritime transport community in the eighteenth century may tentatively be explained as the result of two parallel processes. In part, it may have been the result of migration of shipmasters from other ports in ZDZ (and probably beyond) to Amsterdam. To a certain extent, however, the rise of Amsterdam's maritime transport community may also have resulted from changing registration and declaration practices in the Sound. Indeed, several historians have noted the ambiguity in domicile registrations in the Danish Sound toll registers. As said, this issue will be dealt with more extensively in Chapter 4. For now, suffice it to say that the standardized population of Amsterdam's maritime transport community had the following estimated composition: 75% of all passages had Amsterdam as dominant domicile and 25% of all passages had Amsterdam as alternative

16 De Vries, *De economische achteruitgang*.

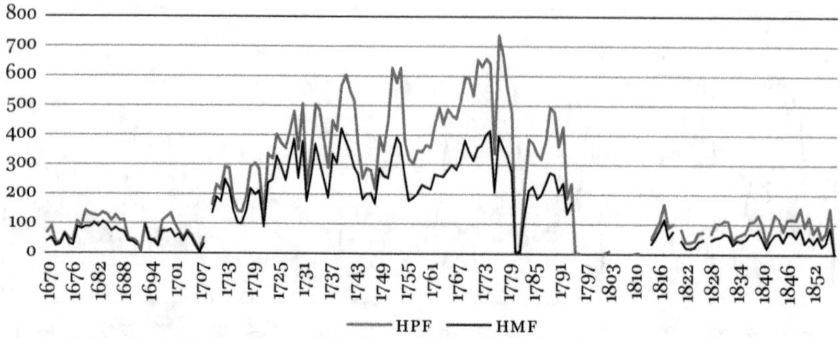

GRAPH 7 *Homeport and home-market frequency of Amsterdam, 1670–1850.*
BASED ON STRO.

domicile. This is in line with the median figures and standard deviations, that were calculated for thirty-two Dutch maritime transport communities (see Chapter 4 for details) and shows a slight overestimate of the size of the maritime transport community of Amsterdam. After the Fourth Anglo-Dutch War, Amsterdam's maritime transport community partially recovered, but the Napoleonic Wars struck a final blow at the end of the eighteenth century. After 1812, the size of Amsterdam's maritime transport community was thrown back to pre-1700 levels. Due to almost unchanged port frequencies in an increasingly competitive Baltic market, where Dutch ports no longer played a major role, Amsterdam lost its status as a large gateway at the end of the eighteenth century. Throughout the first half of the nineteenth century, Amsterdam was a small gateway, at least insofar as can be assessed on the basis of STRO. It had a maritime transport community for which the home-market occupied up to 60% of the services supplied.

The third major development in the port system of ZDZ is that of the relative and absolute decline of the supply of maritime transport services after 1670. Both find expression in the classification of ports itself (see appendix 2.1.) and in the statistics of port, homeport and home-market frequency. At the aggregated level, ZDZ's relative and absolute decline are best visualized using the calculated values for home-delta frequency (left axis on Graph 8), on one hand, and home-delta share of transport services supply and transport-trade-ratio (right axis on Graph 8), on the other. Both calculations are analogous to the homeport frequency and the home-market share, but observations are made at the aggregated level of the delta.

Graph 8 visualizes the aggregated number of ship movements by masters domiciled in one of the ports of ZDZ as well as the total number of movements by these shipmasters going to or departing from ports in ZDZ. The graph shows that until 1740, the home delta was predominant in maritime transport services supplied by shipmasters from ZDZ. About 80% of all maritime transport movements were destined to or departed from the home delta. This pattern started to change drastically around 1750, when suddenly only 60% of all ship movements by masters from ZDZ had a direct link to the home delta. The operational patterns that underlie this structural change, which can, again, be understood as a sign of transport market integration, will be discussed in detail in Chapter 3.

The first signs of absolute decline appeared in the late 1770s and were intensified by the Fourth Anglo-Dutch War. After 1783, the maritime transport sector in ZDZ never fully recovered. After yet another long period of crisis between 1795 and 1810, the maritime transport sector of ZDZ had shrunk to about one sixth of its mid-eighteenth century size. In the same period, transport market integration stagnated.

The severe blow of the Napoleonic era to the maritime transport sector of ZDZ also becomes clear in ZDZ's transport-trade ratio. Before 1795, ZDZ's maritime transport sector could typically accommodate between 80% and 120% of all shipping in the delta. The size of the supply was in balance with maritime transport demand and there was little or no excess supply of maritime transport services. The reality of the operational patterns, however, was more complex than the balance between maritime transport services supply and demand at the level of the delta suggests. As will be explored later on, maritime transport services were increasingly rendered between regions lying outside the home delta, especially after 1750 (see Chapter 3). The balance between maritime transport supply and delta-based demand was lost after 1795. The home delta was no longer in a position to supply the maritime transport services needed in ZDZ. The structural change in ZDZ was the result of a long process of geographical restructuring in the regions surrounding the delta, where former maritime transport communities in North Holland disappeared from the international scene, as 'new' locations in Friesland emerged. As discussed in Chapter 1, these 'new' maritime transport communities were located inland rather than directly on the coastline.

As indicated in Chapter 1, the development of the maritime transport sector in LDE was in the opposite direction. Insofar as they can be traced in STRO, prior to the eighteenth century, maritime transport services were supplied by shipmasters from Dokkum, Groningen and sometimes Delfzijl. Before 1740, these maritime transport communities were insignificant on an international scale. The rise of the maritime transport sector in LDE started roughly around 1750 and is largely undisputed in historiography (see Chapter 1). The first observation that must be made with regard to the port structure of LDE after 1750 is that it had, and continued to have, a profoundly transport services orientated structure. Insofar as ports were categorized as hubs, entrepôts or gateways,

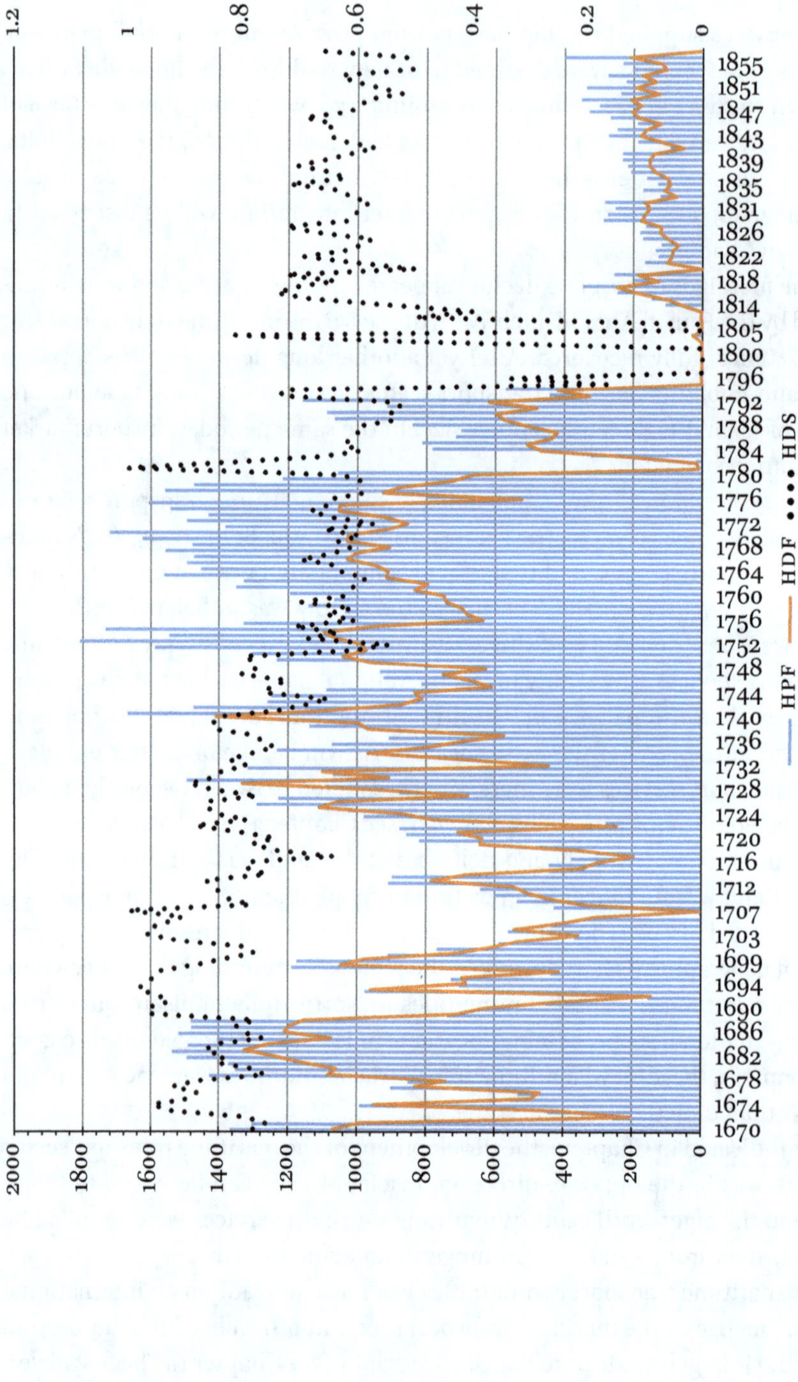

GRAPH 8 *Home-delta frequency and home-delta share in* ZDZ, *1670–1850.*
BASED ON STRO.

they were always insignificant and there were never more than four.[17] At the same time, the number of marinas increased from four to nine between 1750 and 1780. In 1792, this number reached a peak, when shipmasters from thirteen different maritime transport communities in LDE were registered in STRO. Contrary to the development of the port structure in ZDZ, neither the Fourth Anglo-Dutch War nor the Napoleonic Wars seem to have had a severe negative impact on the maritime transport sector in LDE. After 1815, the number of marinas stabilized at between eight and ten; most of them were insignificant, two or three were small. At the same time, the maritime transport sector of LDE became an increasingly important supplier of transportation services between the Baltic and ZDZ (and SMR) (for more details, see Chapter 1). This particular type of transport market integration, in which one delta (LDE) provided maritime transport services to the other Dutch Deltas, rather than maintaining a focus on the own delta (as was the case with the maritime transport sectors of SMR and ZDZ), provides evidence of the emergence of a maritime transport services economy in LDE.

From Graph 9, it is clear that LDE never had a large home-market of its own, at least not insofar as interaction with the Baltic was concerned. Home-market frequency was always low, and for most of the time, a few decades in the mid-eighteenth century notwithstanding, the maritime transport sector was almost entirely orientated to ports located outside of the home delta. ZDZ was clearly dominant throughout the entire period, but SMR emerged in the late eighteenth century as a secondary departure area. Taking into account the specific structure of the maritime transport sector of SMR, which, like ZDZ, was predominantly home-delta orientated, it may be suggested that the maritime transport sector of LDE provided the other Dutch Deltas with much needed services. The fact that, contrary to the fate of the maritime transport sector in ZDZ, the communities in LDE seem not to have experienced any negative long-term impact from the Fourth Anglo-Dutch War or the Napoleonic Wars, raises questions about their survival strategies during these periods. In Chapter 1, it was suggested that changes in the regional economic structure and access to international trade and transport markets were decisive elements in the development of a maritime transport sector. Both elements are clearly present in the port structure of the Dutch Deltas, most obviously when LDE is concerned, but also when the large number of marina's in ZDZ is taken into consideration. In Chapter 3, an in-depth survey of routes sailed and cargoes carried will show that shipmasters from communities in LDE operated with more flexibility

17 The only exception is 1785, when five ports in LDE were categorized as insignificant gateways.

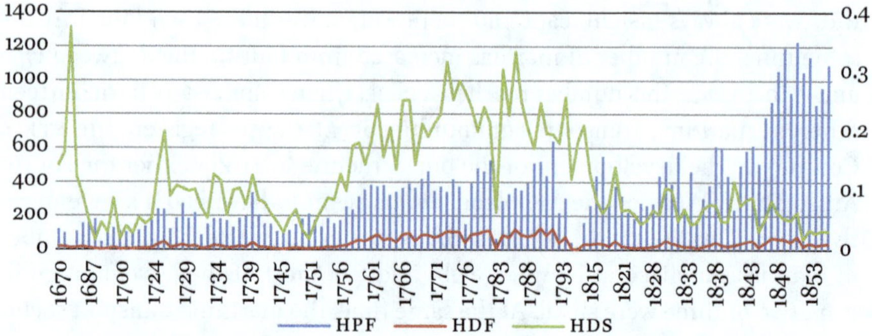

GRAPH 9 *Home-delta frequency and home-delta share of* LDE, *1670–1850.*
BASED ON STRO.

than their counterparts in ZDZ. Finally, in Chapter 4, a particular aspect of the survival strategies of shipmasters from LDE, namely 'virtual migration', will be highlighted. Shipmasters from LDE 'migrated' in significant number to neutral places such as the Duchy of Arenberg during the Napoleonic wars, which put them in a better position to continue their business than their colleagues in ZDZ.

The remaining delta, SMR, needs to be treated with caution, since it is known from historiography, that the main ports in SMR—mostly Rotterdam, Dordrecht and Middelburg—always had a different geographical orientation than the ports in ZDZ or LDE. Moreover, very little is known about the maritime transport sector of SMR, a few case-studies about individual communities notwithstanding.[18] The analytical perspective of the historiography of SMR, which seems to be mostly concerned with labour markets, international (slave) trade and urban development, does not reveal much about the presence of a maritime transport sector in SMR's cities and villages.[19] The situation seems to be even more precarious when the ports along the southern border of SMR—Gent, Antwerpen and Brugge—are taken into account. Only few studies pay attention to their maritime trade before the (late) nineteenth century; an insight into the maritime transport sector of these ports and their surrounding areas is almost completely absent. This makes it particularly worthwhile to have a closer look at STRO for observing changes in the port structure of SMR, even though the observations cannot easily be contextualised and interpreted. The classification of ports in SMR, based on STRO, returned results that differ fundamentally from the classifications of ports in ZDZ and LDE. Ports in SMR were insignificant at almost any point in time between 1670 and 1850. Moreover, their number was low and rarely ever exceeded ten. There were very

18 De Wit, *Leven, werken en geloven.*

19 E.g. Brusse, *Gevallen Stad*; Paesie, *Lorrendrayen*; Van der Schoor, *Stad in aanwas.*

few marinas in SMR, at least insofar as STRO informs us. During the 1670s, one of these marinas was categorized as small and occasionally this would happen again in the course of the eighteenth century. After 1795, the role of marinas became even less significant. Whereas until 1795 shipmasters from one to six marinas annually were active in the Baltic, their appearances would become sporadic in the first half of the nineteenth century. At the same time, however, a modest rise can be witnessed in the other port functions, especially from the late 1760s onwards, when one, sometimes two ports functioned in Baltic- North Sea traffic as small entrepôts or small gateways. This pattern was continued for some time after 1815, but it came to an end rather abruptly in 1833. From that moment, none of the (very few) ports in SMR would play any significant role in Baltic-North Sea traffic. This, however, does not mean that the ports in SMR were small. On the contrary, in the development of the port structure of SMR between 1670 and 1850, as far as it can be reconstructed on the basis of its trade and transport activities in the Baltic Sea, the long-term effects of the rise of the Atlantic economy accompanied with the rise of the British foreland, as well as the decline of the Dutch economy proper and its reorientation towards its immediate German hinterland,[20] seem to find confirmation in the absence of SMR in Baltic-North Sea traffic. It is indeed remarkable, that so little information could be found about the size of the maritime transport sector in SMR, while the rise of Rotterdam as a port in the Altantic economic system is undeniable. Perhaps Dutch shipmasters did not play that big a part in the rise of Rotterdam as a port of international trade? It seems plausible that the carriers responsible for Rotterdam's rise were British and American rather than Dutch shipmasters, but the data to substantiate such claims have yet to be gathered and interpreted.

One of the questions that remains to be answered with regard to the observed changes in the port structure of the Dutch Deltas as well as with regard to processes of transport market integration in the Dutch Deltas is: were these developments unique, or did similar processes occur in different European regions? This topic will be briefly touched upon in the final section of this chapter.

The Port System of the Dutch Deltas in a European Context

The goal of this section is twofold. First of all, it is an attempt to find an answer to the question of whether the Dutch maritime transport system was unique or not. At the same time, this section aims to provide a description of the functions of different regions in the European maritime transport

20 De Vries, *De economische achteruitgang*.

sector. Specifically, two topics are dealt with. First, the ratio between home-region frequency and home-market frequency, which provides relevant insights into the functional orientation of the maritime transport sector, from which it is argued that a maritime transport sector (be it 'regional' or 'national') that primarily served its home-market differed from a maritime transport sector that provided most of its services to 'third parties'. A particular question that will be raised in addition to the ratio of home-region vs. home-market frequency is that of the categories of ports that constituted the maritime transport sector in different European regions and their relative importance. Both topics will be dealt with at the level of large geographical entities, such as the Dutch Republic, England or France, and will then be compared with the regional level of the Dutch Deltas.[21] In the final part of this section, a tentative answer is formulated about the uniqueness of the Dutch maritime transport system.

The ratio of home-region frequency (HRF) to home-market frequency (HMF), which calculates the home-region share of the supply of maritime transport services, reveals structural differences between the various geographical entities that were active in Baltic shipping and trade.
For the regions defined as Norway and the Netherlands, which corresponds, depending on the time period that is observed, to the Dutch Republic, the northern Netherlands or the Dutch Kingdom, the average HRF:HMF(region)-ratio diminished slightly over time, from more than 0.8 at the end of the seventeenth century to 0.7 in the first half of the nineteenth century (see Graph 10). The average ratios for England and France were even higher and remained almost constant throughout the entire observed period.

The development of the home-region share was quite different for other regions and two more groups can be distinguished. The ratios for Ostfriesland and the Danish Duchies became very low in the course of the eighteenth century and their decline was clearly related to an increase in the number of Sound passages executed annually by shipmasters from these regions (see Graph 11). The ratios for Scotland, Mecklenburg, the German States (including Bremen and Hamburg) and Pomerania are more ambiguous. For as long as their respective number of Sound passages was low, the ratio remained high, indicating that most shipping departed from or was destined to the home region of the shipmaster. Once the number of Sound passages started to rise, however, a decline in the home-region share can be observed, from about 0.8

21 The categorization of these large geographical entities is based on Johansen, *Shipping and trade*, Welling, *The Prize of Neutrality* and Welling, *Reconstruction*. The same classification of 'big regions' and 'small regions' is now incorporated in the search engine of the electronic database of STRO.

to about 0.4 (see Graph 12). This means that the maritime transport services of shipmasters from these regions were almost equally divided among the home region and other regions of departure or destination.

Perhaps the differences between the various 'big regions' can be explained as the outcome of different regional economic structures that entailed different transport demand structures. England, for example, seems to have needed almost all of its available maritime transport supply to fulfil its own, increasing needs that were caused by population growth, rapid urbanisation and the take-off of the industrial revolution. Consequently, the increase of Baltic shipping by shipmasters domiciled in England did not have the same effect as it had on Scotland, Mecklenburg, the German States or Pomerania. On the contrary, England's home-region share even increased during the first half of the nineteenth century. The same is true for France, although Baltic shipping only becomes significant in the late 1830s. The evolution of the home-region share of Norway is similar to that of England and might also be explained as the outcome of processes of early industrialization,[22] but the frequency of its Baltic shipping was significantly lower than England's and the ratio started to decline somewhat in the 1830s. The situation in the Dutch Deltas was different. Contrary to the development of Baltic shipping in England, France and Norway, which all experienced an increase in the late eighteenth and the first half of the nineteenth century (France's admittedly only after 1830), the Dutch Deltas' involvement in Baltic shipping declined rapidly after 1780 and would not fully recover in the first half of the nineteenth century. This may be explained both by the absence of population growth and stagnating urbanization and by the absence of an industrial revolution—three elements that were different from the processes observed in England, and to a lesser degree Norway and France.

The emergence of regional maritime transport services sectors for which the home region was significant but not dominant. The supply of maritime transport services to regions different from the home region was an essential part of overall transport services supply, as was the case for Scotland, Mecklenburg, the German States and Pomerania, perhaps may be explained as the outcome of emerging processes of industrialisation in less densely and less urbanised regions. Probably, these regions benefited from the continued professionalization of the maritime transport sector as well as advances in communication and information exchange, which made it easier to provide maritime transport services to third parties and to increase shipping income on these grounds.

---

22 See extensively in: Hutchison, *In the doorway to development.*

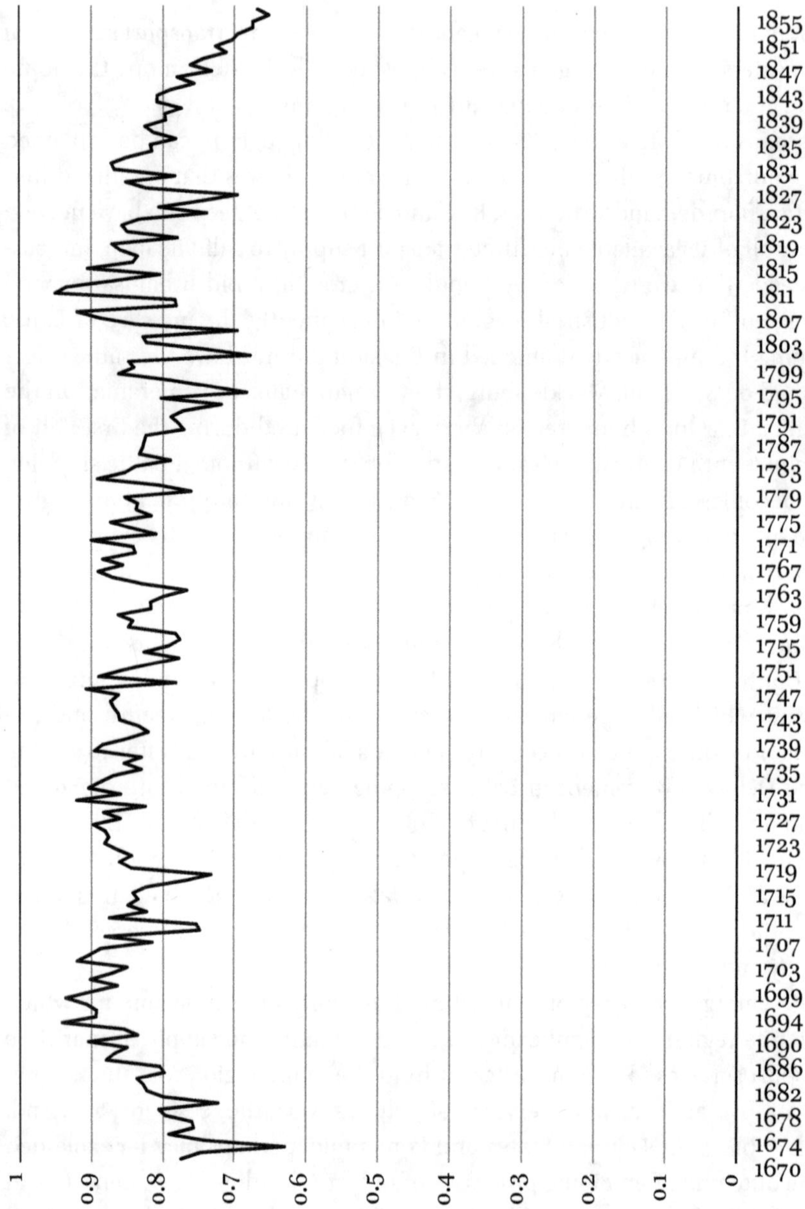

GRAPH 10 *Average home-region share of Norway, Dutch Republic, England and France, 1670–1857.*
BASED ON STRO.

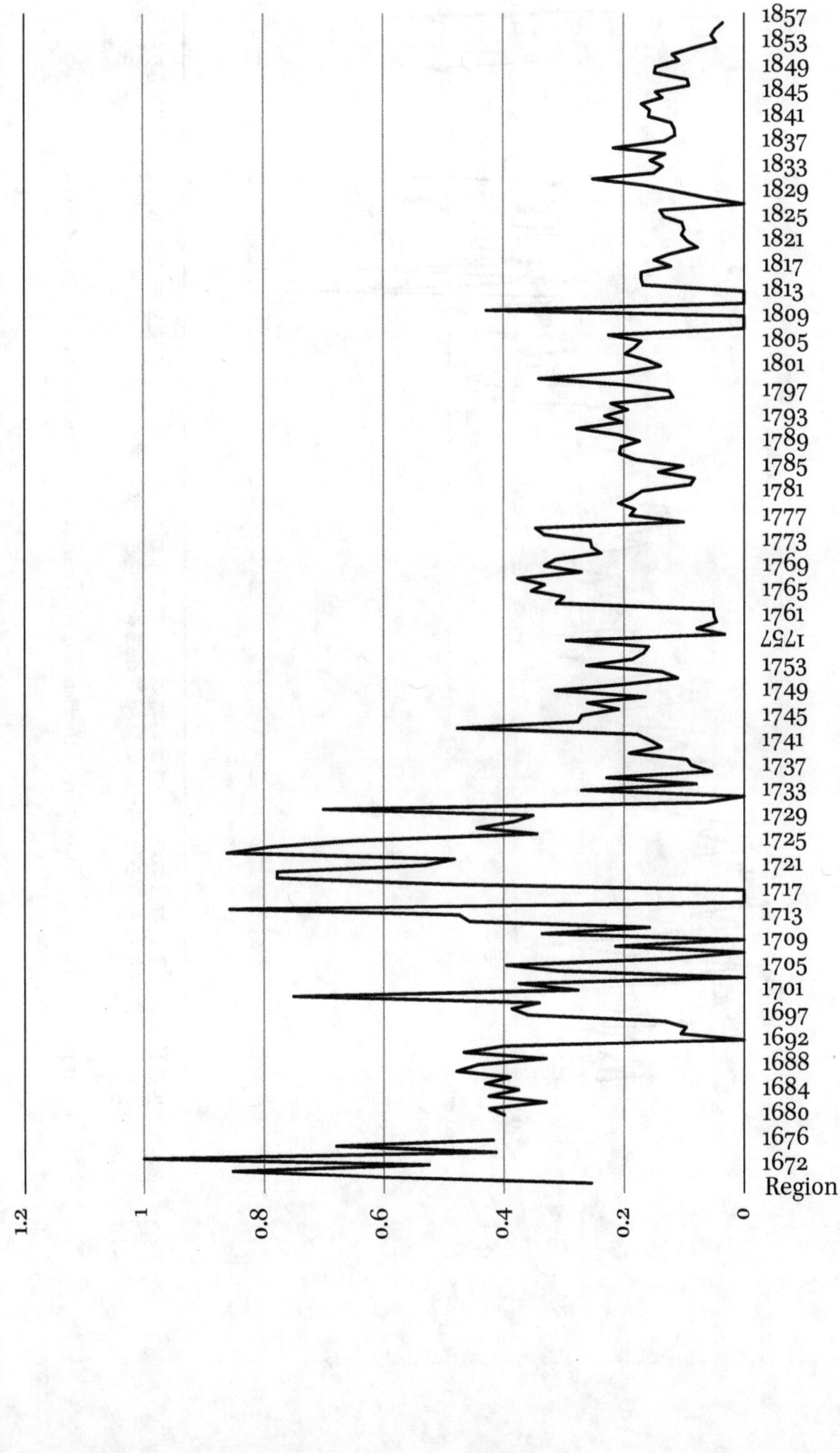

GRAPH 11 *Average home-region share of Danish Duchies and Ostfriesland, 1670–1857.*
BASED ON STRO.

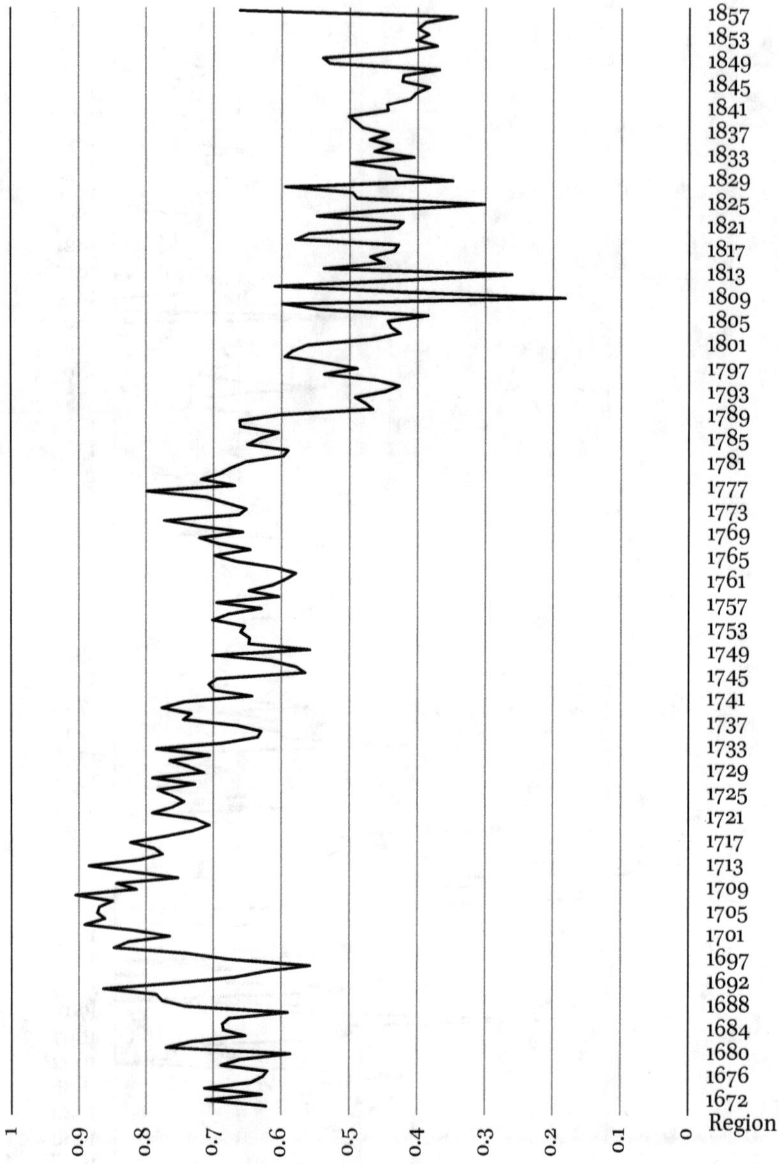

GRAPH 12 *Average home-region share of Mecklenburg, Pomerania, Scotland and the "German States", 1670–1857.*
BASED ON STRO.

Finally, the development of the maritime transport sectors of Ostfriesland and the Danish Duchies requires a different explanation. Here, the supply of maritime transport services was orientated almost exclusively towards third parties; the home region of shipmasters from Ostfriesland or the Danish Duchies only played a marginal role in their transport services production. Ostfriesland and the Danish Duchies were largely rural areas, with very limited urbanization, a small population and (almost) no signs of industrial development. In this respect, both regions are very similar to LDE. It is conceivable that the development of the maritime transport sector in Ostfriesland and the Danish Duchies have followed roughly the same path as that of LDE. However, it should not be ignored, that the maritime transport sector of ZDZ was also largely composed of by marinas on the Wadden Islands and in Friesland, which had hardly any homebound shipping and were bound to Amsterdam as their main port of origin and destination. Of course, aggregated levels of observation, such as the delta or the 'big regions' employed above, make it impossible to identify such nuances. Classification of ports, however, may overcome such issues of taxonomy and provide more detailed information regarding the presence of transport services regions within the boundaries of larger geographical entities.

The observations summarized in Graph 13 suggest that the argument of Chapter 1, in which the emergence of regional transport services economies were considered to be the result of a dual process of regional economic development in the Dutch Republic and integration in the European economy through access to the Dutch Republic's centres of international trade, may be valid for other European regions as well. In that sense, the Dutch maritime transport sector was not unique. For example, the contribution of marinas to total shipping of LDE was similar to that of Ostfriesland. The size of the contribution of marinas in the Dutch maritime transport system and its periodization, however, makes it different from its counterparts in most other European regions, where the emergence of a maritime transport sector in the second half of the eighteenth century coincided with population growth, urbanisation and industrial development, leaving less room for rural coastal communities to participate in international maritime transport services than in the Dutch Deltas.

Conclusions

In this chapter, a method for the classification of ports participating in maritime transport in northern Europe was elaborated and applied to the primary data source of foremost importance regarding Dutch maritime transport communities in the early modern period: the Danish Sound toll registers. The proposed

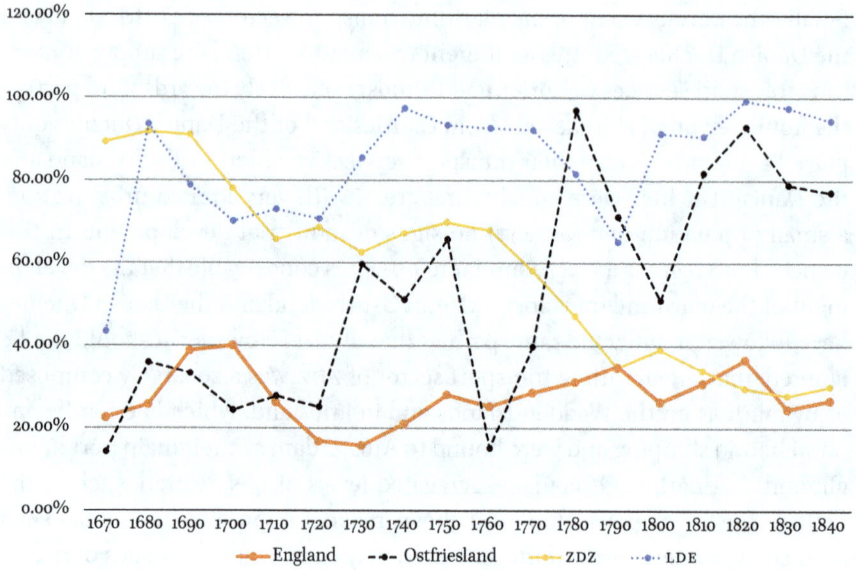

GRAPH 13 *The contribution of marinas in the production of maritime transport services, 1670–1850. 10-year averages of home-region or home-delta frequency.*
BASED ON STRO.

method was based on the functions and frequencies of ports in registered ship movements. The classification was built around five parameters—frequency of ports in their function of ports of departure or destination, frequency of ports registered as shipmasters' domiciles, frequency of the home-market, home-market share of supply of maritime transport and the transport-trade ratio of the port. A distinction was made between gateway, entrepôt, hub, marina and insignificant ports. This classification allowed the capture and assessment of long-term developments in the organisation of maritime transport as a whole, but also addressed the level of distinct geographical regions, like the delta, or of individual homeports.

In this chapter the functions of ports in the Dutch Deltas were analysed. The next chapter will be devoted to the production of maritime transport services in communities in the Dutch Deltas. In this way, the economic-geographical perspective of Chapter 1 and the functional perspective of this chapter are complemented with a long-term quantitative analysis of the routes that were sailed and the cargoes that were carried, so as to obtain additional insights into the structure of the Dutch maritime transport system between 1670 and 1850.

CHAPTER 3

Production

This chapter analyses the relation between trade and transport from the perspective of the production of maritime transport services by shipmasters domiciled in the Dutch Deltas. The chapter starts with a case study about grain transports through the Sound between 1670 and 1850, in which it is established that structural changes in supply and demand led to the emergence of new transport routes, provoking a shift from a bilateral trade pattern to a diversified pattern with many locations involved but none strictly dominant. In the second section, an attempt is made to explain the impact of these changes on the structure of the Dutch maritime transport sector. This chapter contributes to the traditional issue of route vs. cargo specialisation, which was considered extensively by Christensen.[1] But how did Dutch maritime transport services providers react to changes in the demand and supply of goods? Did they move swiftly from one route to the other, depending on the circumstances? Or was their range of activity limited and bounded by external constraints? These questions essentially deal with the impact of changes in the European economy on the Dutch maritime transport sector. Therefore, demand for maritime transport is analysed from an economic-geographical perspective, arguing that changes in hinterland and foreland had reciprocal effects on the spatial structure of maritime transport. In very basic terms, this chapter attempts to answer the following questions: (1) Where were maritime transport services needed? And (2) Who fulfilled these needs?

The Production of Grain Transport Services

The bulk of European commercial exchange was executed by merchants operating as individuals or in small businesses, far from monopolies, exclusive rights and privileges. These entrepreneurs exploited descent and religion to establish commercial networks from which they expected to benefit in one way or another. It is in this context that we situate a case study on the production of grain transport services through the Sound at Elsinore between 1670 and 1850. The market for Baltic grain was highly competitive. There was no single producer and demand varied from year to year. Throughout the early

1 Christensen, *Dutch trade*; Scheltjens, "The influence," 11.

DOI 10.1163/9789004273078_005

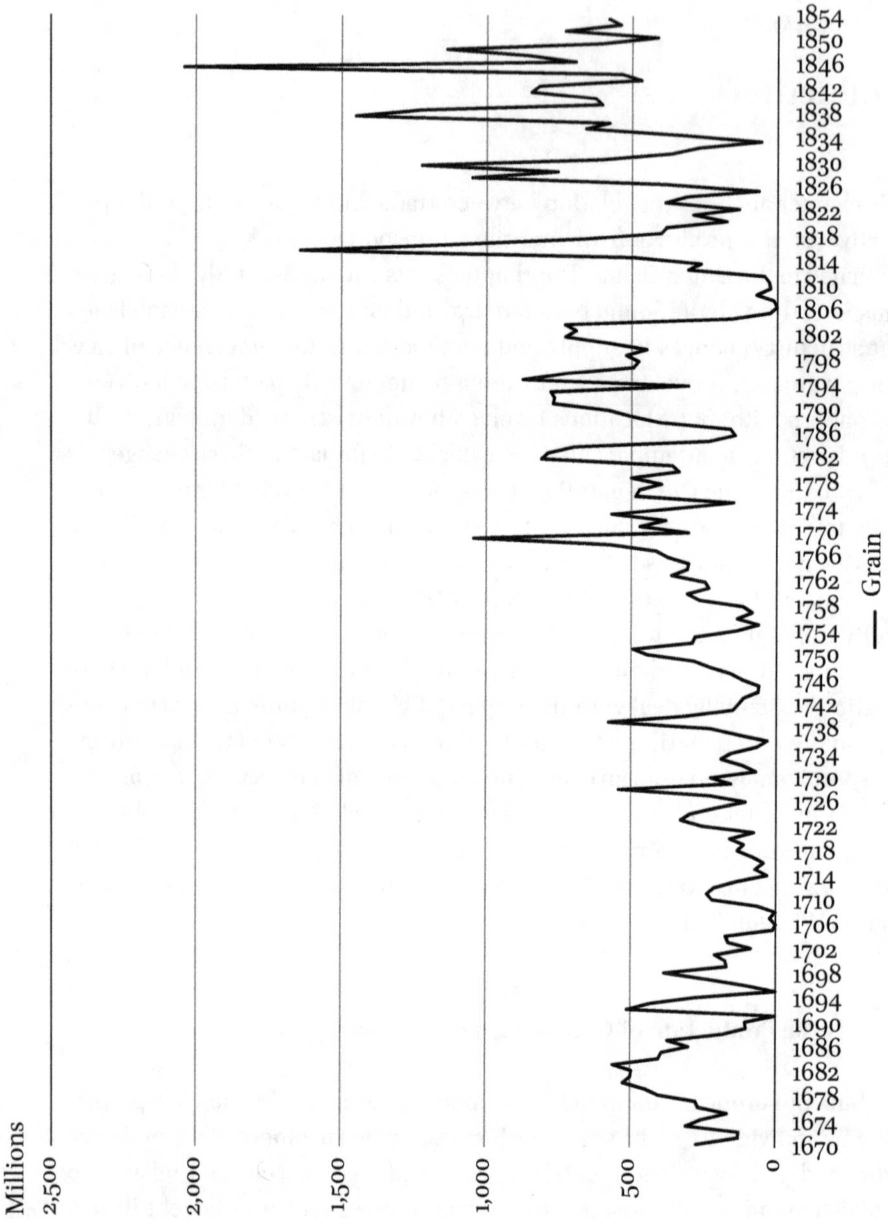

GRAPH 14 *Total Baltic grain transport production, expressed in million tonkm, 1670–1850.*
SOURCE: STRO.

modern era, the Dutch Baltic grain trade was known by contemporaries as the 'Mother of all trades' or *moedernegotie*, indicating its vital importance for the early modern Dutch economy.[2]

The following analysis of the production of grain transport services through the Sound between 1670 and 1850 comprises a statement about its volume, followed by a survey of the changing spatial structure of the grain trade and its main routes. Demand for maritime transport services was, strictly speaking, a merchant's issue. He was the one who decided where he would purchase a cargo, and where this cargo would have to go. While there is no doubt that the rationale behind his decisions was subject to change in time and place, our survey will be restricted to the visible results of a merchant's business decisions, particularly, the routes and directions of grain transports through the Sound.

The central part of the analysis deals with the question of where did merchants look to fulfil their needs for maritime transport and why there precisely? What is reviewed here is the geography of *Dutch* maritime transport services insofar as the shipment of loads of grain through the Sound at Elsinore is concerned. We focus on Dutch supply of maritime transport, but take into account both Dutch and non-Dutch demand, thus highlighting the potential importance of 'third party'-shipping in the changing structure of grain transports through the Sound.

As can be seen in Graph 14 above, grain transports through the Sound were subject to large annual fluctuations, which were due, in most cases, to disturbances of war, or to harvest failures at the supply or demand side.[3] Despite these fluctuations, the volume of grain transports through the Sound, expressed in tonkm, grew from less than 500 million tonkm before 1770 to more than 2 billion tonkm in the exceptional year 1846. Since the measure applied to these estimates (i.e. tonkm) comprises both a volumetric and a spatial component, including both the tonnage transported and the distance along which this was done, the spectacular growth of grain transports through the Sound was probably due to larger volumes being transported over longer distances. In the following sections, the changing structure of Baltic grain transport production and its main routes will be examined more closely.

2 Van Tielhof, *The Mother of all Trades*, 4. The term 'mother of all trades' was coined for the first time by Johan de Witt in 1671 to designate trade with the Baltic. See: Lindblad, "Nederland en de Oostzee," 8.

3 An extensive literature of case-studies on these topics can be found in many historical journals.

TABLE 8 *Routes of grain transport production, 1670–1699*

	Departure	Destination	Production	%
Main routes	Danzig	Amsterdam	4.262.888[4]	48%
	Königsberg	Amsterdam	1.210.203	14%
	Riga	Amsterdam	748.644	8%
	Reval	Amsterdam	509.481	6%
Routes with a share between 2% and 5%	Libau	Amsterdam	162.435	2%
	Pillau	Amsterdam	146.470	2%
Routes with a share between 1% and 2%	Stettin	Amsterdam	102.530	1%
	Elbing	Amsterdam	73.739	1%
	Danzig	Bremen	64.895	1%
	Danzig	Hamburg	62.562	1%
	Windau	Amsterdam	61.729	1%
	Danzig	Harlingen	57.711	1%
	Danzig	Bergen	46.110	1%
Number of other routes with a share of less than 1%				879

Source: STRO. Production values expressed in 1.000 tonkm.

Before 1700, grain transports through the Sound were highly concentrated on a few routes, of which Danzig-Amsterdam and Königsberg-Amsterdam were by far the most important (see Table 8). About 60% of all grain transport production through the Sound between 1670 and 1699 occurred on these two routes. Additionally, much smaller production was realised where grain was exported to Amsterdam from Riga, Reval, Libau, Pillau, Stettin, Elbing and Windau. The only routes with a share in total grain transport production of more than 1%, which did not link the Baltic to Amsterdam, went from Danzig to Bremen, Hamburg, Harlingen and Bergen (for details, see Appendix 3).

With 82% of all Baltic grain transports arriving at Amsterdam between 1670 and 1699, and 71% originating from Danzig and Königsberg, it is fair to say that the structure of Baltic grain transports during this period was almost monopolized by these three ports. Interestingly enough, however, STRO reveals that grain was transported on 893 different routes during this period, which connected

4 Including 170.778.000 tonkm from Danzig to Unknown destinations.

eighty-seven ports of departure with 207 ports of destination.[5] Grain transport production varied between 287 million tonkm on the route Danzig-Amsterdam in 1684 and 41 million tonkm in 1691, which was even lower than the 60 million tonkm in the 'disaster year' of 1672).[6] Similarly, grain transport production on the route Königsberg-Amsterdam was the highest in 1682, when total production amounted to 91 million tonkm and the lowest in 1673, when a mere 7.5 million tonkm were produced. Grain transport production was much smaller on the other routes, although it should be noted that in 1694 an estimated 128 million ton-km were produced on the route Riga-Amsterdam, thus exceeding the production on all other routes in that particular year.

The heyday of the Baltic grain trade was the first half of the seventeenth century and the dominance of Amsterdam was still overwhelming in 1670–1700.[7] The eighteenth century, however, announced itself as a difficult period, disturbed by war in western Europe (Spanish War of Succession, 1702–1713) and in the Baltic (Great Northern War, 1700–1721). A period of depression started in the late 1690s, lasted until about 1720 and was followed by a period of stagnation accompanied by great instability between 1720 and 1760.[8] Annual fluctuations were enormous throughout this period, but despite a few peaks in the volume of grain shipped through the Sound (1713, 1729, 1740),[9] average grain transport production would remain fairly constant until the mid-1760s[10] (see Graph 14). The Amsterdam grain market mostly served its internal market during this period of stagnation.[11] Three factors have been put forward by Van Tielhof as possible explanations of this enduring state of stagnation in the first half of the eighteenth century. Grain prices declined after 1662, population growth began to slow down or even decline in many regions in Europe, while

5 The large amount of data that underlies these numbers cannot be reproduced in this monograph.

6 1692 and 1696 excluded as data incomplete. Here and in the remainder of this chapter, tonkm estimates for individual routes and years are based on STRO and *Distances between ports, Pub. 151*.

7 Van Tielhof, *The Mother of all Trades*, 50.

8 Morineau, "Le Commerce de la Baltique," 36–39; Van Tielhof, *The Mother of all Trades*, 51.

9 For details, see: Scheltjens, "The Changing Geography," appendix 1 and 2.

10 Van Tielhof, *The Mother of all Trades*, 57.

11 De Vries and Van der Woude, *Nederland 1500–1815*, Ch. 10.2.1. Similarly, Morineau argued that the importance of re-exports of grain from Amsterdam should not be exaggerated: "A [l']allure épisodique de la réexportation [des céréales à partir d'Amsterdam] s'opposent, en partie, une constance rémanente des besoins de l'approvisionnement intérieur", See: Morineau, "Le Commerce de la Baltique," 33–34.

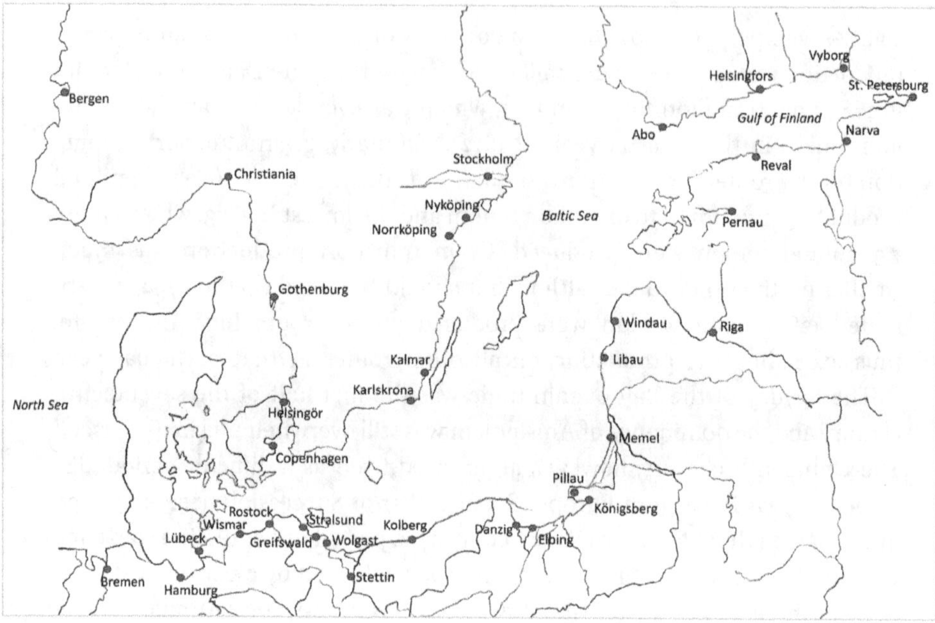

MAP 2 *Map of the Baltic Sea. Based on the map "Denmark, Sweden and Russia on the Baltic", 1847.*

grain cultivation intensified, and political and economic issues at the supply side sometimes severely obstructed the grain trade.[12]

The structure of Baltic grain trade in the first half of the eighteenth century was fairly simple. Demand for Baltic grain was concentrated in the Netherlands, with small shares being distributed to the ports of Bremen and Hamburg, to Norway (primarily Bergen), the Swedish west coast and Scotland (Firth of Forth). Demand in England, France and Portugal was subject to external circumstances, like crop failures, and did not have the same regularity as the aforementioned destinations. The emergence of England as a grain-exporter in the first quarter of the eighteenth century is important and certainly had a negative effect on the size of Amsterdam's market for Baltic grain.[13] Between 1710 and 1760, about 50% of all grain imported to Amsterdam came from England.[14]

---

12 Van Tielhof, *The Mother of all Trades*, 52–54; De Vries and Van der Woude, *Nederland 1500–1815*, Ch. 10.2.1.

13 De Vries and Van der Woude, *Nederland 1500–1815*, Ch. 10.2.1; Ormrod, *The Rise of Commercial Empires*, Ch. 7.

14 De Vries and Van der Woude, *Nederland 1500–1815*, Table 10.2. Even so, Ormrod argues that England's grain surplus accounted only for about 8% of its total domestic production. See: Ormrod, *The Rise of Commercial Empires*, 207.

The supply of Baltic grain continued to be dominated by Danzig and Königsberg, and was complemented by Russia's new possessions on the eastern shores of the Baltic (Riga, Reval, Pernau and Arensburg), two ports in the Kingdom of Prussia (Memel and Pillau), and two ports in the Duchy of Courland and Semigallia (Windau and Libau) (see Map 2).

The participation of Russia's ports in the Baltic, primarily Riga and Reval, was the subject of a series of Russian tax regulations and domestic economic policies, which in the long run proved to be beneficial to these ports.[15] A tax rule of 1731, in which the level of taxation in Riga was made equal to that of Danzig and Königsberg, marked the breakthrough of Riga as a grain supplier. The same rule was applied to Pernau and Reval in 1732 and 1733 respectively.[16] Novel practices in agriculture, like the use of highly fertile land in newly settled regions, were another stimulus to the breakthrough of Russian ports in Baltic grain exports.[17] Still, the Russian government had to deal with crop failures on various occasions, forcing them to prohibit grain exports in 1742, 1744, 1745–1752 and in 1756.[18] Swedish Pomerania held a particular position in the supply of grain, the region had a relatively large number of small grain exporting ports from where shipments were carried out through the Sound on a regular basis. It is likely that Swedish Pomerania catered for the Swedish home-market first and that Sound traffic was only an additional activity.

As a result, grain transport production during the period 1700–1760 was generally lower than in the closing decades of the seventeenth century. The high concentration of grain transports on the routes Danzig-Amsterdam and Königsberg-Amsterdam remained intact during a major part of this period. Between 1700 and 1729, 68% of all grain transport production was realised by carrying grain from Danzig and Königsberg to Amsterdam. The same additional routes as during the previous period completed the picture, although, again, it should be noted that the total number of routes for Baltic grain was much larger (733 different routes in the period 1700–1729) (for details see Table 9).

Only eighteen routes had a share of more than 1% of total grain transport production between 1700 and 1729, connecting ten different ports of departure and eight different ports of destination (see Table 9). In this sense, continuity prevailed in the spatial structure of the Baltic grain trade in the first decades of the eighteenth century. In fact, the disturbances of war, which led to a contraction in the production of grain transport services, resulted in an even stronger

15 Van Tielhof, *The Mother of all Trades*, 60.

16 Semenov, *Izuchenie istoricheskikh svedenii*, 147.

17 Van Tielhof, *The Mother of all Trades*, 60.

18 Semenov, *Izuchenie istoricheskikh svedenii*, 147–148.

TABLE 9 *Routes of grain transport production, 1700–1729*

	Departure	Destination	Production	%
Main routes	Danzig	Amsterdam	2.381.594	52%
	Königsberg	Amsterdam	740.102	16%
Routes with a share between 2% and 5%	Pillau	Amsterdam	76.290	2%
Routes with a share between 1% and 2%	Libau	Amsterdam	58.857	1%
	Reval	Amsterdam	54.465	1%
	Riga	Amsterdam	53.473	1%
	Danzig	Hamburg	53.075	1%
	Danzig	Bremen	48.128	1%
	Stettin	Amsterdam	43.497	1%
	Danzig	Bergen	42.076	1%
	Danzig	Bordeaux	34.385	1%
	Danzig	Emden	31.460	1%
	Danzig	Harlingen	30.999	1%
	Stralsund	Amsterdam	28.875	1%
	Windau	Amsterdam	27.342	1%
	Wolgast	Amsterdam	27.117	1%
	Königsberg	Bremen	26.871	1%
	Danzig	Nantes	24.154	1%
Number of other routes with a share of less than 1%				715

Source: STRO. Production values expressed in 1.000 tonkm.

concentration of Baltic grain transports in a few ports east and west of the Sound. Most clearly, the additional grain routes, linking Riga, Libau and Pillau to Amsterdam suffered a severe blow in the first decades of the eighteenth century, with their share in total grain transport production dropping from 12% in 1670–1699 to only 4% in 1700–1729. The first changes in the spatial structure of Baltic grain transport production would become apparent in the next 30-year period (1730–1759). In particular, three developments stand out.

At the supply side, the recovery of the additional grain routes is remarkable, even though total grain transport production was still significantly lower than in the late-seventeenth century. This, however, is also true for Danzig and

Königsberg, whose grain transport production stayed at roughly the same level as during the period of war that marked the beginning of the eighteenth century. At the demand side, Amsterdam's predominance in Baltic grain transport production started to weaken after 1730. Grain transport production destined to Amsterdam was more than halved in comparison to the final decades of the seventeenth century and there was no sign of recovery after 1730. Moreover, the share of Amsterdam in total Baltic grain transport production declined steadily from 82% in 1670–1699 to 78% in 1700–1729 and 63% in 1730–1759. At the same time, the number of places involved in Baltic grain transport production on either side of the Sound progressively increased.

Moreover, despite the absolute decline in the volumes transported, the number of ports involved in grain exports rose from 87 in 1670–1699 to 104 in 1730–1759, an increase of about 16%. The dominance of the main suppliers, however, hardly experienced any pressure: the ports of Danzig, Königsberg, Riga, Pillau and Libau largely maintained their previous positions. In 1730–1759, grain transport production from these ports amounted for 82% of the total, whereas in the closing decades of the seventeenth century, their share was about 85%.

Finally, the number of ports receiving grain from the Baltic rose from 207 in 1670–1699 to 273 in 1730–1759, an increase of 24%. The increasing participation of ports east and west of the Sound in the Baltic grain trade did not miss its effect on the routes followed by this trade. The number of routes rose to almost 1,000 and for the first time, the traditional routes Danzig-Amsterdam and Königsberg-Amsterdam accounted for less than 50% of all grain transports (see Table 10).

The number of ports involved in the Baltic grain trade and the number of routes sailed would explode in the next decades, which witnessed both a revival of Baltic grain and a structural change in its spatial structure. The era of overwhelming dominance by only a few ports was over.

In the 1760s a period of renewed growth started as a consequence of rising grain prices provoked by renewed population growth.[19] This relatively steady period was disrupted only by the Fourth Anglo-Dutch War (1780–1784) and crop failure in France in 1788.[20] But while in most European countries population grew and prosperity increased in the second half of the eighteenth century, this was not the case for the Netherlands.[21] The total volume of grain transported through the Sound rose quickly after 1760, but the volume of grain

19 Van Tielhof, *The Mother of all Trades*, 58–60.

20 Van Tielhof, *The Mother of all Trades*, 58–59; Faber, "Shipping to the Netherlands," 100–102.

21 Faber, "Structural Changes," 93–94.

TABLE 10 *Routes of grain transport production, 1730–1759*

	Departure	Destination	Production	%
Main routes	Danzig	Amsterdam	2.097.447	37%
	Königsberg	Amsterdam	508.505	9%
Routes with a share between 2% and 5%	Riga	Amsterdam	252.098	4%
	Reval	Amsterdam	141.540	3%
	Danzig	Bergen	129.295	2%
	Libau	Amsterdam	125.977	2%
	Pillau	Amsterdam	102.279	2%
Number of routes with a share between 1% and 2%				18
Number of other routes with a share of less than 1%				973

Source: STRO. Production values expressed in 1.000 tonkm.

transported to the Dutch Republic stayed at roughly the same level as during the first half of the eighteenth century. After 1760, the share of the Dutch Republic in total grain transports from the Baltic declined rapidly.[22]

The main routes of the Baltic grain trade were still Danzig-Amsterdam and Königsberg-Amsterdam, but their share had now diminished to only 20% of total Baltic grain transport production. Especially the traditional route Danzig-Amsterdam went into absolute decline, with volumes being even lower than during the preceding decades of general crisis. For several reasons, probably most importantly the impact of the first partition of Poland with its adverse Prussian policies designed to demolish Danzig's grain trade,[23] the route Danzig-Amsterdam was unable to benefit from the general revival of Baltic grain in western Europe. The route Königsberg-Amsterdam appeared to be more stable, although it does not seem to have profited particularly much from the forced decline of Danzig. This was different for the so-called additional routes of Baltic grain: Riga-Amsterdam, Libau-Amsterdam and Pillau-Amsterdam, whose share rose to 9% of total grain transport production, or 1.2 billion tonkm (see Table 11).

22 Ormrod, *The Rise of Commercial Empires*, 231.

23 Rachel, *Der Handels-, Zoll- und Akzisepolitik Preußens*, II, 397–409; Straubel, "Stettin," 136.

TABLE 11 *Routes of grain transport production, 1760–1789*

	Departure	Destination	Production	%
Main routes	Danzig	Amsterdam	1.588.482	13%
	Königsberg	Amsterdam	825.910	7%
Number of routes with a share between 2% and 5%	Pillau	Amsterdam	417.426	3%
	Riga	Amsterdam	407.851	3%
	Libau	Amsterdam	375.762	3%
	Danzig	Nantes	330.408	3%
	Danzig	Bordeaux	275.050	2%
	Libau	Schiedam	188.270	2%
Number of routes with a share between 1% and 2%				24
Number of other routes with a share of less than 1%				1993

Source: STRO. Production values expressed in 1.000 tonkm.

Compared to the previous period, the volumes transported on the additional routes rose by a factor of 2.5. But even so, more fundamental changes occurred after 1760. The total share of the traditional and additional Baltic grain routes diminished sharply, from 72% in 1700–1729 to 54% in 1730–1759 and to only 29% in 1760–1789; a development which coincided with a spectacular rise in the number of ports involved in the supply of grain. Baltic grain was transported on more than 2.000 different routes between 1760 and 1789, marking the end of the era of dominance of Danzig, Königsberg and Amsterdam in this trade (see Table 13). It is true, however, that still about 50% of total grain transport production originated in Danzig and Königsberg, but their share continued to decline. Of greater importance was the gradual rise of grain transport production from Riga, Libau and Pillau, whose volumes rose exponentially after 1760. These three ports, which have been labelled 'additional ports' earlier in this chapter, now accounted for 29% of total grain transport production between 1760 and 1789, or almost 3.6 billion tonkm. These ports benefited from the isolation of Danzig and (to a lesser degree) Königsberg as well as from the fact that Russia began to produce grain surpluses on a regular basis.[24]

24 Van Tielhof, *The Mother of all Trades*, 60.

TABLE 12 *Production of grain transport services to Amsterdam (A'dam) and Rotterdam (R'dam), expressed in million tonkm*

	1730–1759			1760–1789		
	Total	To A'dam	To R'dam	Total	To A'dam	To R'dam
Danzig	2.914	2.097	59	4.008	1.588	115
Königsberg	928	509	42	2.240	826	
Riga	334	252		1.602	407	362
Pillau	165	102		986	417	
Libau	216	126		982	376	321

Note: Rotterdam includes Schiedam
Source: STRO.

Generally speaking, the number of cities creating demand for grain increased significantly. The number of different destinations for Baltic grain rose from 273 in 1730–1759 to 439 in 1760–1789. The share of Amsterdam diminished to 33%, while demand for Baltic grain spread widely along the Dutch coastline, going as far south as Vlissingen and as far east as Delfzijl. Of particular importance was the emergence of a new concentration of demand for Baltic grain in SMR, notably in Schiedam en Rotterdam, where jenever distilleries required more and more grain to fulfil the needs of an increasing number of consumers.[25] The effects of this development are summarized in the table below: the production of grain transport services increasingly shifted towards different Dutch markets than that of Amsterdam (see Table 12). In comparison to the preceding period, the production of grain transport services destined to Rotterdam or Schiedam experienced an eightfold increase after 1760.

Demand for grain also spread along the East Frisian coastline, with Emden as its main port. Along the Swedish west coast and in southern Norway, demand for Baltic grain became widely spread, most likely as a result of the economic prosperity and related population growth that was provoked by the so-called Bohuslän herring period that had started back in 1752.[26] Demand in England and Scotland started to rise in the late 1760s and grew rapidly to a

25 Ormrod, *The Rise of Commercial Empires*, 221–225; Sluyterman, *Drie Eeuwen De Kuyper*; Van Riemsdijk, *Het Brandersbedrijf*.

26 Ljungman, "The Great Bohuslän," 220–239; Corten, *Herring and Climate*; Lilja, "Small towns in the Periphery," 50–76.

wide-spread, steady demand by the early 1780s. Not only ports located on the English east coast were now involved in demand for Baltic grain, Liverpool also started to receive large amounts of grain from the Baltic. Clearly, England and Scotland were no longer capable of fulfilling their own demand for grain.[27] In addition to a process of diversification in the traditional markets for Baltic grain, the emergence of demand concentrations in southern Europe, mostly along the French Atlantic Coast and in Portugal, but also in the Mediterranean, are evidence of a second structural change on the demand side. A third structural change on the demand side occurred in the Baltic proper, where Sweden emerged as a grain importing country, procuring grain mostly from Pomerania and from the Baltic provinces of the Russian Empire. This, in turn, reduced the surpluses available for export out of the Baltic sea region.[28]

The processes of structural change that took off after 1760 became more and more apparent after 1790. The diversification of the distribution of Baltic grain continued to increase. More ports supplied Baltic grain and more ports received it too, leading to 2,244 different routes overall on which grain transport services were provided between 1790 and 1819. The volumes exported through the Sound continued to rise as well, despite the disturbances caused by the Napoleonic wars.

Maintaining a focus on the traditional routes of Baltic grain, Danzig-Amsterdam and Königsberg-Amsterdam, it becomes clear that they had finally lost their leading position, both in absolute and relative terms (see Table 13). Their place was not taken wholly by the additional routes to Amsterdam, which accounted for 7% of total grain transport production, although the highest volumes were in fact produced on the route Riga-Amsterdam. Alongside the 'additional routes', a number of routes connecting the Baltic to SMR emerged and rapidly gained importance. Grain transport production originating in Riga, Pillau and Libau and destined to Rotterdam, Schiedam or 'The Meuse' accounted for 11% of total grain transport production between 1790 and 1819. The production destined for SMR equalled that of Amsterdam in 1790–1819. Amsterdam's share declined to 15% of total Baltic grain production, while the share of Rotterdam, Schiedam and 'The Meuse' rose to 14%, equal to that of London. In absolute terms, the decline of Amsterdam and the rise of new centres of demand for Baltic grain becomes even more apparent. The production of grain transport services destined for Amsterdam halved, while those destined for London, Lisbon and SMR experienced fivefold, threefold and twofold increases respectively (see Table 14).

---

27 Ormrod, *The Rise of Commercial Empires*, Ch. 7; Faber, "Structural Changes," 88.

28 Van Tielhof, *The Mother of all Trades*, 61.

TABLE 13 *Routes of grain transport production, 1790–1819*

	Departure	Destination	Production	%
Main routes	Riga	Amsterdam	529.491	4%
	Pillau	London	462.895	3%
	Danzig	Amsterdam	462.548	3%
	Königsberg	Amsterdam	442.912	3%
	Libau	Schiedam	412.417	3%
	Riga	Schiedam	374.930	3%
	Danzig	London	371.700	3%
	Pillau	Amsterdam	347.152	2%
	Danzig	Lisbon	343.094	2%
	Danzig	Liverpool	294.109	2%
	Riga	Rotterdam	271.369	2%
	Riga	London	258.046	2%
	Riga	Lisbon	253.911	2%
	Riga	De Maas	230.987	2%
	Königsberg	London	227.489	2%
Number of routes with a share between 1% and 2%				24
Number of other routes with a share of less than 1%				2205

Source: STRO. Production values expressed in 1.000 tonkm.

TABLE 14 *The main destinations of grain transport production, 1760–1819, in million tonkm*

Destination of production	1760–1789	1790–1819
Amsterdam	4.149	2.283
Rotterdam, Schiedam, Meuse	1.013	2.186
London	419	2.078
Lisbon	475	1.283
TOTAL	6.056	7.830

Source: STRO.

Structural changes at the supply side were equally drastic. Riga, Pillau and Libau overtook Danzig and Königsberg, while other ports, like Memel, St. Petersburg and, more importantly, a number of ports in Pomerania made clear advances in the production of grain exports. Again, a structure of clear dominance was finally replaced by a new structure that was dispersed, yet capable of producing higher volumes than ever before. In general, the evolution of the spatial structure of Baltic grain transport after 1760 has been accurately described by Faber as a process of "enlargement, dispersal and stabilisation of the [European] market" that made it no longer necessary for the staple function to be confined to the Netherlands.[29] While the Dutch Republic continued to play a major role in Baltic grain transport, its position was no longer one of clear dominance. The shares of England, Scotland and of other European countries increased significantly and, taken together, their volume surpassed that of the Netherlands.

After 1820 the new structure, that had emerged under the influence of the partitions of Poland, the rise of Russia's Baltic provinces and Pomerania as grain exporting regions, the growth of new centres of demand for Baltic grain in SMR, Scotland and England as well as along the Atlantic coastline, and the diversification of routes that was a logical consequence of these developments, was largely consolidated. At the supply side, among 140 different ports exporting grain through the Sound between 1820 and 1849, three clusters can be discerned, each of which had a share of 20–30% in total grain transport production. The first cluster includes the traditional ports of Danzig and Königsberg, which still provided about 27% of total grain transport production. The second cluster comprises the 'additional' ports of Riga, Pillau and Libau, which accounted for 21% of total grain transport production, while the third cluster is constituted by the ports of Stettin, Stralsund, Wolgast, Greifswald and Wismar, whose share in total grain transport production rose drastically to about 30%.

At the demand side, Amsterdam's position continued to decline, both in relative and absolute terms. Only 9%, or 1,747 million tonkm, of total grain transport production was destined for the gateway of ZDZ. Amsterdam's role as the biggest consumer of Baltic grain was now overtaken by London (19%) and other English and Scottish ports such as Liverpool (5%), Hull (4%), Leith (3%) and Newcastle (3%). Unfortunately, the registrations in STRO become less precise after 1820, which explains why 'The North Sea' becomes an important destination for Baltic grain transport production. Its production volumes cannot be attributed to a specific port, which affects the breakdown of routes

29 Faber, "Structural Changes," 94.

TABLE 15 *Routes of grain transport production, 1820–1849*

	Departure	Destination	Production	%
Main routes	Wolgast	North Sea	1.729.566	9%
Routes with a share between 2% and 5%	Stettin	London	731.627	4%
	Danzig	London	695.141	4%
	Danzig	Amsterdam	523.628	3%
	Königsberg	Amsterdam	490.979	3%
	Königsberg	London	338.742	2%
	Pillau	London	305.106	2%
	Riga	Schiedam	284.055	2%
	St. Petersburg	London	281.263	2%
Number of routes with a share between 1% and 2%				30
Number of other routes with a share of less than 1%				2491

Source: STRO. Production values expressed in 1.000 tonkm.

of Baltic grain transport production. The route with the highest production volumes connects Wolgast with the North Sea (see Table 15).

The number of different routes continues to rise in the first half of the nineteenth century and the diversification becomes even more prominent. Whereas a century earlier, two routes were responsible for 60–70% of Baltic grain transport production, the share of the most productive routes (Wolgast-North Sea excluded) now no longer exceeded 4% and only thirty-nine out of 2,560 routes had a share of more than 1%. Together, these routes accounted for about 60% of total Baltic grain transport production between 1820 and 1849.

On an international scale, grain exports to the Dutch Deltas had definitively lost their previous importance in the first half of the nineteenth century. The traditional routes Danzig-Amsterdam and Königsberg-Amsterdam still served the greater part of the Amsterdam market, while the emerging routes from Riga, Pillau and Libau to Rotterdam, Schiedam and 'The Meuse' were equally important in terms of production. Both had a share of 6% in total grain trans-

TABLE 16 *Evolution of the Dutch market for Baltic grain, annual averages in million tonkm*

	ZDZ	SMR	LDE	Rest	Total
1670–1699	245,13	2,03	0,23	47,77	295,17
1700–1729	120,03	0,47	0,63	30,13	151,27
1730–1759	117,83	7,53	0,73	60,77	186,87
1760–1789	143,63	36,1	2,8	232,3	414,83
1790–1819	80,53	75,87	0,37	338,03	494,8
1820–1849	66,37	73,97	0,77	477,87	618,97

Source: STRO.

port production between 1820 and 1849. The routes from Riga, Pillau and Libau to Amsterdam disappeared almost completely after 1820. Grain transport production destined to the Dutch Deltas was a fraction of what it had been before 1790 (see Table 16) and it can, therefore, be concluded that the famous Dutch market for Baltic grain, that once was 'the mother of all trades', had now become a secondary player in an international grain market that continued to expand.

The Spatial Structure of Grain Transport Services

Inevitably, the changes in the spatial structure of Baltic grain transport services had profound effects on the routes on which these services were produced. In the previous section, we have gained insight in the evolution of the Baltic grain market, focussing explicitly on the production of grain transport services and the structural changes it underwent both at the supply and at the demand side. The relative and absolute decline of the 'mother of all trades' was discussed, highlighting in particular the declining share of Amsterdam in total grain transport production via the Sound. Additionally, it was emphasized that changes in the economy of the Dutch Republic, in particular the rise of SMR as a new centre of trade and commerce, which was already described in Chapter 1, led to the emergence of new routes for Baltic grain. These routes complemented and later partially replaced both the traditional routes of Danzig-Amsterdam and Königsberg-Amsterdam and the additional routes connecting Riga, Pillau and Libau to Amsterdam. Of course, grain transport services were

TABLE 17 *Selection of Dutch maritime transport communities*

ZDZ	SMR	LDE
Ameland, Vlieland, Terschelling, Amsterdam, Hoorn, Harlingen, Workum, Woudsend, Stavoren, Lemmer, Hindeloopen, Makkum, Heerenveen, Texel, Broek, Warns, Molkwerum, Edam, Terhorne, Zaandam, Enkhuizen, Joure, Schellinkhout, Venhuizen	Rotterdam, Dordrecht, Schiedam	Veendam, Groningen, Pekela, Dokkum, Delfzijl

produced on many other routes, but until well into the nineteenth century, they remained of secondary importance to the spatial structure of total grain transport production to the Dutch Deltas.

This section is devoted to the reaction of the Dutch maritime transport sector to the changes in volume, spatial structure and geography of distribution routes that were described above. To this end, the behaviour of communities of shipmasters on the traditional, additional and emerging routes of Baltic grain is examined, focusing on turnover, evolving market shares and operational patterns. Evolving market shares are necessary to gain insight in the demand for Dutch maritime transport on the traditional and emerging routes of Baltic grain, while turnover and operational patterns serve as indicators of the presence and strength of path dependence in the behaviour of Dutch shipmasters. The analysis of the behaviour of maritime transport communities in the Dutch Deltas is executed using a selection of thirty-two different places. The criterion for selection was a more than 1% share in the production of grain transport services on traditional, additional or emerging routes during at least one thirty-year period. The maritime transport communities are geographically distributed as follows (see Table 17). ZDZ is highly represented with twenty-four communities, whereas SMR and LDE are represented with three and five places, respectively.

It has been noted in historiography (and explained from an economic-geographical perspective in Chapters 1 and 2), that a concentration of maritime transport services related to the Baltic area is found north and north-east of Amsterdam, in West Friesland, Friesland and the Province of Groningen.[30] This is reflected in the spatial distribution of the thirty-two maritime transport communities that are subject to analysis in this section.

For the Dutch communities that were involved in Baltic grain transport production, their degree of participation was subject to changes in the spatial structure of the grain trade and the operational strategies adopted by the shipmasters belonging to these communities. A first distinction must be made between routes on which Dutch shipmasters had the dominant share in the transportation of grain, routes on which they were a minority and routes where their share was negligible. The latter were mostly routes that did not have a 'Dutch' component, i.e. routes linking places in the Baltic Sea with places along the coastline of the North Sea. Dutch shipmasters had no share whatsoever in routes to England, Scotland, Norway and Bremen. Many of these routes, except those to Norway, were dominated by shipmasters originating from the port of destination of Baltic grain. For example, between 1770 and 1795, when England and Scotland became large importers of Baltic grain, the average share of shipmasters from Newcastle on grain transports between Danzig and Newcastle exceeded 90%. On the other hand, routes that emerged later in the eighteenth century, like Danzig-Liverpool, were dominated by shipmasters from Danzig (average share of 72% of all shipments). Similarly, the grain routes on which Dutch shipmasters had a major share connected ports in the Baltic Sea to ports in the Dutch Deltas. In the previous section, it was established that the market share of these routes underwent dramatic changes in the course of the eighteenth and early nineteenth centuries. The traditional and additional routes were complemented and then gradually replaced by emerging 'new' routes.

Our questions regarding the behaviour of Dutch shipmasters examine the balance shift from traditional and additional to emerging routes. Were the same communities active on the traditional, additional and emerging routes? If yes, did Dutch maritime transport communities change routes in the course of the eighteenth century? If not, where did the shipmasters on the emerging routes come from? What happened if maritime transport communities did not move away from declining traditional routes? Did they develop a new cargo specialisation maybe? Who replaced them on the traditional routes when they did move away?

A first indication of different operational strategies being applied by different maritime transport communities can be found in the strength of the correlation coefficient between the participation of individual Dutch maritime transport communities in the production of Baltic grain transport services and

30 Van Der Woude, *Het Noorderkwartier*, 384; Van Royen, "Varend volk," 86–87.

total grain transport production on the traditional, additional and emerging routes (see Table 18).

The correlation coefficients reveal substantial differences between the maritime transport communities selected (see Table 18). Three groups can be distinguished. The first group comprises maritime transport communities that had a high positive correlation with the traditional routes and a negative correlation with the emerging routes. This group consists of the maritime transport communities of Ameland, Enkhuizen, Harlingen, Hoorn, Molkwerum, Terschelling, Vlieland, Warns, Woudsend and Zaandam. The second group contains maritime transport communities that had a lower positive correlation on the traditional routes (less than 0.5) and a negative correlation on the emerging routes. The maritime transport communities that belong to this group are Amsterdam, Heerenveen, Hindeloopen, Joure, Makkum, Schellinkhout and Texel. The third group comprises maritime transport communities that returned correlation coefficients below 0.5 on all types of routes, including negative correlations on one or two of them. This heterogeneous group is composed of the maritime transport communities of Broek, Delfzijl, Dokkum, Dordrecht, Edam, Groningen, Lemmer, Pekela, Rotterdam, Schiedam, Stavoren, Terhorne, Veendam, Venhuizen and Workum.

In the following analysis of these maritime transport communities attention will be paid to the two main determinants of the operations of maritime shipmasters: the routes they sailed and the cargoes they carried. Rather than going into detail regarding either of these determinants, a general analysis based on broad categories is pursued. A first distinction is made between grain, timber and other cargoes. The category grain comprises thirty-six different cargo descriptions in STRO, of which rug (rye), hvede (wheat), byg (barley), ærter (peas), malt (malt), mel (flour), gryn (groats), boghvede (buckwheat), rug og byg (rye and barley) and hirsegryn (groats of millet) are the most frequent. The category timber comprises twenty-four different cargo descriptions of which dehler (deals), staver (staves), bjælker (balks), brædder (boards), planker (planks), lægter (laths), sparrer (spars), master (masts), spirer (wooden spikes) and bohler (bowls) are the most frequent.[31] The other goods comprises about a 1,000 different products exported from the Baltic on Dutch ships, most importantly hemp, hempseed, flax, flaxseed, potash and iron; their full range can be reconstructed using STRO.

A second division is made between 'traditional', 'additional' and 'emerging' grain and timber routes. The former have been introduced in the previous sec-

31 The translations of the original Danish cargo descriptions are taken from a preliminary multilingual list of products in STRO, which is available at www.soundtoll.nl.

TABLE 18 *Correlation coefficients of selected Dutch maritime transport communities on traditional, additional and emerging grain transport routes, 1670–1850*

Group	Domicile	Traditional	Additional	Emerging
Group 1	Ameland	0,76	0,04	-0,39
	Enkhuizen	0,52	0,02	-0,35
	Harlingen	0,86	0,32	-0,21
	Hoorn	0,58	0,23	-0,28
	Molkwerum	0,73	0,35	-0,27
	Terschelling	0,81	0,13	-0,17
	Vlieland	0,73	0,36	-0,31
	Warns	0,69	0,53	-0,24
	Woudsend	0,62	0,21	-0,26
	Zaandam	0,59	0,26	-0,15
Group 2	Amsterdam	0,41	0,39	-0,07
	Heerenveen	0,48	-0,06	-0,37
	Hindeloopen	0,44	0,38	-0,24
	Joure	0,35	0,29	-0,14
	Makkum	0,40	0,21	-0,16
	Schellinkhout	0,28	0,09	-0,42
	Texel	0,49	0,40	-0,19
Group 3	Broek	0,43	0,41	0,23
	Delfzijl	0,26	0,37	0,12
	Dokkum	0,13	0,16	-0,04
	Dordrecht	0,09	0,40	0,45
	Edam	0,38	0,39	0,03
	Groningen	-0,03	0,22	0,23
	Lemmer	0,44	0,47	0,07
	Pekela	0,12	0,08	0,30
	Rotterdam	0,21	0,37	0,13
	Schiedam	0,62	0,28	0,06
	Stavoren	0,32	0,22	0,04
	Terhorne	0,53	0,40	0,17
	Veendam	-0,01	-0,16	0,11
	Venhuizen	0,69	0,61	0,24
	Workum	0,21	0,31	0,09

tion; the latter selection of routes is based on an analogical quantitative survey of Baltic timber transport production between 1670 and 1850. The routes Narva-Amsterdam and Vyborg-Amsterdam were labelled 'traditional timber routes'. The 'additional' timber routes were Danzig-Amsterdam and Riga-Amsterdam. Finally, the 'emerging' timber routes connected Vyborg, Narva and Riga to Dordrecht and Middelburg. Moreover, a further distinction is made between routes to the Dutch Deltas (ZDZ, SMR and LDE), routes to Bremen, Emden or Hamburg (further referred to as: BEH) and routes that are commonly depicted in historiography as *voorbijlandvaart* (VBL). A major historiographical issue related to the increasing complexity of the spatial structure of Baltic grain transport in the course of the eighteenth century is that of the importance of the so-called *voorbijlandvaart*, routes that bypass Amsterdam or other Dutch ports to connect places in southern Europe directly with places in northern Europe.[32] If, indeed, the Dutch maritime transport sector managed to secure its market position by establishing direct connections bypassing the Dutch Deltas, this could be seen as the successful adaptation of the maritime transport sector to new economic circumstances. The growing number of routes involved in Baltic grain transport and the dispersal or decentralisation of Amsterdam's economic functions indicate that there is no single route that succeeded the formerly dominant Danzig-Amsterdam and Königsberg-Amsterdam. Indeed, a comparison of the main routes and their respective average volumes and shares of total grain transports in a number of given years makes clear that the Baltic grain market became an increasingly complex market in which many parties were involved, both on the supply and on the demand side of business.

The rationale for including VBL and BEH routes in the analysis is that if it can be sustained that the Dutch maritime transport sector managed to secure its market position by establishing direct connections bypassing the Dutch Deltas, this could be seen as the successful adaptation of the maritime transport sector to the new economic circumstances which had led to the dispersal and decentralisation of international commodity markets and the increasing complexity of the maritime transport market. If this proves to be impossible, we may have to conclude that the Dutch maritime transport sector was not the first modern transport system after all.

Group 1

Estimates of total transport production of the first group of maritime transport communities, expressed in tonkm, reveal that the character of their occupations was rather straightforward. Grain transport continuously dominated

32 Christensen, *Dutch trade*, 246–248.

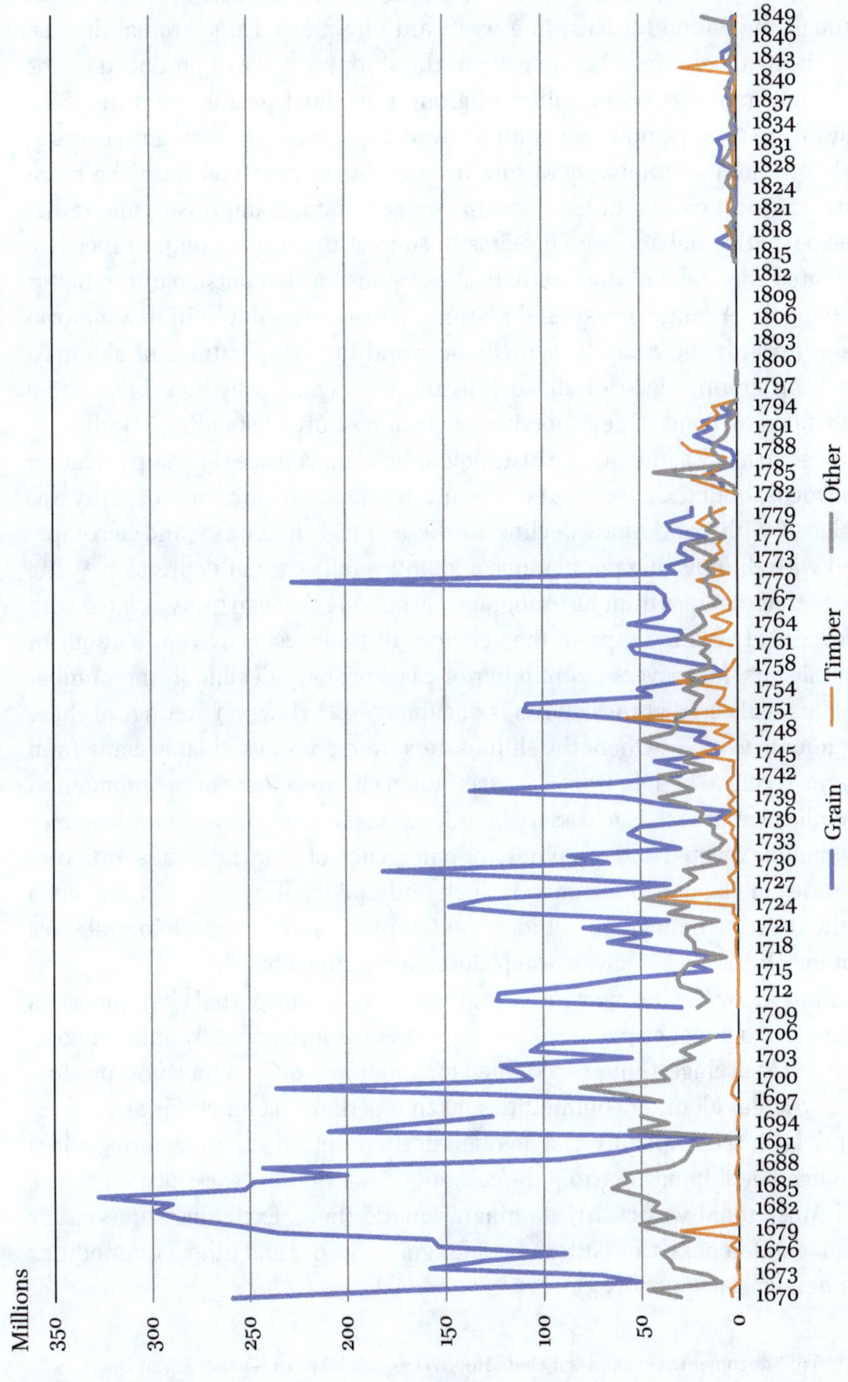

GRAPH 15 *Estimates of total transport production by the first group of Dutch maritime transport communities, expressed in million tonkm.*
BASED ON STRO.

their operations, leaving little room for the transportation of other goods through the Sound (at least in a westward direction). Large annual fluctuations notwithstanding, the intensity of the shipping services produced by the maritime transport communities of group 1 declined steadily from the 1680s onwards to disappear almost completely at the end of the eighteenth century. Only on one exceptional occasion, in 1770, did shipmasters from the maritime transport communities of group 1 make a sudden, impressive and rather unexpected comeback, which seems to suggest that those communities had relocated the scope of their activities to regions located outside of the Baltic, perhaps the Atlantic. For several reasons, however, it is doubtful that a movement 'towards the Atlantic' actually occurred in the operations of shipmasters from group 1. First of all, such assumption can hardly be substantiated with facts. Second, a few notable exceptions notwithstanding (Harlingen, Woudsend and Zaandam), the historiography of most places in group 1 is rather equivocal about the absence of economic development: former prosperity was replaced with continuous decline in the eighteenth century and accompanied with diminishing populations and an overall sense of depression.[33] The image that emerges from historiography is not one that can be associated with a successful 'Atlantic turn' in the activities of their respective populations of shipmasters. Moreover, in a number of cases, a sharp decline in the number of ships registered at these ports is documented.[34] However, neither of these arguments explains where the shipmasters from group 1 suddenly came from in 1770, trying to benefit from the exceptional circumstances of the moment. A microlevel approach is necessary in order to make a step towards understanding the unexpected and temporary re-emergence of maritime transport communities in decline, unfortunately, such undertaking lies outside of the scope of the present monograph, although in Chapter 4 some methodological tools that may bring such analysis a step closer are outlined briefly.

Grain was by far the most important commodity transported by shipmasters from group 1 and, between 1670 and 1850, it was not replaced by another commodity. On average, timber accounted for about 10% of total transport production, whereas all other commodities taken together accounted for about 25% (for details, see Appendix 3). The main destinations of the transport services produced by shipmasters from the communities in group 1 were located in ZDZ and Amsterdam was clearly dominant among them. Excluding some minor relative differences, this pattern held for grain, timber and other commodities (for details, see appendix 3).

33 Well-documented examples include Hoorn (Lesger), Wadden (Draper & Van Duijl).

34 For example in: Lesger, *Hoorn*, 151–178.

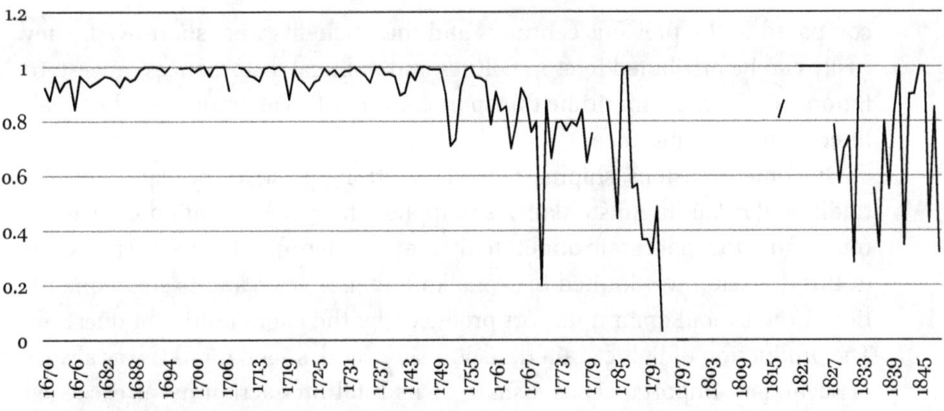

GRAPH 16 *Share of* ZDZ *as destination of grain transport services produced by shipmasters of group 1.*
BASED ON STRO.

The predominance of ZDZ as a destination for grain transport services produced by shipmasters from group 1 was of an exceptional degree during most of the eighteenth century, but it started to decline somewhat after 1760. The direction of grain transports produced by shipmasters from group 1 became increasingly diverse after 1780 (see Graph 16). It should be noted, however, that the latter development coincided with a sharp decline in total grain transport production by shipmasters from group 1.

The level of engagement of shipmasters from these communities in *voorbijlandvaart* was limited, both in terms of volume and duration. VBL was insignificant between 1670 and 1750, with shares of total grain production by shipmasters from group 1 never exceeding 6%. Between 1750 and 1795, VBL obtained a slightly more regular character, with shares in individual years rising up to 30% in the early 1750s and the early 1780s. During the same period, two exceptional episodes are recorded, in 1770 and between 1785 and 1790. In 1770, VBL accounted for almost 79% of total grain transport production by shipmasters from group 1; in 1785–1790 the average share of VBL was 27%. There was a fundamental difference between these two episodes, however: in 1770, the total production of grain transports by shipmasters from group 1 was estimated at more than 227 million tonkm, whereas in the late 1780s, their production did not exceed 10 million tonkm in a single year. After 1794, VBL again lost its significance, which can be explained by the continued decline of transport services production by the maritime transport communities in group 1. Only during a few years, like 1817 or 1829–1832, could modest upswings in the production of grain transports be witnessed, but these were very small compared to the previous centuries and their benefits were short-lived. They easily can be attributed to crop failures and political disturbances in western Europe. By 1795, the maritime transport communities of group 1 had lost their international significance.

On some occasions, shipmasters with a strong positive correlation on the traditional routes to Amsterdam also supplied their transportation services on routes from the main grain outlets to Bremen, Emden and Hamburg. The extent of this diversion was limited to a small number of years and to small quantities of the overall grain transport produced by the communities in question. The production of Baltic grain transport services destined for BEH was always of secondary importance, as a share of its total production never exceeded 10% per year. Grain transport production destined to BEH never obtained a truly regular character. The largest uninterrupted series of grain transports to BEH were 1679–1689 and 1719–1726. In absolute terms, total production of grain transport services never exceeded 9 million tonkm (in 1688, accounting for 3.9% of total grain transport production in that year).

Sporadically and, again, in small quantities, the shipping communities of the first group were engaged in shipments of grain to other places in the Dutch Deltas, besides Amsterdam. The impact of such transport services was limited at first but from the late 1760s onwards, shipments to SMR obtained a more regular character, with a production of transport services destined for SMR that accounted for between 5% and 20% of total grain transport production by shipmasters from group 1. In the early 1790s, roughly the same level of production was achieved, accounting now for up to 50% of total grain transport production by shipmasters from group 1. Diminishing volumes transported to ZDZ as well as the relative decline in the overall production of grain transports by shipmasters from group 1 are important elements in the explanation of these changes. Following the interruptions caused by the Napoleonic Wars and the continental blockade, a continuation of the transport production pattern of the 1790s can be observed, albeit in ever diminishing volumes. SMR complemented ZDZ as a destination of grain transports and occasionally (in 1832, 1837 and 1842) it even surpassed ZDZ. However, the transport production level was low and internationally insignificant.

The production of timber transports and the transportation of other goods never overtook the role of grain transports but, in light of the declining production of the latter and the relatively stable production of timber and other transports, represented an increasingly large proportion of shipmasters' total transport production in group 1. This, however, does not mean that they presented a full-blown alternative for the production of grain transports. If that had have been the case, the shipmasters communities' absolute decline (fewer and fewer shipmasters in the community) would not have been as large as it was, nor would the total production have declined to the extent that it did. Looking ahead, the decline will become clear especially in comparison with the production patterns of shipping communities in groups 2 and 3. The dominant destinations for grain, timber and other goods transports from the Baltic were similar and their relative importance developed in a similar way too. In the course of the eighteenth century, VBL and SMR emerged as secondary destinations for the production of transport services by shipmasters from group 1. But all of this happened during a continuous process of declining production.

The main routes on which shipmasters from group 1 were active were the 'traditional' and 'additional' grain routes to Amsterdam as well as the 'additional' timber routes, which also had Amsterdam as their destination. Not surprisingly, the 'additional' timber routes overlapped with some of the grain routes to Amsterdam, Danzig-Amsterdam and Riga-Amsterdam. Did a change of product occur on these routes? Not at all, as can be seen on the graph below (see Graph 17; 1790 excluded).

The decline in the production of grain transports was not compensated with a rise in the production of timber transports. Similarly, the decline in the production of grain transports to ZDZ, which was its most important destination, as we have seen before, was not compensated with a rise in the production of the transportation of other goods (see Graph 18), although the relative importance of the other goods did indeed increase in the course of the eighteenth century.

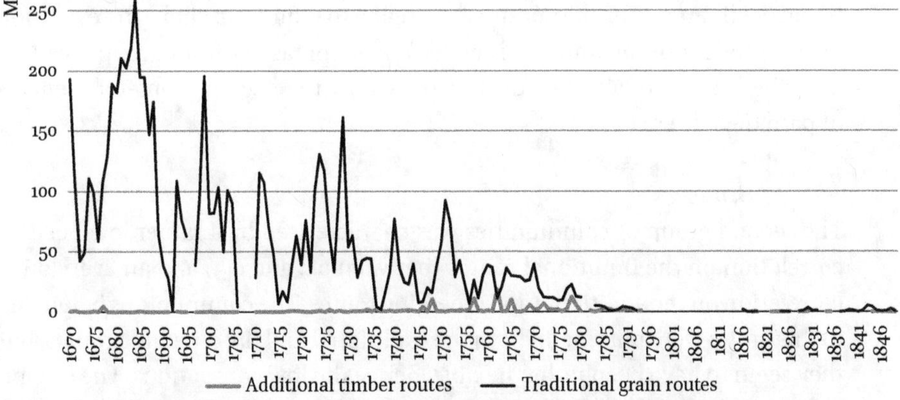

GRAPH 17 *Comparison of transport services production by shipmasters of group 1 on traditional grain routes and additional timber routes, expressed in million tonkm.* BASED ON STRO.

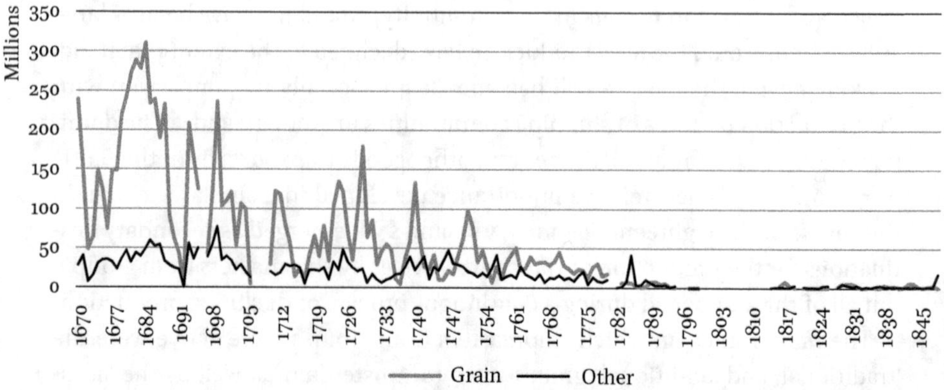

GRAPH 18 *Comparison of grain and 'other' transport services production by shipmasters of group 1, expressed in million tonkm.*
BASED ON STRO.

No structural change occurred in the production of transport services by shipmasters from group 1, at least not from the perspective of the goods they transported. Neither did such change take place in the routes sailed by shipmasters from group 1. The emerging routes for grain and timber remained insignificant at all times between 1670 and 1850. VBL continued to be of limited importance throughout the entire period, even though the second half of the eighteenth century shows some development towards increasing integration of distant destinations into the Dutch Deltas' transport system. Finally, the decline of Amsterdam resulted in a general downturn in the activities of shipmasters from group 1; they did not manage to manifest themselves on transport markets in the other Dutch Deltas. Ports in SMR or LDE never replaced the traditional destinations for shipmasters belonging to group 1. In short, the evolution of the production of transport services by shipmasters from group 1 can be described as one of absolute decline caused by the negative long-term effects of path dependence.

Group 2

The second group of communities was characterised by a rather low positive correlation on the traditional routes (between 0.4 and 0.5) and an even lower correlation on the additional and emerging routes. The communities belonging to this group were present on both the traditional and the emerging routes, but they seem to have been rather flexible, hence the low correlations. The second group consists of the communities of Amsterdam, Heerenveen, Hindeloopen, Joure, Makkum, and Texel, all located in ZDZ. Additionally, the maritime transport

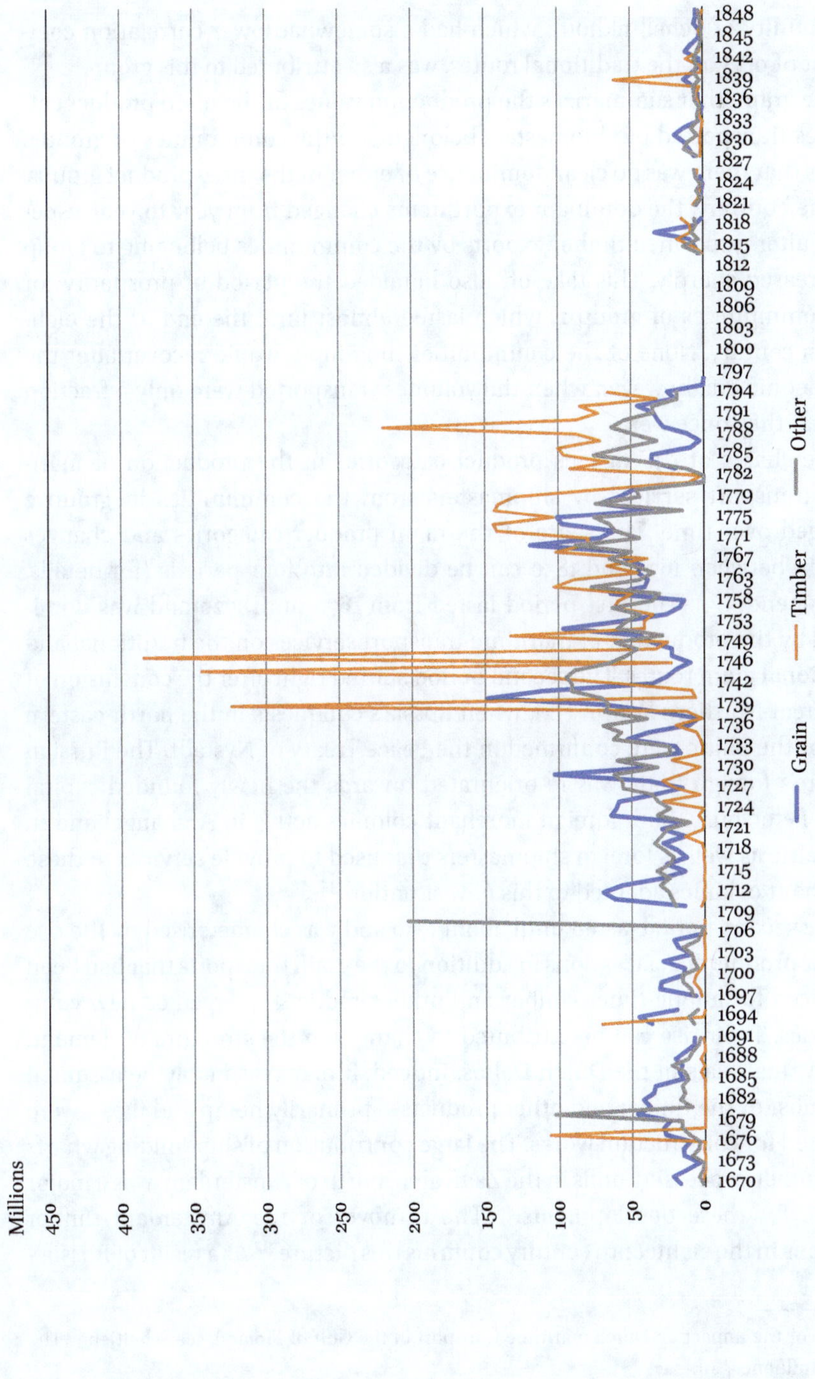

GRAPH 19 *Estimates of total transport production by the second group of Dutch maritime transport communities, expressed in million tonkm.*

BASED ON STRO.

community of Schellinkhout, which had a somewhat lower correlation coefficient of 0.28 on the traditional routes, was also attributed to this group.

The graph that summarizes the production values of the main product categories transported by shipmasters belonging to the communities of group 2 shows that there was no clear dominance of either of the three product groups. On the contrary, the dominant export items changed from year to year, especially after 1720, when timber exports by the communities belonging to group 2 increased sharply. This take-off also heralded the period of prosperity for the communities of group 2, which lasted almost until the end of the eighteenth century. None of the communities in group 2 would recover after the Napoleonic era, however, when the volumes transported were only a fraction of what they once were.

The shares of the various product categories in the production of maritime transport services by shipmasters from the communities in group 2 changed over time. The share of the main product categories and changes therein between 1670 and 1850 can be divided into four periods (for details, see appendix 3). The first period lasted from 1670 until 1720 and was dominated by the production of maritime transport services on the traditional and additional grain routes. The second period started right after the conclusion of the Great Northern War in 1721, when Russia's conquests in the north-eastern part of the Baltic were confirmed in the Peace Treaty of Nystadt. The Russian Empire's foreign trade was re-orientated towards the newly founded capital of St. Petersburg. Both foreign merchant colonies active in Archangel and in the Baltic as well as foreign shipmasters that used to provide services to these merchant colonies adapted to this new situation.[35]

The second period lasted until about 1750 and was characterised by the rise of new product specialisations in addition to the grain transports that had been practiced for a long time. Timber and other products emerged as new commodities. Their rise can be attributed to changes in the structure of demand for raw materials in the Dutch Deltas. Indeed, it may reasonably be assumed that most of the timber and other products—primarily hemp and flax—were destined for construction works. The large contribution of shipbuilding wharfs and timber-processing mills in the Zaan area, north of Amsterdam, was a major engine for these developments.[36] The turnover of the Amsterdam timber auctions in the eighteenth century confirms this picture.[37] As a result of Russia's

35 For the impact on Dutch maritime transport in the Gulf of Finland, see Scheltjens, "The Influence," 115–147.

36 See: Moore, "Amsterdam," 200–201.

37 Middelhoven, "De Amsterdamse veilingen," 86–114.

foreign economic policies, all foreign imports were directed to St. Petersburg (and to a lesser extent Riga), while exports of timber became the exclusive function of secondary ports in the Gulf of Finland, like Narva and Vyborg.

Between 1750 and 1795, more changes occurred in the production of maritime transport services by shipmasters from group 2. Grain transports on the traditional and additional routes were gradually complemented by grain transports destined towards SMR as well as towards destinations along the Atlantic coast (VBL). The balance, however, would not shift until the late 1770s. In the late 1760s and during the 1770s, there were attempts to change the routes of grain transports but the emerging grain routes never fully replaced existing patterns and after the Fourth Anglo-Dutch War, the emerging routes were not able to replace the rapidly declining volumes transported on the traditional and additional routes. On the contrary, despite a rapid recovery after the war, the overall level of the production of maritime transport services declined in the 1780s. Grain transports were hit hardest by the decline but other goods, which took over the role of grain for a short period of time between 1783 and 1790, could not bring a lasting solution. After 1750, the production of timber transports took a radically different course. Following a gradual rise in the 1720s and 1730s, the production of timber transports literally exploded in the 1740s, clearly exceeding the production of grain and other transports in single years. The predominance of timber transports in the operational patterns of shipmasters from group 2 can be explained by the length of the routes and the size of the ships they used.[38] Quite different from the production of grain transports, which continued to be realized mainly on the traditional and additional routes, the production of timber transports took place on routes linking the Gulf of Finland, where St. Petersburg emerged as an additional export port in the second half of the eighteenth century, with ZDZ as well as with destinations along the French, Spanish and Portuguese Atlantic coast. Thus, the average distances over which timber transports were produced was more than twice the distances of grain and other transports. At the same time, the volumes of timber transported on these routes were much higher than the volume of grain and other products transported on the traditional routes.

Until 1795, timber transports would maintain their dominant share in the production of maritime transport services by shipmasters from group 2 but its size declined rapidly after the Fourth Anglo-Dutch War, falling back to about half of the production in the 1760s and 1770s (see appendix 3 for details). This decline was of a general nature and was not limited geographically to one or

38 For estimates of the size of ships involved in timber transports from the Baltic, see Knoppers, *Dutch trade*.

another demand area. From 1795 onwards, the production of maritime transport services by shipmasters from group 2 did not differ much from that of shipmasters from group 1. After the Napoleonic era, destinations along the Atlantic coast were largely lost; the size of the production of maritime transport services to ZDZ was a far cry of what they had used to be and their replacement by other destinations—be it one of the Dutch Deltas or some 'new' destinations (in appendix 3 denoted as 'rest')—never occurred. Maritime transport communities from group 1 and group 2 relied very strongly on a limited set of routes for their services production and structural changes that occurred on either end of these routes could have a devastating effect. Since these communities were attached all too strongly to the routes that they knew, they were unable to change direction when necessary.

Group 3

The third group of maritime transport communities was rather heterogeneous and composed of communities from all three Dutch Deltas that had a relatively low positive correlation on traditional, additional and emerging routes. These communities in ZDZ were Broek in Waterland, Edam, Lemmer, Stavoren, Terhorne and Workum; in LDE: Delfzijl, Dokkum, Groningen, Pekela and Veendam; In SMR: Dordrecht and Rotterdam. Generally speaking, the production of maritime transport services by shipmasters that were members of these communities, insofar as Baltic shipping is concerned, can be summarized as follows. Between 1670 and 1720, the production of maritime transport services was insignificant. Most of the services produced in this period were grain transports. After 1720, a steady rise of the production level began, which lasted until 1795. A few peaks in the production of timber transports notwithstanding, the services were predominantly orientated towards the transportation of grain and some other goods from the Baltic. After the disruption caused by the Napoleonic Wars, the production of maritime transport services continued in the nineteenth century, following by and large the same grain-dominated pattern insofar as goods are concerned but without experiencing further growth. Since the composition of group 3 is rather heterogeneous, comprising shipmasters' communities from all three Dutch Deltas, it is worthwhile to take a closer look at the production patterns of the communities in each of these Deltas.

Interestingly, the communities in ZDZ, which were located in North Holland (Broek in Waterland, Edam) and south-west Friesland (Lemmer, Stavoren, Terhorne and Workum) make up some of the oldest towns in the area, including Edam, Stavoren and Workum. Unlike their counterparts in the other groups, Hindeloopen, Hoorn, Enkhuizen and Amsterdam to name but

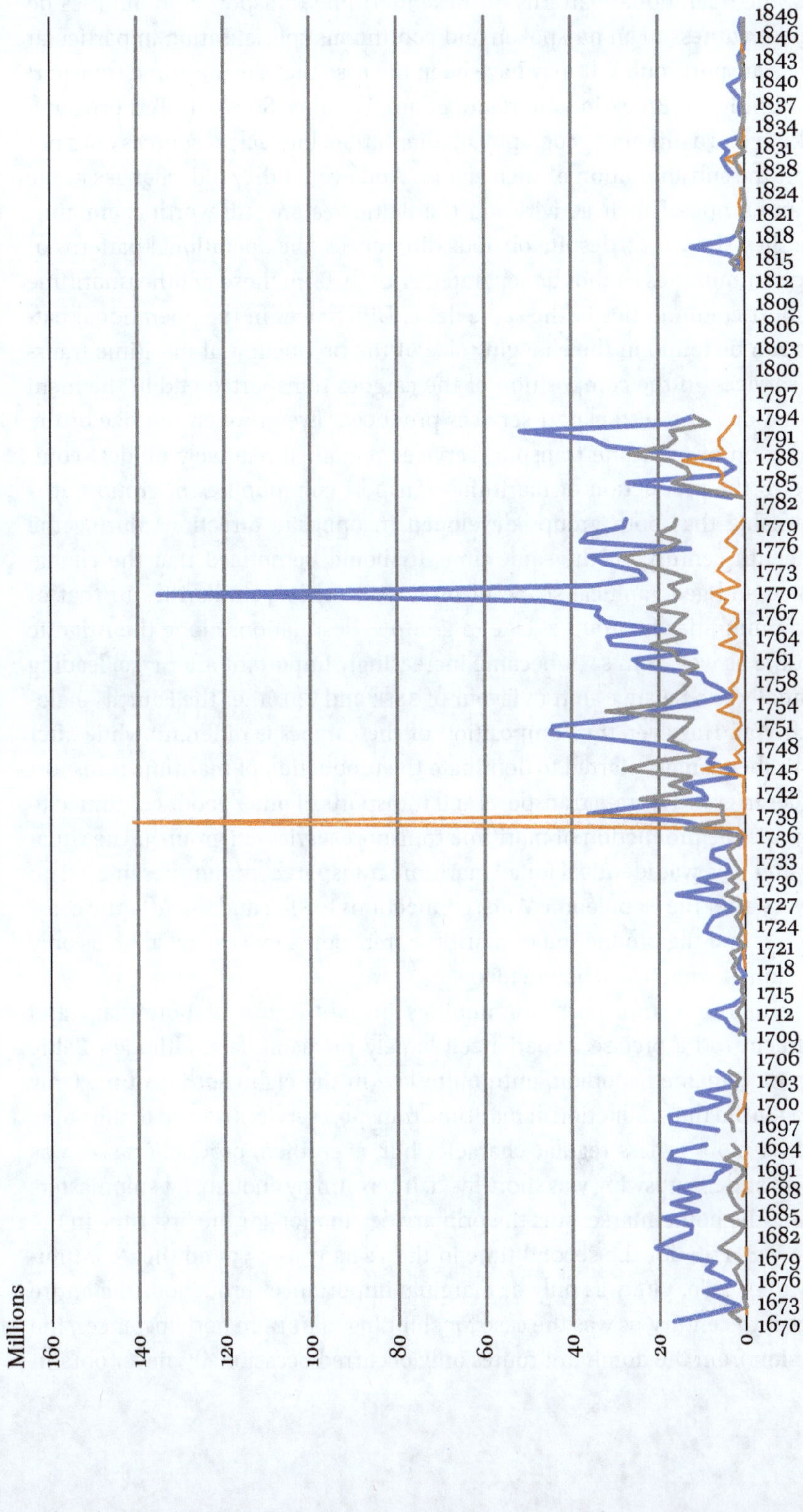

GRAPH 20 *Estimates of total transport production by a selection of the third group of Dutch maritime transport communities, located in* ZDZ, *expressed in million tonkm.*

BASED ON STRO.

a few, the operational patterns of these maritime transport communities do not bear witness to an outspoken and continuous specialisation in particular Baltic transport routes. It may have been the case that the maritime transport communities of Broek in Waterland, Edam, Lemmer, Stavoren, Terhorne and Workum had a different geographical orientation, but lack of sources does not permit the substantiation of such claims. And even if they did, changes in the size and scope of their activities in the Baltic Sea are still worth examining, since they show that, despite obvious differences, the operational patterns of these communities cannot be separated entirely from those of other maritime transport communities in the same delta. Differences in the operational patterns can be found in the changing size of the production of maritime transport services, in the composition of the cargoes transported and in the main destinations of the transport services produced. Even though the size of the production of maritime transport services remained relatively modest compared to the production of maritime transport communities in group 1, it is remarkable that both groups developed in opposite directions during the eighteenth century. At the same time, it should be noticed that the changing size and geographical scope of production shows similarities to that of the communities in group 2. Like in group 2, destinations along the Atlantic coastline as well as in SMR became increasingly important after 1750, leading eventually to a balance shift in favour of SMR and VBL after the Fourth Anglo-Dutch War. However, the composition of the cargoes is different: while after 1750 timber cargoes started to dominate the production of maritime transport services in group 2; grain transports and transports of other goods continued to dominate the production of maritime transport services in group 3. The situation after 1795 was identical for all maritime transport communities in ZDZ: no recovery after the Napoleonic Wars, connections to SMR and the Atlantic coast were lost and the production of maritime transport services ended up as only a fraction of what it was before 1800.

The maritime transport communities in SMR, those of Rotterdam and Dordrecht to be precise, experienced largely the same fate, although Baltic shipping remained sporadic until quite late in the eighteenth century. Only after 1760 did the production of maritime transport services related to the Baltic obtain a more or less regular character but, even then, production size was modest and the upswing was short-lived. Interestingly enough, for shipmasters of SMR, the home-market was the primary destination for the first time in the years 1738–1744, for the second time in the years 1751–1755 and then continuously after 1760. VBL was only of marginal importance throughout the entire eighteenth century, as was the case for shipping to BEH. In both instances, the diversion from the dominant routes only occurred occasionally, never obtain-

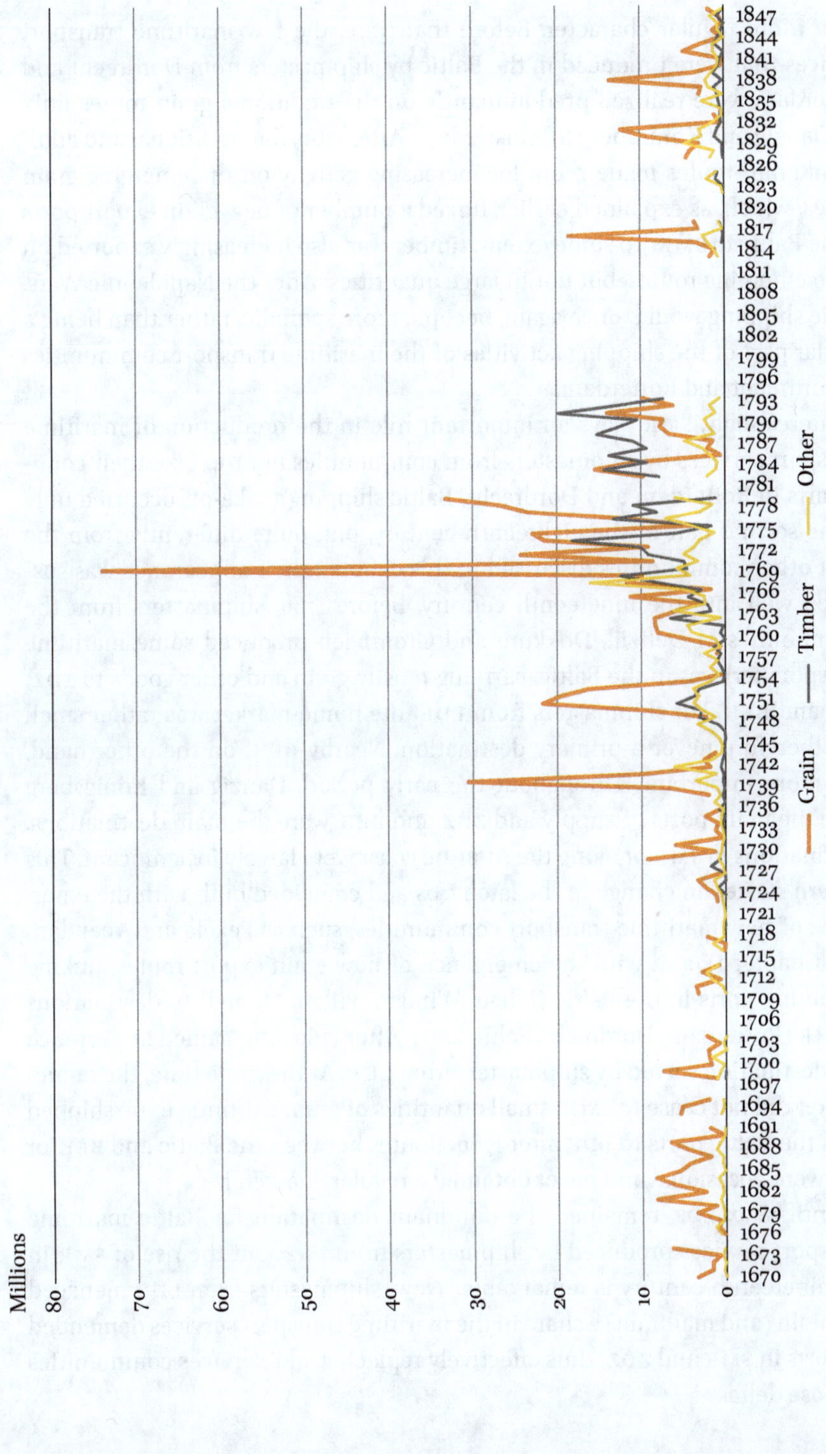

GRAPH 21 *Estimates of total transport production by a selection of the third group of Dutch maritime transport communities, located in SMR, expressed in million tonkm.*

BASED ON STRO.

ing a more regular character. Before that time, the few maritime transport services that were produced in the Baltic by shipmasters from Dordrecht and Rotterdam were realized predominantly on the traditional grain routes linking Danzig and Königsberg to Amsterdam. After 1760, the traditional and additional grain routes made room for increasing activity on the emerging grain routes, which, as explained earlier, linked a number of new grain export ports in the Baltic to SMR. To some extent, timber was also increasingly exported on these emerging routes, but not in large quantities. After the Napoleonic Wars, Baltic shipping would, once again, become more sporadic, rather than being a regular part of the shipping activities of the maritime transport communities of Dordrecht and Rotterdam.

Interestingly, SMR plays an important role in the production of maritime transport services by shipmasters from communities in LDE. Like their counterparts in Rotterdam and Dordrecht, Baltic shipping's take-off occurred only in the second half of the eighteenth century, but, quite differently from the most other communities discussed in this chapter, their appearance was sustained well into the nineteenth century. Before 1760, shipmasters from the communities of Delfzijl, Dokkum and Groningen produced some maritime transport services in the Baltic, carrying mostly grain and other goods to ZDZ, LDE and BEH. For shipmasters from LDE, the home-market was rather small and therefore never a primary destination. Nearby BEH, on the other hand, was more important. Throughout the early period, Danzig and Königsberg were the main ports of supply and ZDZ and BEH were the main destinations. Destinations in SMR or along the Atlantic coast were largely insignificant. This pattern started to change in the late 1740s and coincided both with the emergence of new maritime transport communities, such as Pekela and Veendam (see Chapter 1), and with the emergence of new grain export routes, linking upcoming ports in the Baltic (Libau, Windau, Pillau, Memel) to destinations in SMR (Rotterdam, Dordrecht, Schiedam). After 1760, ZDZ gained importance as a destination served by shipmasters from LDE. At the same time, the home-market did not cease to exist; small quantities of grain continue to be shipped from the Baltic ports to LDE after 1760. Routes between the Baltic and BEH or VBL were occasional and never obtained a regular character.

Until 1850, ZDZ remained the dominant destination for Baltic maritime transport services produced by shipmasters from LDE, but the rise of SMR in the nineteenth century is remarkable. 'New' shipmasters from LDE managed to obtain (and maintain) a share in the maritime transport services demanded by ports in SMR and ZDZ, thus effectively replacing 'old' services communities in those deltas.

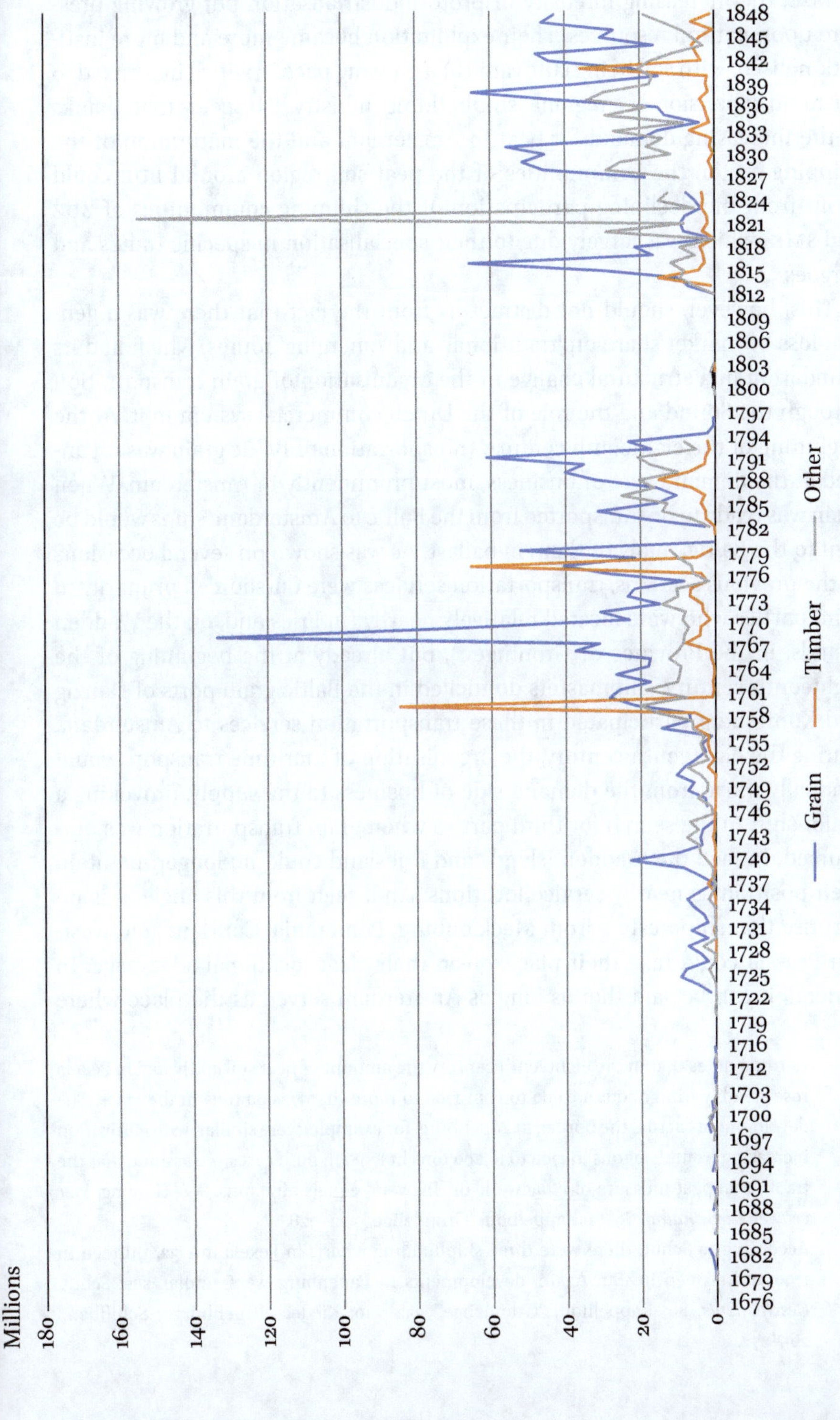

GRAPH 22 *Estimates of total transport production by a selection of the third group of Dutch maritime transport communities, located in* LDE, *expressed in million tonkm.*

BASED ON STRO.

In LDE, the increasing intensity of proto-industrialisation put growing pressure upon natural resources. Their exploitation became more and more institutionalised, with soil being cultivated at a growing pace,[39] which in turn led to the rapid expansion of a regional shipbuilding industry.[40] It seems that, thanks to the increasing demand for peat in Amsterdam and the maturation of the shipping sector, the communities of the peat soil region around LDE could profit from the declining participation of the shipping communities of ZDZ and SMR, which was largely due to their specialisation in specific routes and cargoes.

This, however, should not distract us from the fact that there was a general loss of market share on 'traditional' and 'emerging' routes, which had its foundation in a structural change in the organisation of grain transportation through the Sound and the role of the Dutch commercial system in it. At the beginning of the eighteenth century, transportation of Baltic grain was organised at the demand side of business; most prominently in Amsterdam. When grain was ready to be transported from the Baltic to Amsterdam, ships would be sent to the Baltic, many of them in ballast. As was shown on several occasions in the previous sections, transportation services were outsourced primarily to third parties who were located relatively nearby (in Friesland, on the Wadden Islands, in the Province of Groningen), but already at the beginning of the eighteenth century, shipmasters domiciled in the Baltic grain ports of Danzig and Königsberg participated in these transportation services to Amsterdam. During the eighteenth century, the organisation of maritime transport would gradually move from the demand side of business to the supply, provoking a major shift in the search for third parties when grain transportation was outsourced. In fact, the Wadden Islands and Friesland could no longer maintain their position as nearby service locations. And, seen from this angle, it is no surprise that shipmasters from Mecklenburg, Pomerania, Courland and western Prussia could take their place, given their clear locational advantage. In general, it can be said that as long as Amsterdam served as the place where

39 Gerding notes that in the eighteenth century the amount of peat soil cultivated in Pekela rose steadily from about a 9,000 tons in 1700 to more than 72,000 tons in the 1790s. The developments across the border, in Papenburg, for example, were similar, with cultivation increasing from 1800 tons in 1700 to 16,200 tons in 1798. In both cases, we assume that the traditional peat measure of 1 'dagwerk' or 'Tagwerk' equals nine tons. See: Gerding, *Vier eeuwen turfwinning*, 69–71; Kappelhoff, "Grundzüge," 322–328.

40 According to Schut, there were three shipbuilding wharfs in Pekela in 1732, thirteen in 1790 and sixteen in 1811. Again, developments in Papenburg were similar. see: Schut, *Geschiedenis*, 86; Kappelhoff, "Grundzüge," 332–340; Kiedel, "Papenburger Schiffbau," 265–272.

grain transports were organised, the operational knowledge offered by Frisian shipmasters and those from the Wadden Islands was easily accessible. This, however, changed dramatically when the supply side of the grain trade started to take care of the organisation of transportation.

Not surprisingly, this line of reasoning is in agreement with developments in the structure of the Dutch commercial system in the eighteenth century and, more precisely, the role of Dutch merchant colonies in it. While until the beginning of the eighteenth century, Dutch merchant communities abroad were "integral to the mechanism of Dutch world-trade primacy",[41] their role started to change under the influence of the "postal revolution", leading to a situation where travelling or settling abroad was no longer required.[42] For a large part of their business, merchants could rely on local partners with whom intense correspondence was established. Local merchants were also capable of taking care of business and, when maritime transport needed to be organised, it was logical that these merchants would rely increasingly on shipping services that were supplied by communities that were easier to access than the distant traditional shipping services regions of the Dutch Republic. Both in the case of VBL and on the 'emerging' routes it can be observed that these routes attracted more and more attention as the decline of Amsterdam's staple market continued. However, when the shares and origins of Dutch shipmasters active on these routes are compared with the routes to Amsterdam, it becomes clear that VBL was not a replacement of the traditional routes to Amsterdam. Clearly, Dutch shipmasters did not manage to stay in business through engagement in VBL. Its character was too sporadic for this strategy to be successful in the long run.

Conclusions

In this chapter, the relation between trade and transport was analysed from the perspective of the production of maritime transport services by shipmasters domiciled in the Dutch Deltas. The chapter started with a case study about grain transportation in the period from 1670 until 1850, in which it was established that structural changes in supply and demand led to the emergence of new transportation routes, provoking a shift from a bilateral trade pattern to a

41 Veluwenkamp, "Merchant colonies," 162.

42 Morineau, "Le Commerce de la Baltique," 41–42; Veluwenkamp, "International Business Communication Patterns," 121–134; Veluwenkamp, "Het Nederlandse Handelsstelsel," 70–72.

diversified pattern with many locations involved, but none strictly dominant. In the second section, the impact of these changes on the structure of the Dutch maritime transport sector was explained. It was explained that market entry constraints, path dependence, positive and negative lock-in and access to operational knowledge also had an impact on the structure of the Dutch maritime transport sector and on the rise and decline of maritime transport communities in the Dutch Deltas. In the next chapters, the perspective of the production of maritime transport services by shipmasters domiciled in the Dutch Deltas is combined with an analysis of the behaviour of the populations of maritime transporters that constituted these communities.

CHAPTER 4

Domiciles

In the preceding chapters attention has been paid to the locations of maritime transport communities, their functions in the maritime transport system of the Dutch Deltas and the structure of their services production. In this chapter, the populations of maritime transporters, that constituted the communities examined in the preceding chapters, take centre stage. In the first part of this chapter, the focus lies on the issue of domicile ambiguity, which has been introduced briefly in Chapter 2. Additionally, some attention will be paid to a phenomenon described as 'virtual migration', which stands for the 'administrative' relocation to places with neutral flags in times of trouble.

Domicile Ambiguity

A recurring phenomenon in research based on the Sound toll tables, which were compiled and published by Bang and Korst in the first decades on the twentieth century,[1] is the attention to the problem of ambiguity in the registered domiciles of shipmasters. One shipmaster could be found in the original registers with different homeports. Usually, such ambiguity is interpreted as a sign of the unreliability of the registers. Additional research found irregularities between the domicile registered in the Sound and in other registers, such as the *galjootsgeldregisters*.[2] Logically, these research results have led to significant scepticism regarding the reliability of domicile registrations in the Sound toll registers. It should be noted, however, that the unreliability of the domicile registrations in the Sound was never the subject of a structural analysis nor have there been attempts to interpret or contextualise the observed ambiguity. So far, the so-called 'hjemsted'-problem has been analysed at best in an impressionistic way.

The 'hjemsted'-problem can be summarized as follows. The domicile of shipmasters registered in the Sound may differ from year to year and in some cases even from voyage to voyage; therefore these registrations are unreliable. Often it is assumed that (either) false declarations or wrong registrations at customs in Helsingör lie at the basis of these ambiguous registrations. Various

1 Bang and Korst, *Tabeller over skibsfart*.

2 For an overview of relevant literature, see: Scheltjens, *De invloed*, 73–96.

 | DOI 10.1163/9789004273078_006

scholars have touched upon the 'hjemsted'-problem for different time-periods and have provided proof of their unreliability, using different sources for comparison with the Sound toll registers. De Buck and Lindblad provided evidence from the *galjootsgeldregisters* of Amsterdam to support their argument. They found that Amsterdam was mentioned as domicile far more often in the Sound toll registers than in the *galjootsgeldregisters*, and concluded that the reliability of domicile registrations in the *galjootsgeldregisters* is higher than that of the Sound toll registers.[3] Boon has compared the domiciles of West Frisian shipmasters registered in the Sound with local sources, reaching the conclusion that these local sources provided more detailed information about shipmasters' domiciles.[4] In itself this is valuable information, but it does not mean that the data in the Sound toll registers is unreliable. It seems that the domicile data in the Sound toll registers is often of a more general nature, registering, for example, Hoorn rather than one of the very small villages in its immediate surroundings as the domicile of the shipmaster. Contrary to previous claims about the reliability of declarations or registrations made by clerks in Helsingör, my opinion is that the rationale behind this ambiguity may simply be that the further one is away from 'home', the less significant detailed geographical information becomes.[5] What remains relevant is a certain point of reference, a 'bigger' place with a name that people know and recognize. Obviously, one such place may have been Amsterdam, but regional denotations such as Friesland, Holland or Zeeland may have fallen into the same category. Domicile ambiguity may thus be interpreted as a variation of the cognitive strategy known as *totum pro parte* (the whole for a part), which is a kind of synecdoche that is quite common in geographical references.

This, however, is not the whole story. In the Sound toll registers shipmasters are registered with domiciles that do not fit into the above-mentioned scheme. Many, ambiguous registrations do involve Amsterdam or a regional denotation as an 'alternative' domicile, but not all the ambiguous domicile registrations in the Sound toll registers are a combination of a commonly known 'point of reference' with smaller 'real' domiciles. There are two options: ambiguous domicile registrations were the result of randomly chosen declarations of domiciles

3 De Buck and Lindblad, "Navigatie en negotie," 27–48.

4 Boon, "West Friesland and the Sound," 171–189; Boon, *Bouwers van de zee*, 83–89.

5 One may compare this to everyday situations, like in the following example. Among Flemish people, when asked, the author says that he comes from Mortsel, because he can reasonably assume that Flemish people will know where Mortsel is—some 8 kilometers from Antwerpen. Among foreigners, when asked, he says that he comes from Antwerpen, since Mortsel is not a place that a foreigner usually knows.

or of an unknown rationale that may explain the differences. Either way, examining combinations of registered domiciles for large groups of individual shipmasters over a long period of time appears to be the only way to test these assumptions.

Apart from the fact that the electronic database of the Sound toll registers only became available in 2013, there is one other major factor that explains the lack of knowledge about domicile ambiguity: techniques for its examination in a structural way are still largely non-existent in the field. Of course, attempts to create algorithms for standardizing nominal data in large databases have been around for several decades and some efforts have been made to establish their reliability for historical databases of maritime shipping[6] but unfortunately, none of these algorithms is capable of dealing in an acceptable way with the vast amounts of spelling differences in the electronic database of the Sound toll registers.

Domicile ambiguity, however, is an issue of vital importance to the advancement of maritime history as a discipline, especially since ambiguity (in general) is a recurrent issue in many historical databases. In this chapter, domicile ambiguity was analysed in the registrations of shipmasters from thirty-two different ports in the Dutch Deltas between 1670 and 1850, which corresponds to a total of 227,705 Sound passages. Several preparatory steps were taken, before these records were ready for analysis. Most importantly, spelling and other variations in the names, first names and patronymics (if any) of the shipmasters in the data set had to be removed as much as possible. Given that the 227,705 registrations in the dataset account for about 12.000 variations of first names and about 35,000 variations of last names and patronymics, it is clear that removing variation was a painstaking task that could only be completed partially. As before (see introduction), the frequency of each variation was taken as a guide for limiting the number of variations that had to be dealt with, while at the same time achieving the highest possible efficiency during the process. In other words, the variations with the highest frequency were dealt with first and the process was cut off at a minimum frequency of three. Upon completion of the name-matching procedures described in the Introduction, about 13% of all registrations were excluded from further analysis.

The remaining 197,000 registrations constituted the basis for the second step, in which an overview was created of all different combinations of standardised first name, standardised name and standardised patronymic with their respective registered domiciles, the first and last year of occurrence of this name

6 See: Welling, "Price-supply relations," 457–469, For an alternative approach (no explanation of method; intuitive standardisation), see: Dataset of the *Convooigeldregisters*.

TABLE 19 *Ambiguity breakdown for shipmasters domiciled in the Dutch deltas. Based on a standardised sample data set of 168,682 sound passages by shipmasters from thirty-two different domiciles (details in the text)*

	Shipmasters	Passages
1 Delta	29.242 (90,2%)	107.631 (63,8%)
2 Deltas	2.892 (8,9%)	49.469 (29,3%)
3 Deltas	301 (1%)	11.582 (6,9%)
TOTAL	32.435	168.682

combination and its frequency in the Sound toll registers. All combinations for which the difference between the first and last year of occurrence was more than forty years (or the time of one long career) were excluded. Then, domicile ambiguity was analysed in the remaining 168.682 records. Theoretically, it would have been possible to review all remaining records and the sailing patterns attached to them in order to review if shipmasters with slightly different standardised names, were more likely to refer to the same master or had to be regarded as different masters after all. This step, however, was not executed because it would not significantly alter the results that are aggregated below. In the following tables, domicile ambiguity is treated both at the level of the Dutch Deltas and at the level of individual domiciles.

It is clear, that the majority of the Dutch shipmasters were registered in one or more domiciles located in the same delta (see Table 19). About 9% of all masters, however, was registered in STRO with domiciles in two different Dutch Deltas; less than 1% was registered in three different deltas. Interestingly, the share of shipmasters with domicile registrations in two or three of the Dutch Deltas was somewhat larger when expressed in number of passages, that they executed. The shipmasters that were registered with domiciles in one delta can be specified further (see Table 20).

There was a significant amount of ambiguity in the domicile registrations of shipmasters in ZDZ. More than half of the passages through the Sound by shipmasters domiciled in ZDZ were executed by masters that were found registered with two or more different domiciles in STRO. Such ambiguity is much less prominent for SMR and LDE (where the number of different domiciles was also smaller). In the following breakdown, shipmasters' domiciles in two different deltas are reviewed (see Table 21).

TABLE 20 *Ambiguity breakdown for shipmasters with domiciles located in one delta. Based on a standardised sample data set of 168,682 sound passages by shipmasters from thirty-two different domiciles (details in the text)*

	ZDZ	SMR	LDE
1 Domicile	13.204 (34.413)	1.087 (2.354)	4.976 (16.235)
2 Domiciles	2.701 (27.026)	29 (161)	276 (3.264)
3 Domiciles	779 (14.109)		15 (289)
4+ Domiciles	348 (9.780)		

TABLE 21 *Ambiguity breakdown for shipmasters with domiciles located in two deltas. Based on a standardised sample data set of 168,682 sound passages by shipmasters from thirty-two different domiciles (details in the text)*

	ZDZ+SMR	ZDZ+LDE	SMR+LDE
2 Domiciles	434 (2.851)	977 (9.856)	95 (1.253)
3 Domiciles	231 (3.332)	490 (9.372)	17 (307)
4+ Domiciles	225 (6.394)	422 (15.540)	1 (24)

Even though time is not included in the breakdown of domicile ambiguity for shipmasters registered in two different Dutch Deltas, it is clear that the observed patterns can be related to the structural developments described in Chapters 1 and 3 (see Table 22). The dominant combinations of registered domiciles are found in ZDZ+SMR, on one hand, and ZDZ+LDE, on the other, indicating a significant degree of transport market integration between these two deltas. It may be assumed that the observed dominance reflects the dominant position of Amsterdam and the emergence of Rotterdam as the Low Countries' central transport markets, where shipmasters could easily register in order to obtain the necessary documents for executing return voyages to the Baltic. Moreover, as was highlighted in Chapter 1, Amsterdam and Rotterdam were probably also attractive outlets for the domestic produce of Frisian and Groningen villages, which makes the combination of domicile registrations in two different deltas even more self-explanatory.

TABLE 22 *Ambiguity breakdown for shipmasters with domiciles located in three deltas. Based on a standardised sample data set of 168.682 sound passages by shipmasters from 32 different domiciles (details in the text)*

	ZDZ+SMR+LDE
3 Domiciles	67 (1.091)
4+ Domiciles	234 (10.491)

Finally, some shipmasters were registered at domiciles in all three Dutch Deltas, but their number is not very large (301) and their share in the total number of passages in the sample is limited to about 7%.

Based on the detailed matrix of domicile combinations and their respective frequencies (see Appendix 4), that were calculated on the basis of the standardised sample of 168,682 passages executed by Dutch shipmasters from thirty-two different domiciles, the observed ambiguity can now be assessed from two different perspectives. In order to streamline the subsequent analysis of the standardized Sound passages, two additional calculations were made: the lowest (min) and the highest (max) frequency among the different domiciles with which a shipmaster in the standardised data set based on STRO was registered. They allow a relevant distinction to be made between two perspectives on ambiguity: the first perspective, where the analysed domicile has the highest frequency and is complemented with alternative domiciles with lower frequency; the second perspective, where the analysed domicile has the lower frequency and serves as alternative to another predominant domicile. In the former case, domicile ambiguity is estimated from the perspective of the domicile with the highest frequency. In the latter case, domicile ambiguity is estimated from the perspective of the alternative domicile. The matrix in Appendix 4 comprises two different instances of domicile ambiguity. The columns contain a breakdown of the alternative domiciles for all shipmasters with the port mentioned in the column heading as their most frequently registered domicile. In the column 'Hindeloopen', for example, it can be read that shipmasters with Hindeloopen as their primary registered domicile (i.e. the frequency of Hindeloopen = max) executed 10,490 Sound passages between 1670 and 1850. Of them, 8,267 were executed with Hindeloopen as their registered domicile. A further 1,175 passages were executed with Amsterdam as registered domicile. Other domiciles appeared next to the shipmasters' name occasionally, as can be seen in the matrix. Summarizing, in the case of the maritime transport community of Hindeloopen, it can be said that they executed almost 79% of all their voyages indicating their proper domicile in STRO (it is assumed, therefore, that the domicile with the highest frequency in the shipping pattern of an individual shipmaster is also his actual domicile). In the row 'Hindeloopen', on the other hand, it can be read how often Hindeloopen occurred as an alternative domicile for shipmasters with another dominant domicile. The sum of each row gives an impression of the role of Hindeloopen as an alternative domicile for other maritime transport communities in the sample. In this case, Hindeloopen was registered as an alternative domicile in 1,667 passages, which corresponds to a share of about 14% of all Sound passages executed by shipmasters in the sample. The example of Hindeloopen is representative for all thirty-two domiciles in the sample (see graphs below). As dominant domiciles, each domicile had a median share in the total number of passages by its community members of 75.7% with a standard deviation of 7.1%. As alternative domiciles, each had a share of 16.9% (standard deviation of 6.2%) in the total number of passages.

From a bottom-up perspective, in which the names of shipmasters are standardised first and observations regarding domicile ambiguity are only made on the basis of a standardised data set, the 'hjemsted'-problem becomes much less problematic, than it was previously assumed. Observations made using the domicile as a key variable in quantitative analyses, for example, in the previous chapters of this book, capture about 75% ± 7% of all passages of shipmasters with this domicile as their main domicile while, on the other hand, they also capture about 17% ± 6% of all passages with this domicile as their alternative domicile. As a result, it is unlikely that the use of the domicile as a key variable, without prior execution of name standardisation in order to reconstruct in detail the population of a specific domicile, significantly distorts the results.

In short, the ambiguity in domiciles registered in the Sound was not a random ambiguity; not just any domicile was mentioned instead of the 'real' one. In most cases, a relation can be assumed between the primary, secondary and tertiary domiciles mentioned in STRO. This relation was probably of a purely economic nature, when the domicile ambiguity is the result of the shipmaster's transport activities and the registered 'alternative' domiciles are most likely to have been the places where the shipmaster received his most recent sea letter (*zeebrief*). As an example, we may refer to the domicile ambiguity that occurs in the shipping pattern of Dooitse Eelkes Hinxt, who is registered in STRO with three different domiciles: Dokkum (seven times), Harlingen (five times) and Amsterdam (three times). The activities of Dooitse Eelkes Hinxt are extremely well documented; a journal of his travels during the years 1763–1776 has been preserved.[7] Dooitse Eelkes Hinxt was a shipmaster from Dokkum.

7 Nationaal Archief, 1.01.47.09: Admiraliteitscolleges XIII Hinxt.

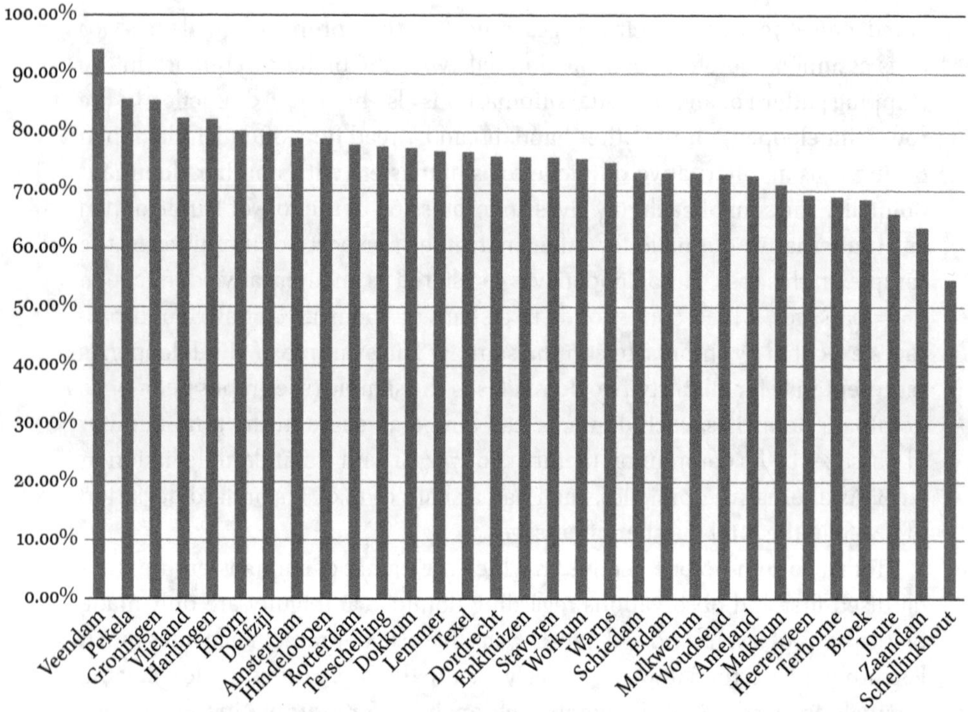

GRAPH 23 *Share of the main domicile in total number of passages by a community of shipmasters.*
BASED ON STRO.

The accountant responsible for the ship lived in Leeuwarden. In the journal, the entries for the years 1772–1776 show that Dooitse Eelkes Hinxt made several calls at the ports of Amsterdam and Harlingen. He appears to have started a number of voyages through the Sound from Harlingen. Moreover, in April 1773, the ship's journal states that Dooitse Eelkes Hinxt had purchased a new set of sea passes from the secretary (no name given) in Harlingen. Combination of the evidence from the ship's journal with the three different domiciles Hinxt is registered with in the Sound leads to the conclusion that domicile ambiguity results from the fact that Dooitse Eelkes Hinxt probably purchased his sea passes in Amsterdam, Dokkum or Harlingen. Where and when exactly he purchased a new sea pass depended on the course of his professional activities. Arguably, the efficiency of the Dutch administrative system made such pragmatism possible. It may be concluded that the ambiguity was due either to a cognitive strategy called synecdoche (*totum pro parte*, in particular), which

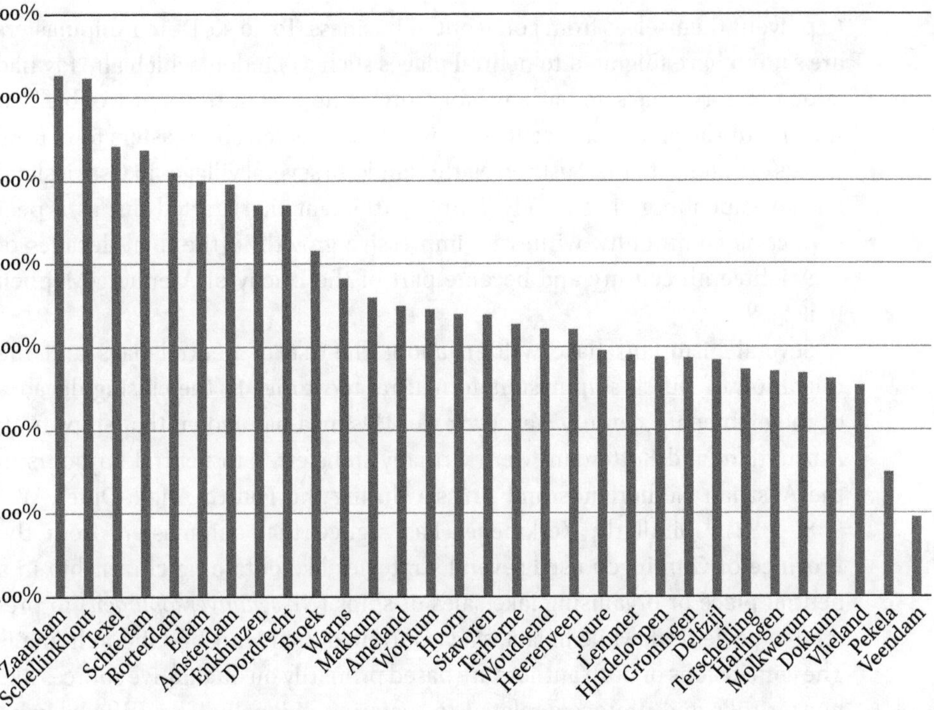

GRAPH 24 *Share of registrations as alternative domiciles in total number of passages by a community of shipmasters.*
BASED ON STRO.

was employed by shipmasters or clerks during the clearing of ships in the Sound, or to a pragmatic attitude of shipmasters towards administrative procedures like acquiring a new sea pass. In the next section, such pragmatism is explored further.

Virtual Migration

Dutch shipmasters did not disappear from international waters during periods of political disturbance. They pursued their operations under a neutral flag. After 1772, when northern Europe entered a long and exhausting period of political turbulence,[8] obtaining a neutral flag was the only way for shipmasters

8 See: Kirby, *The Baltic World 1772–1993*.

to prevent themselves from going out of business. To do so, Dutch shipmasters are said to have migrated to neutral places such as Emden, which already had a long history as a 'safe harbour' for Dutch shipping in times of trouble.[9] At the end of the eighteenth century, migration of Dutch shipmasters to neutral places expanded to the Austrian Netherlands, to several villages in Ostfriesland and to Papenburg, a relatively young settlement that started out as a peat extracting community, witnessed impressive growth in the final decades of the eighteenth century and became part of the Duchy of Arenberg-Meppen in 1803.[10]

Several historians have written about the use of neutral flags and the migration of Dutch shipmasters to neutral territories in the closing decades of the eighteenth century. Van Eyck van Heslinga has shown that ships from Amsterdam and Rotterdam were virtually transferred to neutral harbours in the Austrian Netherlands and Prussia during the Fourth Anglo-Dutch War (1781–1784).[11] Similarly, Kerkmeijer has argued that shipmasters from the Province of Groningen used several strategies like obtaining citizenship in a neutral place or organising fake sales of ships (Nl. *schijnverkopingen*) to pretend to be neutral, while in fact, they continued their work and life 'at home'.[12] The conclusions of both authors are based primarily on qualitative sources like muster rolls and protocols related to seafaring. Kappelhoff and Kiedel, who have observed migration patterns from the viewpoint of one of the neutral places—Papenburg—also contend that the migration of Dutch shipmasters should not be mistaken for 'real' migration, pointing out to the fact that in many cases their migration was coordinated by what may be described as 'neutral passport brokers' while in other cases false passports could also be purchased.[13] More recently, Scheltjens and Veluwenkamp, have made a first attempt to approach the issue of virtual migration of Dutch shipmasters to Papenburg using STRO.[14] So, the case of Papenburg is well-documented and will be introduced briefly below as a typical example of a virtual community of shipmasters.

---

9 Kernkamp, *De handel op den vijand.*

10 For an economic history of Papenburg, see: Kappelhoff, "Grundzüge," 319–476. More general works about the history of Papenburg and the Duchy or Arenberg include: Krömer, *Kleine Wirtschaftsgeschichte Ostfrieslands und Papenburgs*; Buck, *Die Arenberger in Emsland.*

11 Van Eyck van Heslinga, "De vlag dekt de lading," 102–113.

12 Kerkmeijer, "Groningers onder neutrale vlag," 149.

13 Kappelhoff, "Grundzüge," 358–360. About the false passports, see Kappelhoff, "Grundzüge," 455, note 237.

14 Scheltjens and Veluwenkamp, "Sound Toll Registers Online," 322–325.

TABLE 23 *The early development of the maritime transport community of Papenburg, 1772–1784. Based on Kappelhoff, 332–340. Schmackschiffe and Kuffschiffe are categorized as seaworthy; Schüttschiffe, Mutten and Flottmutten are categorized as coastal ships. The data for 1772–1774 are probably incomplete. The figures are based on membership lists of the great skippers' guild of Papenburg*

	Seaworthy ships	Coastal ships	Total
1772			13
1773			24
1774			24
1775	9	72	81
1776	12	72	84
1777	13	73	86
1779	15	69	84
1780	18	64	82
1782	30	60	90
1783	36	61	97
1784	35	52	87

A maritime transport community started to emerge in Papenburg in the second half of the eighteenth century. In 1771 a Great Skippers' Guild was founded in Papenburg.[15] Membership lists of this guild allow tracing of the early development of maritime shipping in Papenburg. The maritime transport community of Papenburg experienced a first boom in the early 1780s (for details, see Table 23). During the Fourth Anglo-Dutch War, Dutch shipmasters migrated to neutral places like Emden and Papenburg to continue their operations. At the same time, the local business community in Papenburg tried to leverage the situation, ordering the construction of at least forty-one ships at their local shipyards between 1781 and 1783.[16] After 1784, the maritime transport community, i.e. those shipmasters who sailed seaworthy ships like smacks, koffs and tjalks, maintained its 1783 status. For at least a decade, their number would not change significantly. The maritime transport community of Papenburg consisted of about fifty masters.[17] From 1795 onwards, Papenburg

15 Kappelhoff, "Grundzüge," 332.
16 Kappelhoff, "Grundzüge," 337.
17 For details, see Kappelhoff, "Grundzüge," 339.

TABLE 24 *The development of maritime shipping in Papenburg, 1800–1810. Based on Kappelhoff, 358–361 and Staatsarchiv Osnabrück, Nr. 7294 Hzgl. Arenbergisches Archiv zu Meppen, Dep 62 b: Liste der 1807 in der Großen Schiffersgilde zu Papenburg verzeichneten Schiffe. The figures are based on membership lists of the great skippers' guild of Papenburg. The figures for seaworthy and coastal ships in 1803 and 1806 are Given by Kappelhoff (p. 360)*

	Seaworthy	Coastal	Total
1800			216
1801			200
1802			74
1803	Appr. 60		100
1804			290
1805			406
1806	Appr. 80	Appr. 50	350
1807			356
1808			360
1809			395
1810			422

experienced a second boom; once again political circumstances forced Dutch shipmasters to start sailing under neutral flag.

In 1800, 216 shipmasters were registered as members of the great skippers' guild of Papenburg. Their number dropped in 1801–1803, but rose to unprecedented heights when Papenburg became part of the Duchy of Arenberg (for details, see Table 24). One of the reasons for the rapidly increasing popularity of Papenburg as a domicile in the beginning of the nineteenth century lies in the fact that acquiring a neutral pass in Papenburg was much cheaper than in Emden.[18] Another reason is that in the early years of the development of maritime shipping in Papenburg, local masters from Papenburg made use of the Dutch flag and Dutch passports to find their way into international

18 Kappelhoff, "Grundzüge," 340: "Die sehr hohen Forderungen, die in Emden mit dem Erwerb des Bürgerrechts verbunden waren, ließen Papenburg als Alternative zunehmend attraktiv erscheinen, denn hier genügte die Stellung eines Bürgen und die mit der pflichtgemäßen Eintragung ins Gildebuch verbundene Zahlung einer relativ geringen Summe Geldes".

transportation.[19] In this sense, the hospitality of Papenburg towards Dutch shipmasters may be interpreted as mutually beneficial.

Soon, there were many more shipmasters that claimed to be from Papenburg than a few years earlier. According to data gathered by Kappelhoff and additional data found in Staatsarchiv Osnabrück their number easily exceeded 200. Of them, only about eighty can be identified as 'actual' shipmasters from Papenburg. Similar figures are available for 1803 and 1806. Overall, the number of maritime transporters from Papenburg remained stable at about eighty; the other members of the great skippers guild were either involved in coastal shipping or they were 'virtual' members.[20] Local shipmasters had become a minority in their own community.

The virtual community was twice as large as the actual maritime transport community, which was soon felt as a threat by local ship owners and masters, although the benefits were also recognized.[21] On one hand, there were so many shipmasters that claimed to be from Papenburg, that their neutrality became more and more questioned and England would try to seize ships from Papenburg arguing that they were not really neutral. On the other hand, the financial benefits for the local community were quite clear.[22] Even though their migration was 'paper migration' which did not involve moving to Papenburg, these so-called *Schein-Papenburger*, that made their appearance in the Sound thanks to the relative ease with which they could obtain passports and sea passes in the Duchy of Arenberg, affected the local maritime transport community in a very real way.

The argument that I would like to make in the remainder of this section is that the virtual migration of Dutch shipmasters to neutral places like Papenburg in times of political disturbance did not have a random character. Neutral passes were obtained in places that were within reach of Dutch maritime transport communities and regional trade and transport connections defined which neutral places were viable outlets and which were not. In other words, I assume that the spatial structure of regional economies (i.e. the location of its main areas of economic interaction) accounts for the differences in 'virtual migration' between shipmasters from various parts of the Dutch Republic. This argument is in line with Kerkmeijer's conclusion that shipmasters of the Province of Groningen used neutral Prussian flags in great numbers because of

19 Kiedel, "Papenburger Schiffbau," 270–271.

20 Kappelhoff, "Grundzüge," 360.

21 Kiedel, "Papenburger Schiffbau," 270–272.

22 Kappelhoff, "Grundzüge," 359.

Prussia's geographical proximity,[23] but the source which I will use to substantiate my claims differs from Kerkmeijer's source base in type, size and scope. Kerkmeijer's (as well as Van Eyck van Heslinga's, Kappelhoff's and Kiedel's) sources are suitable for a qualitative analysis; the data set employed in this section lends itself primarily to a quantitative analysis. Moreover, Kerkmeijer's and Van Eyck van Heslinga's surveys are focused primarily on ship ownership, whereas my analysis is concerned with the masters who conducted the ships.

Between 1772 and 1814, the 'normal' course of maritime shipping through the Sound was severely disturbed during the Fourth Anglo-Dutch War of 1781–1784 and after 1795, when the Dutch Republic became a part of Napoleon's Empire. Graph 25 visualizes when the change to a domicile under neutral flag occurred among shipmasters registered in STRO with two different domiciles: one in the Dutch Deltas and one in a 'neutral' place. The effects of warfare on the domiciles of Dutch shipmasters can clearly be discerned. Besides, the downward trend of Dutch shipping, be it under neutral flag or not, in the closing decades of the eighteenth century is obvious. Taking its continuation under neutral flag into account, it becomes clear that there was not a complete rupture during the Fourth Anglo-Dutch War of 1781–1784 or during the Napoleonic Wars. It was only between 1807 and 1813 that Dutch maritime shipping came to an almost complete stop.

Conclusions

As in the case of domicile ambiguity, virtual migration was driven by the regional economic structure of various parts of the Dutch Republic. Shipmasters from the Province of Groningen used neutral places in Lower Saxony, such as Oldenburg, Emden, Norden, Leer, Weender, Papenburg and others to continue their maritime transport activities. They did not turn to the Austrian Netherlands in search for neutrality. Obviously, geographical proximity must have played an important role in this respect, but the almost complete absence of the Austrian Netherlands during the Fourth Anglo-Dutch War is nevertheless remarkable. It seems that shipmasters from the Province of Groningen just did not have the connections that were necessary to turn to the Austrian Netherlands to obtain a neutral pass. The situation for shipmasters from Friesland and North Holland was different, even though they also found their way to places in Lower Saxony,

23 Kerkmeijer, "Groningers," 150–153. Kermeijer's table of Groninger shipmasters and the Prussian places of which they possessed citizenship in 1795–1813 (pp. 152–153) is based entirely on sources kept as part of the Judicial Archives of Sappemeer, Oude and Nieuwe Pekela and Delfzijl.

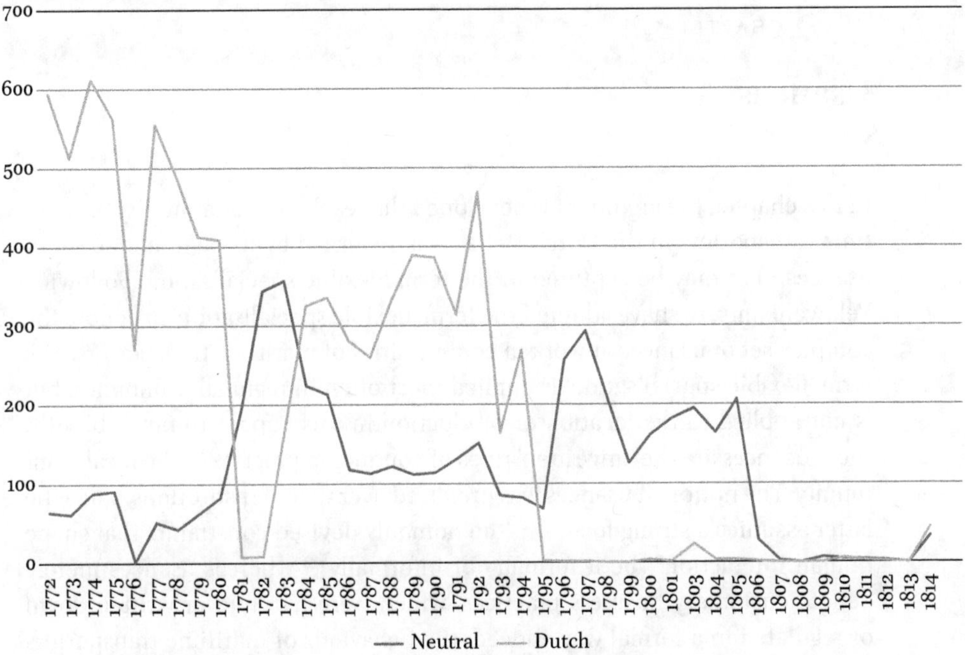

GRAPH 25 *Shift of Dutch shipmasters registered in* STRO *with two different domiciles to neutral territory, 1772–1814.*
BASED ON STRO.

both during the Fourth Anglo-Dutch War and during the Napoleonic Wars. In this respect, their strategies do not differ from those of shipmasters from the Province of Groningen. But rather than being focused *a priori* on places in Lower Saxony, Frisians and Hollanders also turned to the Austrian Netherlands during the Fourth Anglo-Dutch War. In this respect, their virtual migration strategies cannot be compared to those of the Province of Groningen. The difference may perhaps be explained by the relatively strong economic ties between Friesland/North Holland and the Austrian Netherlands.[24]

---

24 Unfortunately, it is impossible to substantiate this claim based on the existing literature on the commercial ties between the Austrian Netherlands—primarily Oostende—and the Dutch Deltas. Even though several scholars have studied aspects of Oostende's maritime trade in the eighteenth century in detail, they have, if at all, only paid attention to the local carriers of this trade: the shipmasters that were domiciled in Oostende itself. The contribution of Dutch shipmasters to Oostende's maritime trade remains to be established. For relevant literature, see: Parmentier, *De maritieme handel*; Houtman-De Smedt, "De Zuidelijke Nederlanden," 105–119; Farasyn, *De 18de eeuwse bloeiperiode*.

CHAPTER 5

Usances

In this chapter, the argument is developed that each local community of maritime transporters in the Dutch Deltas was governed by its own unique set of usances, that may be captured in the term 'flexible specialisation'. Following Veluwenkamp, we have adopted the term 'flexible specialisation' to denote the complex set of usances at work in communities of maritime transporters. The term 'flexible specialisation' is applied most often in regional economics, but is not applied to the location of production in this paper.[1] To put it bluntly, these usances are the 'unwritten' rules of conduct in practice in the local community. The notion of usances was preferred over that of institutions, since the latter assumes a strong focus on "the humanly devised constraints that shape human interaction" (be it formally or informally),[2] whereas usances merely denote customary business practices, which were hardly ever written down or regulated in a formal way. Indeed, the behaviour of maritime transporters did not have a random character; specialisation was at the core of the shipmaster's economic actions. Moreover, maritime transporters did not operate in a vacuum without constraints; any business decision made by individual shipmasters fit into a set of business practices defined by the community of which the shipmaster was part.

The notion of 'flexible specialisation' helps to describe the behaviour of maritime transporters. Flexible specialisation was not only apparent at the individual level, but also at the level of local communities of maritime transporters. While each local community had its own set of usances, which resulted in a sort of monopoly on a limited range of practices, their rules of conduct were also regionally embedded. The community shaped the actions of its members as much as the members themselves shaped the community. Together, they created, shaped, supported and preserved what is possibly described best as a local business culture. Precisely because of its central role, the community could exert a positive or negative impact on the local development of the maritime transport sector. The domicile of the shipmaster was the billboard of the local business culture with its corresponding usances. At the same time, the domicile of the shipmaster expressed his regional 'origins' and referred to the region in which the local business culture was

1 Veluwenkamp, *Archangel*.

2 North, *Institutions*, 3–4.

 | DOI 10.1163/9789004273078_007

embedded. The spatial structure of the Dutch maritime transport sector and changes therein are perhaps best understood as the result of the emergence, prosperity and decline of regionally-embedded local business cultures.

In order to gain insight into the usances of Dutch maritime transport communities, the population of shipmasters of the community of Hindeloopen was analysed, employing a set of tools for semantic enrichment of spatial data resulting from economic behaviour.[3] Relevant for this chapter are the semantic annotation of the status of individual economic entities within a collective of entities and the analysis of the operational patterns of individual economic entities. In our historical case, these results are achieved through application of the entity dynamics tool and operational pattern classification to historical ship movement data. Entity status classification is based on three parameters: new, re-appearing and known entity supply (see Table 25). During classification, all three parameters are related to the total supply (i.e. the size of a population during a certain time interval), both in absolute numbers and as relative shares. The underlying assumption for entity status classification is that a route continuously needs new supply in order to develop and avoid stagnation. However, a route cannot survive without stability (i.e. a share of total supply that is ascertained for a number of consecutive years). Through comparison of the share of new supply and the share of 'known participants' on a certain route at a certain point in time, various stages in the development of populations can be addressed and analysed. It allows a determination of when stagnation becomes a problem; in other words it allows a determination of when a route is starting to decline.

Entity status classification covers the micro-meso domain. It is a type of semantic annotation in which the behaviour of individual entities (micro) is treated in the context of a dynamic population of entities. The status of an entity can be annotated semantically on the basis of its individual trajectories (see Table 25). Semantic annotation, therefore, addresses micro-economic behaviour.

The second of the mentioned types of semantic annotation is involved with operational patterns of individual economic entities and provides breakdowns of the individual behavioural patterns that economic entities adopted in their activities on one route or a set of routes. Operational pattern classification is a horizontal type of classification based on entity dynamics (see Table 26). Distinction is made between specialized participation, regular participation, sporadic participation and incidental participation. Additionally, entity

3 Scheltjens and Dopfer, "Unified Rule Approach," 227–249; Scheltjens, "The operational structure," 357–389.

TABLE 25 *Summary of entity status classification*

Class	Definition
New	Moving entities that appear for the first time at a given location or set of locations during the examined time-period
Re-appearing	Moving entities that re-appear at a given location or set of locations after having been absent for at least one pre-defined time-moment within the examined time-period
Known	Moving entities that have been registered in the immediately preceding pre-defined time-moment or time-moments at a given location or set of locations during the examined time-period

TABLE 26 *Summary of operational pattern classification*

Participation frequency	Pattern density rate	Type of participation
More than once	PDR < BM/2	Specialized participation
	BM/2 ≤ PDR ≤ BM*3/2	Regular participation
	PDR > BM*3/2	Sporadic participation
Once	N/A	Incidental participation

dynamics data (in our case: historical ship movement data) can be semantically enriched with a pattern density rate (PDR), which is a calculation of the time frame of a particular operational pattern divided by the number of voyages executed within this time frame. The closer the pattern density rate comes to the benchmark (BM) (i.e. the empirically established average pattern density rate), the smaller the time frame in which the shipmaster carried out an increasing number of voyages.

The outcome of operational pattern classification provides a breakdown of the geographical trajectories of a moving entity, including their frequency and the time-frame of execution of these trajectories within a pre-defined time-period. Based on these data, the density of the participation of the moving entity on each of these geographical trajectories is calculated. Next, the benchmark value for these geographical trajectories is calculated on the basis of trajectory data of the entity during a certain time-period. The pattern density rate of the moving entity is compared with this benchmark value and the classification criteria described in Table 26 are applied. Thus, the spatial trajectories of a moving entity can be classified according to the participation density. After execution of entity status classification and operational pattern classification, individual trajectory data is annotated semantically in two ways. First, the position of an entity in a dynamic population has been defined. Second, a typology of the entity's participation pattern during a pre-defined time-period has been given.

In a spatio-temporal context, an economic entity cannot be isolated from other economic entities and their respective movements. Collectives and sub-collectives with variable membership can quite easily be identified through simple operations. For example, the sub-collective of moving entities sharing the same route during a certain time period is identified through selection of this route and the consecutive visualization of its members. More complicated is the identification of sub-collectives based on kinship relations, because names are insufficient as identifiers of such relations. In this chapter, we will focus on sub-collectives sharing routes exclusively. One way to achieve semantic annotation of collectives of economic entities is to aggregate the results of entity status and operational pattern classification. This is done in two relatively simple procedures. In the first case, entity status classification is aggregated to provide semantic data about the collective as a whole. Following this procedure, the collective of moving entities can be described and visualized in terms of its composition, with possible distinction between incidental participants (only once member of a collective) and non-incidental participants (more than once member of a collective), but also between new, known and re-appearing participants. In the second case, the pattern density rate for the entire geographical footprint of all trajectories can be calculated and used as benchmark value to provide operational pattern classification for the collective as a whole.

In the following example, the rise and decline of the maritime transport community of Hindeloopen, a small Frisian town that was famous for its large population of shipmasters, is analysed from the perspective of its population of shipmasters. Throughout the early modern period, shipmasters from Hindeloopen were the Dutch Republic's most famous timber transporters.[4] They operated mostly on behalf of merchants and entrepreneurs in North Holland (predominantly Amsterdam and the Zaan area). Already in the sixteenth century, shipmasters from Hindeloopen transported large quantities

4 Lootsma, "De Zeevaart van Hindeloopen," 218–296.

of timber from Norway to Amsterdam.[5] By then, the maritime transport community of Hindeloopen, with timber and northern Europe as its main features, had completed its initial formation process. From that point onwards, the reputation built up by the local community of shipmasters triggered further specialisation in ship type, route and cargo. Scattered evidence from the end of the sixteenth century reveal that a community of shipmasters from Hindeloopen was engaged in timber transports from Norway on a regular basis.[6] In fact, of the seventy-five arrivals of shipmasters from Hindeloopen registered in the 1580 *convooigeldregisters* of Amsterdam,[7] only three were not returning from Norway. Similarly, Bosse provides an overview of Dutch ships arriving in Tönsberg (also in Norway) at the end of the 1580s, which shows that shipmasters from Hindeloopen were the preferred transporters in that area as well.[8] Of the eighty-six registered arrivals, seventy were Dutch ships, of which the shipmasters of forty-eight were domiciled in Hindeloopen. Half a century later, not much seems to have changed, as Kernkamp's data amply reveal.[9]

For a significant period of time, the actions of the community had a positive impact on Hindeloopen's maritime transport sector: shipmasters from the community were considered to be 'the best in the business'; their knowledge and skill was recognised far beyond community borders. As a rule, the reputation of a local population of shipmasters was based on a small set of very specific (repeated) actions. The primary effect of their successful implementation in daily operational practice was one of expansion: newcomers, adhering to local institutions, made the population grow. Two elements are important to stress in this context: first, growth was accomplished within a framework of action that was locally provided and endorsed and, second, local institutions were regionally embedded. With regard to the latter, we may point to the fact that the maritime transport community of Hindeloopen was not the only one of its kind. Several communities with a strong specialisation in a limited range of activities were discussed in Chapter 3.

In general, within a local population of shipmasters, the chances to successfully develop an alternative range of maritime transport activities, parallel to the one in which the community had obtained its reputation, were slim. Indeed, both the process of emergence of a local community and that of its further development were highly path dependent. This, in turn, made local

---

5 Bosse, *Norwegens Volkswirtschaft*, 87–94.

6 Bosse, *Norwegens Volkswirtschaft*, 87–94; Dataset of the *Convooigeldregisters*.

7 Dataset of the *Convooigeldregisters*.

8 Bosse, *Norwegens Volkswirtschaft*, 89–90.

9 Kernkamp, *De handel op den vijand*.

communities vulnerable to changing circumstances in commodity streams or in the community of shipmasters itself. Besides these two factors, there is a whole range of changes that could have an impact on the growth or decline of communities of shipmasters, like changing import and export regulations, wars, capacity constraints at the shipmasters' homeport, international competition and changes in the location of production or consumption. However, when dramatic changes occurred in the supply of timber from Norway[10] and new timber outlets were sought in the Baltic in the last decades of the seventeenth century, shipmasters from Hindeloopen were able to respond to this geographic supply shift and became the dominant transporters on the routes between the eastern Baltic (Narva, Vyborg, Riga, etc.) and the Dutch Republic in the first half of the eighteenth century.[11] Apparently, at that time, their reputation and skill was strong enough to deal with such significant changes.

Inevitably, at a certain moment, a point of saturation was reached. For the shipmasters of Hindeloopen this happened in the late 1730s. There was less and less room for newcomers (see Graph 26) and specialisation became a negative factor for the maritime transport sector in the local community. The commercial environment in which the shipmasters community of Hindeloopen operated was one in which each community of shipmasters had some sort of monopoly on a very specific range of activities. It was hard to break with previous experience, accumulated as a community, and internalised as a local institution. The path dependent specialisation process had a negative impact on the local maritime transport sector: there was no way out. Such phases of decline could be of a temporal nature, like a fluctuation, but in the long run, they could also become a trend, leading towards the obsolescence of the local population of shipmasters. This appears to have been the case for Hindeloopen's maritime transport community in the latter part of the eighteenth century (see Table 27).

Shipmasters' communities provided an informal framework for knowledge acquisition and exchange. Membership of the community was vital. It could have a positive effect, when community membership supported an increase in the benefits of competitive advantage, and a negative effect, when membership led to negative lock-in, leading to the fossilization of the community. Formal or semi-formal extensions, like skipper guilds, mutual insurance boxes or barges, could emerge on the basis provided by existing usances. These extensions did not exert a direct influence on the shipmaster's daily activities; they functioned primarily as quality control or solidarity mechanisms.

---

10 Moore, "Amsterdam is Standing on Norway, part II," 200–201.

11 Lindblad, "Dutch trade," 103–114.

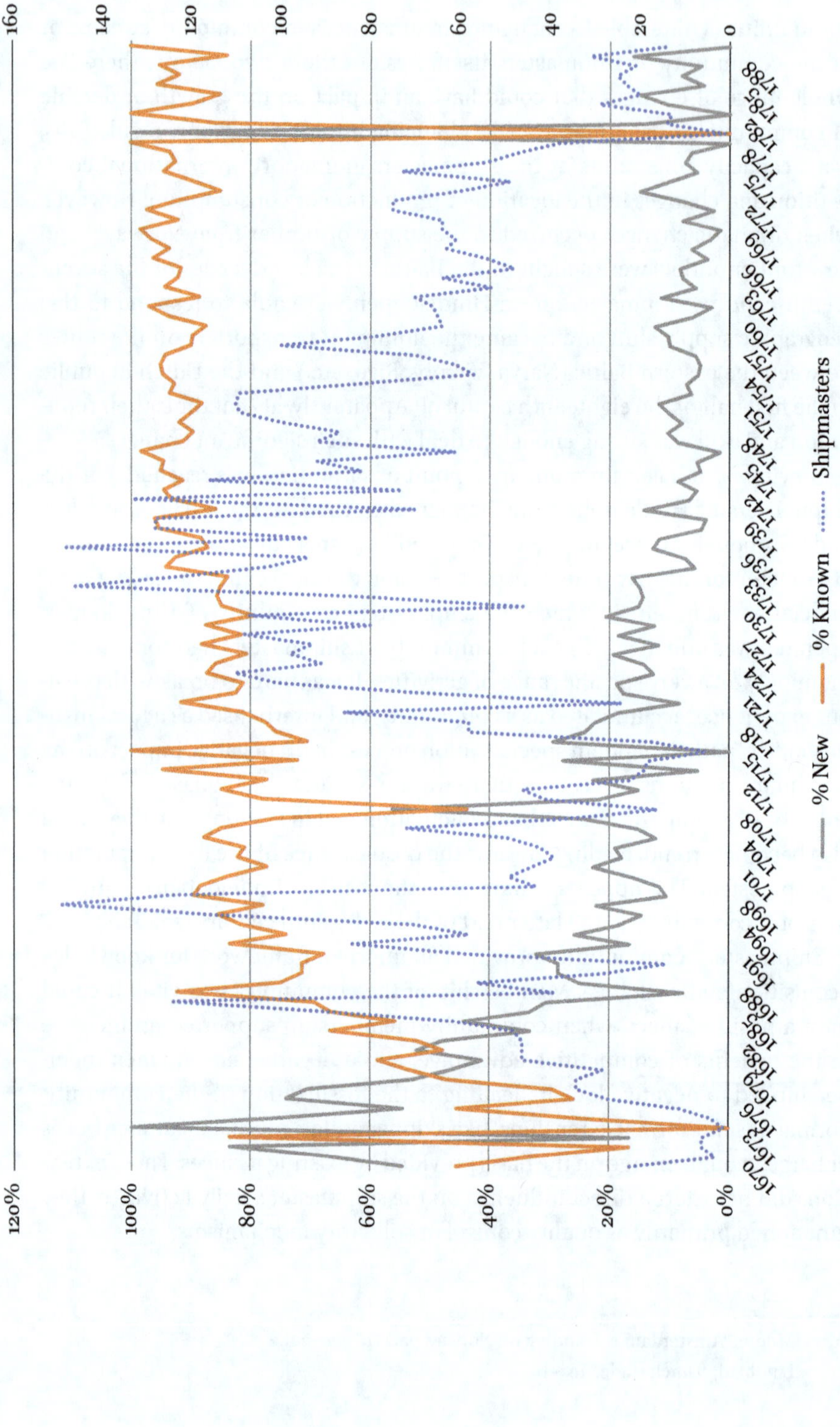

GRAPH 26 *Entity status classification for the maritime transport community of Hindeloopen, 1670–1790. Number of shipmasters on right axis.*
BASED ON STRO.

TABLE 27 *Operational pattern classification of the maritime transport community of Hindeloopen, 1670–1820. Based on STRO*

	Shipmasters							
	Spec	Reg	Spor	Once	% Spec	% Reg	% Spor	% Once
1670–99	48	100	61	238	11%	22%	14%	53%
1700–29	58	48	39	128	21%	18%	14%	47%
1730–59	54	59	33	129	20%	21%	12%	47%
1760–89	17	19	13	55	16%	18%	13%	53%
1790–1819	5	1	1	10	29%	6%	6%	59%
	Passages							
	Spec	Reg	Spor	Once	% Spec	% Reg	% Spor	% Once
1670–99	406	940	219	238	23%	52%	12%	13%
1700–29	1316	527	142	128	62%	25%	7%	6%
1730–59	1680	525	142	129	68%	21%	6%	5%
1760–89	586	219	50	55	64%	24%	5%	6%
1790–1819	14	3	2	10	48%	10%	7%	34%

While the functions of formal organisations are often described in legal sources explicating the formal 'rules of the game', the day-to-day activities of members of the community, which are fundamentally different from the formal rules of the game,[12] largely remain invisible. It is not possible to derive any knowledge about a member's daily actions from a description of the formal rules of a collective. In order to know more about the usances that shaped the community, sources that document the activities of its members have been explored in the previous chapters and empirical evidence in support of 'flexible specialisation' was found at different levels of analysis. At the micro level, specialisation was observed in the operational patterns of individual shipmasters (i.e. the routes they sailed and the cargoes they carried). Operational patterns provided valuable insight into the usances that shaped the shipmasters' actions. The main usances at work were routinization and flexibility. Routinization found

12 See: Meyer and Rowan, "Institutionalized organizations," 341.

expression in the fact that shipmasters usually sailed on a limited number of different routes, often for several years in a row. Flexibility was expressed in occasional divergence from the dominant pattern.

Usances provided structure to communities of maritime transporters in the Dutch Deltas' cities and villages, to the regions to which they adhered and to the maritime transport sector as a whole. At some point, these usances could be subject to institutionalization; a process that could serve various goals, from dealing with technical regulations to securing market access, promoting skills acquisition and establishing solidarity mechanisms. The formal organisations of maritime transporters that emerged in the Dutch Republic had a number of unique features that can be described on the basis of qualitative archival evidence.

In some aspects, formal organisations of maritime transporters were similar to craft guilds, but whereas craft guilds were essentially about the production of goods,[13] formal organisations of maritime transporters were about the production of services. In other aspects, they were similar to merchant guilds and other services-producing guilds, but whereas these services-guilds dealt primarily with the production of services in one particular place,[14] formal organisations of maritime transporters had a much wider geographical scope. A major difference between merchant guilds and other services-producing guilds, on one hand, and formal organisations of maritime transporters, on the other, was that the latter's services were not provided 'on the spot' but 'at a distance', which required a different kind of rules to be established.

Different types of formal organisations of maritime transporters emerged. Evidence of formal organisations of maritime transporters can quite easily be traced in local and provincial archives as well as in primary source editions. In a number of cases, however, the actual guild archives did not survive and the only remaining evidence can be found in legal documents that mention skipper guilds, mutual insurance boxes and barges—the three main types of formal organisations of maritime transporters (see Table 28).[15]

Skipper guilds are the oldest and most traditional. They are equivalent to craft and merchant guilds, even though, as was mentioned above, the location of the services they provided, was different. The first skipper guilds in the Low Countries emerged in the fourteenth century in cities like Elburg, Deventer,

13 Epstein, "Craft guilds," 684–713; Ogilvie, "Guilds," 286–333.

14 Greif, Milgrom and Weingast, "Coordination," 745–776.

15 For an overview of skipper guilds in the Netherlands see: "Database Dutch Craft Guilds", http://hdl.handle.net/10411/10101%20V1. Accessible via: http://www.collective-action.info. A comprehensive overview of barges does not exist.

TABLE 28 *Typology of formal organisations of maritime transporters*

	Crafts guilds	Merchant and services-producing guilds	Formal organisations of maritime transporters
Type of production	Production of goods	Production of services	Production of services
Location of production	On the spot	On the spot	At a distance

Kampen, Amsterdam, Antwerpen, Brugge, Dokkum, Ghent, Groningen, Hoorn, Middelburg, Sluis and Zierikzee. While some of their activities probably were carried out at sea, this was not (yet) their main area of activity in the early years. First and foremost, these skipper guilds dealt with transport services in general, which could be of a local and regional nature. It was only after 1500 that the increasing supply of maritime transport services required some restructuring of skipper guilds that led to the emergence of separate guilds for inland and maritime transportation. Skipper guilds continued to be part of most Low Countries' cities for the entire period covered in this paper.[16]

While skipper guilds emerged in urban centres during the fourteenth to sixteenth centuries, a specialization process unfolded in the Low Countries' rural coastal areas, predominantly in the area north of Amsterdam (Noorderkwartier), Friesland, the Province of Groningen and the Wadden Islands, which led to a sharp increase in the share of villages in the supply of maritime transport services. The diffusion of maritime transport services to rural areas entailed a social dilemma: when the supply of maritime transport services was the main (or even the only) source of income, how could it be secured? The creation of mutual insurance boxes for shipmasters (and their respective households) was probably the most common solution. These so-called 'schipperscompacten' (also called 'guild boxes' or 'mutual boxes')

16 A total of 173 skipper guilds could be retrieved from the database of craft guilds in the Netherlands. See: "Database Dutch Craft Guilds", http://hdl.handle.net/10411/10101 V1. Some of these guilds were continuations under a different name of existing guild structures, some are mutual insurance boxes or barges rather than actual skipper guilds and some may have been involved in domestic and inland rather than maritime transport services, so the actual number of skipper guilds in the Netherlands is probably lower.

TABLE 29 *Typology of communities of maritime transporters*

	Formal	Informal
Types	Skipper guilds (urban) Mutual insurance boxes (rural) Barges (urban & rural)	Community-based populations of maritime transporters
Source base	Direct evidence (mostly legal sources)	Indirect evidence (serial data sources)

emerged primarily (but not exclusively) in areas that lacked a significant international trade of their own but did have the means to supply transportation services to third parties. 'Schipperscompacten' could also emerge in affiliation with urban skipper guilds.[17] Mutual insurance boxes were most prominent in Friesland and Groningen. Their emergence coincided with the maritime expansion of their rural coastal communities' transportation services in the seventeenth and eighteenth centuries (see Table 29).[18]

Guild boxes emerged as early as 1605, when the Great Skippers' Guild (*grootschippersgilde*) of the City of Groningen was founded.[19] Its foundation marked the gradual development of an alternative insurance system in the northernmost parts of the Dutch Republic.[20] Mutual insurance or guild boxes for shipmasters incorporated elements of traditional urban skipper guilds, but

17 Go, *Marine insurance*, 36.

18 Go's dissertation on marine insurance in the Netherlands contains a valuable overview of the literature on mutual insurance boxes in the Province of Groningen. It should be noted, however, that Go's comparative analysis of the marine insurance markets of Amsterdam, Rotterdam and Groningen is rather misleading. The insurance systems that developed in Amsterdam and Rotterdam were concerned with insuring cargoes and/or ships. Quite differently, the mutual insurance boxes that emerged in the Province of Groningen dealt with the shipmasters' risks to life at sea. The insurance systems of Amsterdam and Rotterdam are about the risks of trade; the insurance system of the Province of Groningen is about the risks of transport. Both types of systems cannot and should not be confused. Therefore, in line with the focus of the present monograph on the Dutch maritime transport sector rather than on Dutch trade, our use of Go's study of the Dutch marine insurance system is strictly limited to her survey of the Province of Groningen.

19 Go, *Marine insurance*, 36.

20 Unfortunately, available data on schipperscompacten is scarce and mostly limited to some legal documents.

reserved a more prominent place for regulations concerning life and damage insurance.[21] Moreover, membership was less restrictive (in the case of the guild box organised by the great skippers guild of Groningen it was voluntary)[22] and the geographical scope of the individual members' supply of transport services was taken into consideration when fees were calculated.[23] Often mutual boxes were called skipper guilds, or 'vrije Societeit' or 'willekeur' or 'contract'. Although this complicates their identification at first, the distinction between a "traditional" skipper guild and a 'schipperscompact' is quite clear: the former had restrictive membership, the latter had not; the former was a typically urban phenomenon, the latter was not.

Barges (Nl. *beurtveren*) have a somewhat ambiguous position among the formal organisations of maritime transporters in the Low Countries. Most barges were operational on domestic routes, whereas only a limited number of barges can be attributed to the domain of maritime (sea-going) transportation.[24] The exploitation of domestic barges was organised in separate guilds, for example, the 'Guild of the barge between Workum and Amsterdam'[25] and the several 'métiers des bateliers' in the southern Netherlands.[26] In other cases, barges were part of the activities of local skipper guilds, either as a one-sided organisation, or as a cooperation of two (and sometimes more) places that wanted to operate such regular service together.[27] Besides, Wagenaar stipulated that a great number of barges existed that were not formally organised.[28] Similar organisational differences apply to the international, sea-going barges that appeared in the beginning of the seventeenth century. For example, the international barges between Amsterdam and London, Rouen, St. Valéry, Bremen, Hamburg and (later) Duinkerken were operated under the jurisdiction of so-called commissioners of the Great Skippers' Guild of Amsterdam who were not formally part of the Great Skippers' Guild itself.[29] The barges

21 'Schipperscompacten' should not be confused with 'zeevarende buidels' or 'bootsgezellenbeurzen', even though they have a number of shared characteristics.

22 Go, *Marine insurance*, 36–41.

23 Go, *Marine insurance*, 36–41; Geerdink van der Worp, *De Sociteit Compact*; De Graaff, "De Woudsenders"; Bakker, "Scheepsverzekering".

24 Barges on international riverine routes are not taken into consideration in this chapter. For literature about these barges, see Fuchs, "De Amsterdamse beurtvaart," 30–45; Heirwegh, unpublished; De Peuter, *Brussel in de achttiende eeuw*.

25 Roorda, [*Genealogische collectie Roorda*].

26 Heirwegh, unpublished.

27 Fuchs, *Beurt- en Wagenveren*.

28 Wagenaar, *Amsterdam*.

29 Van Eeghen, *Archief van de Gilden*.

between Rotterdam and Duinkerken, and between Rotterdam and Oostende and Brugge, on the contrary, were organised by the City Council of Rotterdam, but membership of the skippers' guild was stated as a requirement in article III of the ordinance.[30]

Skipper guilds, mutual insurance boxes and barges share two basic functions: preserving quality of service and securing the shipmaster's income. The first function of skipper guilds, mutual insurance boxes and barges was the pursuit of quality of service. Each of the formal organizations approached this issue in its own particular way, but in all cases quality of service was preserved by means of membership regulations (rules for admission or exclusion, membership fees), technical regulations (rules for equipment) and operational regulations (practical rules). The second function was to secure the income of organisation members (i.e. shipmasters) and their respective households. Skipper guilds and barges tended to secure income for their members through exclusivity regulations, whereas mutual insurance boxes treated income-related issues differently.

The difference between the various types of formal communities is not always clear. The so-called skipper guild of Leek, a village located about 15 km west of the City of Groningen, for example, had the same functions as a number of mutual insurance boxes, located roughly in the same region (e.g. Oude en Nieuwe Pekela, Veendam, Wildervank and Oldampt; Woudsend, Makkum, Workum and Hindeloopen; Stavoren, Sneek, Joure and Leek). In Workum, an old Frisian town located on the Zuiderzee, a guild of barge skippers coordinated the exploitation of a fleet of sixteen ships, demonstrating the existence of guilds with a very specific function.[31] To complicate the matter even further, skipper guilds sometimes organised so-called guild boxes for their members, which was a specific feature of skipper guilds in the Province of Groningen.[32] Moreover, within this broad category, institutional developments sometimes led to the split or merger of local skipper guilds into guilds of a more specific or, on the contrary, more general nature. A typical example of such development can be found in the history of the skipper guilds of Amsterdam.[33] Their intrinsic historical value notwithstanding, formal organisations, and the archival documentation about them, only partially shed light on the daily practices of organisation members. Usances remain central to any discussion about collective action in the Low Countries' maritime transport sector, even though

30 Jaerboeken 1770, 898–914.

31 Roorda, [*Genealogische collectie Roorda*].

32 Go, *Marine Insurance*, 41–46.

33 Van Eeghen, *Archief van de Gilden*.

they are harder to grasp, since neither the modalities of informal cooperation between maritime transporters nor the names of those who cooperated were normally written down in legal documents.[34]

In the above survey of communities of maritime transporters in the Dutch Republic, a distinction was made between formal organisations and informal communities of maritime transporters, arguing that both were institutions for collective action that emerged in pursuit of different but related goals. The actions registered in STRO served as the basis for generating knowledge about the principal economic behaviour of collectives of maritime transporters and about the usances that shape this behaviour. Quantitative analysis of historical maritime shipping records like STRO has shed light on the usances at work in the Low Countries' maritime transport sector and has contributed to our knowledge of a vital part of the economy of the Low Countries in pre-modern times. At the same time, evidence from legal sources, which was in it the result of institutionalization, suggested that the community-based usances that shaped these operational patterns were—for the most part—not formalized. Rather, they continued to serve as an informal framework for maritime transport operations. Because of the difference between formal organisations and usances, it would be wrong to equate usances with their formal 'offspring' residing in legal documents of all kinds.

It cannot be denied that a relation exists between formal organisations like skipper guilds, mutual insurance boxes and barges and the usances (i.e. the daily behavioural practices) that we have labelled 'flexible specialisation'. They share a common ground depicted as 'regionally-embedded local business culture'. Informal collectives were the actualization of the regionally-embedded local business culture to which the collective adheres, whereas formal organisations were the result of processes of institutionalization. My viewpoint is that the usances developed by informal collectives are concerned with daily practice, while formal organisations offered ways to secure long-term benefit.

34 There is one major exception: the occasional cooperation of maritime transporters in legal issues. A good example is the case of 92 Makkumer shipmasters who gathered to protest against protectionist regulations installed by the City of Groningen in 1694. While such examples of collective action are numerous in the history of trade and transport, their occasional (non-institutional) nature forced us to exclude them from our survey. To complexify things, the Makkumer shipmasters that joined the action against Groninger city officials called themselves 'schipperscompact', referring to 'compact' in the meaning of 'group' rather than in the more common meaning of 'mutual insurance box'. Information about this 'schipperscompact' of Makkum was kindly provided by Drs. Jelle Jan Koopmans, PhD student at the University of Groningen.

'Flexible specialisation' had local, regional and, logically, also supraregional features; each local collective of maritime transporters had some unique institutions of its own, which were complemented with usances that were shared with other collectives regionally and beyond. Formal organisations shared a common ground with their informal counterpart. Formal organisations of maritime transporters emerged when there was a reason for processes of institutionalization to take off. Among other reasons, formal organisations emerged as a way to organize sanctioning mechanisms and to make the informal collective sustainable.

In the remainder of this chapter, I compare my insights with the literature about the historical emergence of institutions for collective action, and more specifically, with De Moor's review of various forms of corporate collective action in pre-modern western Europe.[35] De Moor calls their emergence both a silent and unnoticed revolution and suggests that the new institutions for collective action that sprung up in pre-modern Europe were essential to the specific development of the European economy. While I am in broad agreement with De Moor's assessment of the role of institutions for collective action, I will use my survey of communities of maritime transporters in the Dutch Republic to support a claim for a more explicit treatment of its usances.

De Moor advances 'corporate collective action' as the concept best suited to describe "the exclusive, self-governed autonomous institutions that formed the core of the silent revolution".[36] The concept of "corporate collective action" highlights "group formation processes accompanied by the design of a set of rules to make collective action work",[37] distinguishes corporate collective action from other types of collective action and is mainly concerned with the governance structure of formal organisations. In my opinion, the concept of corporate collective action presents itself as a promising tool for the description of institutional dynamics, but it fails to capture how these 'corporate' dynamics were shaped by the daily practice of the institution's member populations. There is little doubt that changes in the everyday routines of organisation members eventually found their way into the institutional structure itself. Yet, such institutionalization refers to an advanced stage of the development of collective action, rather than to its starting point.

In my opinion, the study of institutions for collective actions would benefit from making a distinction between daily practice, with its own, mostly *tacit* rules of the game, and the continuous efforts that were made to give coop-

35 De Moor, "The Silent Revolution," 179–212.

36 De Moor, "The Silent Revolution," 192.

37 De Moor, "The Silent Revolution," 193.

eration a sustainable character. The separate study of daily practices, which is most likely to be executed as quantitative analysis, sheds light on features of collective action that would remain invisible if only qualitative evidence were to be explored. I hope to have made this point clear in the above survey of communities of maritime transporters. At the same time, it goes without saying that the study of corporate dynamics or changes in the formal rules of a collective also help to understand apparent changes in the daily routines. In short, there is a reciprocal relation between formal organisations and informal communities. Both strands of institutions for collective action would greatly benefit from a rapprochement between the two, which would also bring us closer to explaining what their contribution to the 'rise of the west' in the pre-modern world actually looked like.

In this context, it may be worthwhile to indicate one promising area of further research. One of the potential motives for corporate collective action was to achieve advantages of scale through 'sticking together'.[38] Advantages of scale helped to explained why maritime transporters active in bulk transports, like the shipmasters from Hindeloopen, operated as communities. It may have been the case that 'communities of merchants' involved in international trade benefited from the fact that they could rely on communities of maritime transporters to carry large amounts of bulk goods like grain or timber. We have seen in the previous sections, that the evidence from STRO, which registered a significant part of the Europe's bulk trade, supports this assumption. Merchants may have been able to get a better freight rate, because they acted as a collective, while shipmasters benefited from such collective deal in different ways: their income was secured, risks were reduced and their social capital was leveraged. It would be of great value to study the impact of collective action on the competitiveness of the Dutch maritime transport sector, expressed in terms of operational efficiency and transport costs. This, however, would require extensive cooperation between the various disciplines within the historical sciences as well as the further elaboration of tools capable of dealing with the dynamics of complex group interaction.

38 De Moor, "The Silent Revolution," 204.

Conclusions

In this book, the delta was advanced as a suitable geographical unit for understanding the regional economic origins of communities of maritime transporters. Three deltas were identified. The first delta includes the mouths of the rivers Scheldt, Meuse and Rhine, with Antwerpen, Middelburg, Dordrecht and Rotterdam as its main gateways. The second delta is that of the Zuiderzee, including Amsterdam, Hoorn, Enkhuizen, the harbours of the mouth of the Ijssel (Kampen, Zwolle), the Frisian harbours and the Wadden Islands Ameland, Texel, Vlieland and Terschelling. The third delta is the Lauwers-Dollard-Ems delta, with Groningen and Emden as its main central places and Dokkum as a humble third.

Maritime transport services could emerge in fishing, peat-extracting, dairy-producing villages, or in settlements that developed into regional (and sometimes supra-regional) market centres. Examples of the former are the maritime transport communities that emerged on the Wadden Islands around 1570, previously involved almost exclusively in fisheries or the maritime transport communities in the fen colonies (Nl. *veenkoloniën*) in the Province of Groningen. Examples of the latter obviously include Amsterdam and Rotterdam, but also secondary cities like Edam, Hoorn or Enkhuizen. Essential for the development of the spatial structure of the Dutch maritime transport sector were changes in maritime trade networks and changes in the structure of regional economies. Both entailed a process of specialisation in the transport sector, which facilitated the emergence of 'professional' (or 'specialised') urban and rural maritime transport communities, on one hand, and the development of an integrated maritime transport market with Amsterdam and Rotterdam as its main centres, on the other.

Before the mid-seventeenth century, the three deltas operated more or less autonomously. For the most part, shipmasters from communities in each delta provided maritime transport services to the central places of their own delta. While initially their existence was based on a set of common features, the functions of maritime transport communities in cities differed from their *counterparts* in rural areas. In the case of rural communities, economic development and access to the maritime transport market were necessary for the development of maritime transport communities. This was different for cities: their own urban economy was initially enough to develop maritime transport communities.

In the second half of the seventeenth century, an important change occurred. A process of centralization of the maritime transport market in one

 | DOI 10.1163/9789004273078_008

single gateway of the delta started, mirroring a similar process of centralization of international trade. Secondary cities lost their autonomy in international trade; their maritime transport communities became dependent on the central maritime transport market of the delta. In the eighteenth century, this process continued and expanded beyond the boundaries of the deltas: the services rendered by shipmasters from one delta became more and more accessible to trade networks in the other deltas. A process of integration between the deltas was initiated, in which the maritime transport communities of LDE owed their activity to the continuing development of the central gateways of Amsterdam (ZDZ) and Rotterdam (SMR). Maritime transport communities in the Fen colonies, but also in Groningen and Dokkum, accounted for part of the shipping to and from Amsterdam and Rotterdam; Groningen and Dokkum also lost their autonomy. At this point, one may speak of a partly integrated Dutch maritime transport space. Partly, because the spatial integration between LDE, on the one hand, and ZDZ and SMR, on the other hand, was one-sided and clearly showed characteristics of a spatial hierarchy with ZDZ at the top. Between SMR and ZDZ one could also distinguish spatial integration in the second half of the eighteenth century, but because of the parallel development of Rotterdam as central *gateway* in SMR there is less hierarchy between these deltas. Changes in the economic structure of coastal regions surrounding the Dutch Deltas and changes in the functions of ports in the maritime transport system of the Dutch Deltas were put forth as the main causes for increasing specialisation and professionalization of maritime transport communities and for the gradual rise of integration between the Dutch Deltas. The dominance of Amsterdam and, later, Rotterdam may seem obvious due to their significance as gateways within the Dutch international trade system. Their role as the central transport markets, however, as was shown in Chapter 1, followed a different, though clearly related course, in which Amsterdam became the central transport market of the Dutch Republic only by the end of the seventeenth century, long after it had been established as the Republic's central gateway. The same applies to the development of Rotterdam in the course of the eighteenth century.

In Chapter 2, a categorization of ports into five categories was elaborated so as to explain how and when the structure of the maritime transport supply in individual places and at delta-level changed. Several developments could be discerned. First of all, the Fourth Anglo-Dutch War of 1781–1783 and the Napoleonic Wars led to a reconfiguration of the port structure of ZDZ. Many marinas disappeared and transport services supply went down. Amsterdam, however, continued to be a large gateway, although it gradually lost its status in the first half of the nineteenth century. Within the maritime transport system of ZDZ, the growth of the maritime transport community of Amsterdam was a remarkable phenomenon, which was explained tentatively as the result of migration of shipmasters from other ports in ZDZ to Amsterdam as well as of changing registration practices in the Sound. A third development in the port system of ZDZ was that of the relative and absolute decline of the supply of maritime transport services after 1670, which found expression both in the classification of ports and in statistics of port, homeport and home-market frequency. As before, the Fourth Anglo-Dutch War and the Napoleonic Wars that followed were identified as causes for the absolute decline of the maritime transport sector in ZDZ.

At the same time, the development of the maritime transport sector in LDE went in the opposite direction. A steadily increasing supply of maritime transport services after 1750 eventually led to processes of integration between the Dutch Deltas, whereas at the same time, the functions of the communities involved in providing these services clearly show the emergence of a particular regional economic structure, which was orientated towards ports located outside the home delta.

In Chapter 3, some key characteristics of the operations of Dutch maritime transporters in the early modern period were revealed. Generally speaking, the production of maritime transport services resulted from specialist operations with a high degree of regularity and predictability, which helps explain the rise and decline of maritime transport communities in the Dutch commercial system as the outcome of complex processes such as market entry constraints, path dependence, positive and negative lock-in.[1] Such a contribution is deemed relevant for a number of reasons. First of all, in his seminal work on the economic decline of the Dutch Republic in the eighteenth century, Johannes de Vries acknowledges that very little is known and can be said about the development of the Dutch maritime transport sector.[2] Second, in Dutch historiography, maritime transport organisation is dominated by two viewpoints. One opinion is that shipmasters were not specialized in certain routes or products, but moved randomly across the seas in search for available cargoes.[3] Elsewhere, it is argued that the opposite was true and that the appearance of new routes coincided with the emergence of new shipmasters' domiciles in the structure of the maritime economy.[4] That both viewpoints are expressed by the same author is not a surprise: operations of maritime

1 See Martin, "Roepke Lecture," 1–27 for definitions.

2 De Vries, *De economische achteruitgang*, 28–29.

3 Van Royen, *Zeevarenden*, 17.

4 Van Royen, "Varend volk," 86–87.

shipmasters have never been subject to a long-term quantitative analysis looking for general developments. Both viewpoints ignore market entry constraints, path dependence, positive and negative lock-in or access to operational knowledge as relevant to the structure of maritime transport. Third, maritime transport is a topic that is usually treated in relation to trade streams, good flows and merchant networks, but only rarely receives more profound attention, almost as if transportation can be assumed to be a fully integrated part of these streams, flows and networks, not requiring specific attention. Indeed, it is commonly acknowledged that maritime transport was an important engine of economic growth in the early modern world. Improved techniques in shipbuilding[5] and navigation[6] facilitated discoveries,[7] but also lead to changes in the structure of international trade[8] and to a shift in the focus of diplomacy and international politics.[9] Maritime transport was an economic activity of great importance, which is reflected in the continuous efforts of cities, regions, and nations to expand and secure their interests through various kinds of economic partnerships, diplomatic agreements and, when deemed necessary, warfare. In long-distance trade, transport services were often an integrated part of commercial exchange, and were either the responsibility of chartered companies exerting exclusive rights on colonial routes to the East and West Indies,[10] or of 'specialist' merchant firms who controlled trade at specific locations.[11] On short and relatively safe distances transportation was outsourced to third parties, which often were established and maintained with the financial support of merchants. This was the case in Baltic trade, where joint stock companies (Nl. *partenrederijen*)[12] and shipmasters-entrepreneurs (Nl. *rederij van één reder*) were prominent providers of maritime transport services already in the sixteenth century.[13] The integration of trade on a global scale, heralded by monopolistic companies and merchant firms leveraging colonial possessions from the early seventeenth century, was paralleled by the

5 Wegener Sleeswyk, *De Gouden Eeuw*; Unger, *Dutch shipbuilding*; Unger, *Ships and shipping*.

6 Davids, *Zeewezen en wetenschap*.

7 Scammell, *The World Encompassed*; Tracy (ed.), *The Rise of Merchant Empires*.

8 Van der Wee, "Structural Changes," 14–33; Steensgaard, "The Growth and Composition," 102–152; Barrett, "World bullion flows," 224–254.

9 Steinberg, *The Social Construction*, Ch. 3 and 4.

10 Den Heijer, *De geoctroieerde Compagnie*, 64–68.

11 See: Tracy, Introduction to *The Rise of Merchant Empires*.

12 Hart, "Rederij," 106–125; Broeze, "Rederij," 106–116; Wegener Sleeswyk, "Omvang en financiering," 52–72; Den Heijer, *De geoctroieerde Compagnie*, Ch.2: Van partenrederij naar handelscompagnie.

13 Hart, "Rederij," 106.

introduction of novel practices in the organisation of short-distance maritime transport. Regular transportation services were established on dense national and international routes, linking centres of trade like Amsterdam, London, Rouen, Hamburg and Bremen.[14]

The production of maritime transport services in the Baltic by shipmasters from the Dutch Deltas seems to have filled a niche between domestic transport and truly oceanic shipping on a global scale. Although theoretically any shipmaster could supply maritime transport services, not being bound by the usual restrictions that applied to the 'domestic' and 'global' destinations, this did not lead to random participation by Dutch shipmasters on the market for transport services to and from the Baltic. On the contrary, it had already been explained that the structure of the regional economy and the position in international trade networks were of vital importance to the maritime transport community.

In Chapter 3, it was shown that some local communities embarked upon a process of specialisation, while others maintain a more diffuse, non-specialist character. These differences were explained as the result of different positions in urban maritime trade networks, on one hand, and differences in the structure of regional economies (of which the local communities were part), on the other. At the macro level, maritime transport communities exerted their influence on the structure of the maritime transport sector as a whole. This impact was positive insofar as it increased the benefits of competitive advantage, but it became negative, when the behaviour of the community aggravated lock-in. In the context of economic-geographical changes in the demand and supply of goods (and their respective effect on the directions and frequencies of good flows between demand and supply areas), group behaviour of maritime transporters in the Dutch Deltas resulted in the rise and decline of communities of maritime shipmasters operating in service of the Dutch Deltas' centres of trade. In the last chapter, the dynamics of group behaviour were captured in the term 'flexible specialisation', which denotes the complex set of usances at work in communities of maritime transporters. It was substantiated that the behaviour of maritime transporters neither had a random character, nor was it without constraints. Any business decision made by individual shipmasters fit into a set of usances defined by the community of which the shipmaster was part.

The analysis of the Dutch maritime transport sector, which was presented in five thematic chapters, has ended with a reference to the study of the dynamics of complex group interaction as a promising area of further research. This

---

14 Fuchs, *Beurt- en Wagenveren*, 226–251; Baasch, *Die Börtfahrt*; *Ordonnantie*, 1663.

should not be a surprise: this monograph should not be seen as an encompassing survey of the emergence, functions and structure of the Dutch maritime transport sector in the early modern period. Its aim was rather to deal with one particular research *desideratum* in Dutch maritime history, namely that of the economic geography of the Dutch maritime transport sector and its long-term development. Consequently, several other important aspects related to maritime transport,—such as shipbuilding, the maritime labour market, maritime insurance, etc.—were taken into account only insofar as they contributed to the main topic of the book. Having said this, it may be a good time to say a few words about the potential contribution of these insights to the discipline of maritime history. The thematic focus on location, functions, production, domiciles and usances which was maintained in this book, and the methodological tools that were elaborated for their analysis, may be a stimulus for maritime-economic historians of other European as well as non-European regions to adopt a long-term, bottom-up perspective to the study of maritime transport systems.

If at all, it is hoped that this monograph has made two things clear. First of all, that a strong and broad quantitative basis for the study of maritime transport systems, in which international trade relations and regional economic development play a central role, is essential for producing novel analytical results with regard to the rise and decline as well as the functions of maritime transport communities in the early modern period. In this context, it should be mentioned that important gaps still remain in our knowledge of the Dutch maritime transport sector, especially with regard to the Scheldt-Meuse-Rhine delta and the presence of a maritime transport sector in the Southern Netherlands. In particular, too little is known about the long-term effects on the Southern Netherlands' maritime transport sector of the Blockade of the Scheldt in 1588. But the lack of published primary sources probably also negatively affected the level of detail with which maritime transport communities in Zeeland and along the Meuse and Rhine could be described.

Second, it is hoped that it has been made clear that the delta is a widely applicable analytical perspective, that combines rural hinterlands and urban centres as well as estuarine, coastal and maritime shipping in a single analytical framework. The delta as an analytical concept may be seen as a geographical unit of analysis that is comparable (but not identical to) the littoral, which is less precisely defined by Pearson as "the coastal sea zone, the beach, and some indeterminate frontier on land".[15] Potentially, the analytical concept of the delta may be applied to coastal areas all over the world, which may stimu-

15 Pearson, "Littoral society," 354.

late the revision of the historiography of single locations in a delta, the production of new research into the emergence, structure and functions of ports in a delta and, last but not least, the comparative analysis of deltas worldwide, for example in the context of examining divergent economic development patterns.

The present analysis of the Dutch maritime transport sector in the early modern period may also serve as a starting point for further interdisciplinary research in the structure of the maritime labour market, a topic that has received considerable attention in recent historiography, but that is still mostly dominated by a bird eye's view on the maritime labour market, in which the sailors are anonymous and their lives are reduced to numbers. Adopting a long-term, bottom-up perspective to the study of the maritime labour markets, in which the locations, functions and production of maritime services are examined at the level of the community first and aggregation to the level of the delta follows afterwards, might reveal significant similarities and differences between the locations and functions of the maritime transport sector (i.e. the shipmasters) and the maritime labour sector (i.e. the 'sailors').[16] How did these two, obviously related groups, interact? Can similar strategies be identified in the maritime labour market? Can similar concentrations of maritime labourers be identified in the Dutch Deltas, or further inland? How might their emergence be explained? What was the place of 'communities of maritime labourers' in the international transport systems? Do similar or different structures become visible in the behaviour of communities of maritime labourers? Did maritime labourers specialise in specific kinds of shipping? Did they interact with specific maritime transport communities? If so, what might have been the underlying rationale of such specialisation? Answering these and related questions may significantly advance our understanding of past economic structures and the role of "ordinary" people therein. The present monograph provides—at its best—a small step in that direction.

16 A first, very modest attempt to compare the structure of the maritime transport sector with that of the maritime labour market, using the electronic database of sailors employed by the VOC in the eighteenth century, has been published in: Scheltjens & Veluwenkamp, "Sound Toll Registers Online," 309–314.

Appendices[1]

Appendix to Chapter 1

APPENDIX 1.1 *Overview of all places in the Dutch Deltas with maritime transport communities between ca. 1300 and 1800, with reference to the corresponding entry in the geographical dictionary of Van der Aa (1839–1851)*

Delta	Class.	Domicile	Ref. (AA)
?	?	Esens	
?	?	Haukynswaard	
?	?	Oldendam	
?	?	Orno	
?	?	Oudebom	
?	?	Ralduch	
?	?	Rinchoove	
?	?	Tarndijk	
?	?	Walls	
?	?	Zuiderdam	
LDE	village	Adorp	I, 53
LDE	village	Ameland	I, 132–139
LDE	village	Anjum	I, 270
LDE	village	Bergum	II, 324–325
LDE	village	Delfzijl	III, 247
LDE	village	Ezumazijl	IV, 298
LDE	village	Farmsum	IV, 301
LDE	village	Holwerd	V, 714
LDE	village	Hoogezand	V, 776
LDE	village	Lange Akkerschans	VII, 41
LDE	village	Leek	VII, 94
LDE	village	Lutjegast	VII, 468
LDE	village	Meeden	VII, 773
LDE	village	Muntendam	VII, 1112
LDE	village	Munterziel	
LDE	village	Nieuwe Zijlen	VIII, 158

1 For an overview of the sources that were excerpted to compile the appendices 1.2.–1.6., see Introduction and Bibliography.

 | DOI 10.1163/9789004273078_009

APPENDIX 1.1 *Overview of all places in the Dutch Deltas (cont.)*

Delta	Class.	Domicile	Ref. (AA)
LDE	village	Nieuweschans	VIII, 147
LDE	village	Nieuwolda	VIII, 176
LDE	village	Oosterhuizen	VIII, 224
LDE	village	Oudega	VIII, 654
LDE	village	Peesens	IX, 15
LDE	village	Pekela	IX, 116
LDE	village	Sappemeer	X, 59
LDE	village	Schiermonnikoog	X, 221
LDE	village	Sloterzijl	X, 450
LDE	village	Statenzijl	X, 670
LDE	village	Termunten	XI, 78
LDE	village	Termunterzijl	XI, 81
LDE	village	Uithuizen	XI, 370
LDE	village	Veendam	XI, 542
LDE	village	Warffum	XII, 98
LDE	village	Weer	XII, 183
LDE	village	Wildervank	XII, 442
LDE	village	Winneweer	
LDE	village	Winschoten	XII, 496
LDE	village	Zoutkamp	XIII, 271
LDE	region	Westerwoude	XII, 320
LDE	city	Dokkum	III, 360
LDE	city	Groningen	IV, 921
LDE	vlek	Appingedam	I, 301–309
LDE	vlek	Kollum	VI, 548
outside	village	Bergen (NH)	II, 289–292
outside	village	Callantsoog	III, 5
outside	village	d'Hoef	
outside	village	Ijzermonde	
outside	village	Kamperduin	
outside	village	Lombartszijde	
outside	village	Yzer	
outside	city	Blankenberge	
outside	city	Duinkerken	
outside	city	Nieuwpoort	
outside	city	Oostende	

Delta	Class.	Domicile	Ref. (AA)
outside	vlek	Den Helder	V, 367
SMR	village	Biervliet	II, 404
SMR	village	Black	II, 457
SMR	village	Cadzand	III, 5
SMR	village	De Lier	III, 212
SMR	village	Domburg	III, 408
SMR	village	Heist	
SMR	village	Hellevoetsluis	V, 388
SMR	village	Hoek	V, 652
SMR	village	Hooge	V, 757
SMR	village	Horssens	
SMR	village	Kats	VI, 334
SMR	village	Katwijk	VI, 340
SMR	village	Kinderdijk	VI, 450
SMR	village	Kortgene	VI, 591
SMR	village	Krimpen	
SMR	village	Langendijk	VII, 48
SMR	village	Lapschuere	
SMR	village	Lek (De)	VII, 167
SMR	village	Maas	
SMR	village	Maasland	VII, 502
SMR	village	Mariakerke	
SMR	village	Meerland	VII, 783
SMR	village	Monnikereede	
SMR	village	Nederveen	VIII, 54
SMR	village	Nieuwehaven	VIII, 138
SMR	village	Nieuwvliet	VIII, 183
SMR	village	Noordwijk	VIII, 297
SMR	village	Oudewater	VIII, 700
SMR	village	Putten	IX, 315
SMR	village	Renesse	IX, 423
SMR	village	Sas van Gent	
SMR	village	Scheveningen	X, 158
SMR	village	Slepedamme	X, 419
SMR	village	St. Anna ter Muiden	X, 455
SMR	village	Temse	
SMR	village	Ter Heide	XI, 72
SMR	village	Westkapelle	XII, 336

APPENDIX 1.1 *Overview of all places in the Dutch Deltas* (*cont.*)

Delta	Class.	Domicile	Ref. (AA)
SMR	village	Wouterkeyne	
SMR	village	Zoutelande	XIII, 266
SMR	village	Zwartewaal	XIII, 412
SMR	region	Vlaanderen	
SMR	region	Zeeland	
SMR	region	Zwin	
SMR	city	Aardenburg	I, 21–26
SMR	city	Antwerpen	
SMR	city	Arnemuiden	I, 322
SMR	city	Axel	I, 394
SMR	city	Bergen op Zoom	II, 297
SMR	city	Brielle	II, 753
SMR	city	Brouwershaven	II, 778
SMR	city	Brugge	
SMR	city	Damme	
SMR	city	Delft	III, 212
SMR	city	Delftshaven	
SMR	city	Dordrecht	III, 454
SMR	city	Gent	
SMR	city	Goedereede	IV, 597
SMR	city	Goes	IV, 616
SMR	city	Gorinchem	IV, 672
SMR	city	Gouda	IV, 705
SMR	city	Hulst	V, 905
SMR	city	Leerdam	VII, 106
SMR	city	Leiden	VII, 163
SMR	city	Maassluis	VII, 537
SMR	city	Mechelen	
SMR	city	Middelburg	VII, 873
SMR	city	Reimerswaal	IX, 406
SMR	city	Rotterdam	IX, 662
SMR	city	Schiedam	X, 170
SMR	city	Schoonhoven	X, 256
SMR	city	's-Gravenzande	
SMR	city	Sluis	X, 456
SMR	city	Steenbergen	X, 694

Delta	Class.	Domicile	Ref. (AA)
SMR	city	Ter Hagen	
SMR	city	Tholen	XI, 124
SMR	city	Tiel	XI, 162
SMR	city	Veere	XI, 612
SMR	city	Vianen	XI, 666
SMR	city	Vlaardingen	XI, 718
SMR	city	Vlissingen	XI, 769
SMR	city	Willemstad	XII, 457
SMR	city	Zierikzee	XIII, 180
SMR	vlek	Roosendaal	IX, 633
SMR	vlek	Westenschouwen	XII, 289
ZDZ	village	Aalsmeer	I, 14–15
ZDZ	village	Aartswoude	I, 33
ZDZ	village	Abbekerk	I, 34
ZDZ	village	Abt	I, 42
ZDZ	village	Akersloot	I, 67
ZDZ	village	Akkrum	I, 70
ZDZ	village	Amerongen	I, 141–143
ZDZ	village	Arum	I, 331
ZDZ	village	Assendelft	I, 370
ZDZ	village	Avenhorn	I, 390
ZDZ	village	Axwijk	I, 401
ZDZ	village	Bakhuizen	II, 46
ZDZ	village	Balk	II, 63–64
ZDZ	village	Barsingerhorn	II, 126–127
ZDZ	village	Beets	II, 236–237
ZDZ	village	Bennebroek	II, 274–276
ZDZ	village	Berkhout	II, 355–356
ZDZ	village	Berlikum	II, 340
ZDZ	village	Bildtziel	II, 428; 435
ZDZ	village	Binnewijzend	II, 451
ZDZ	village	Blokker	II, 495
ZDZ	village	Bovenkarspel	II, 640
ZDZ	village	Broek in Waterland	II, 764
ZDZ	village	Buiksloot	II, 803–804
ZDZ	village	Cadoelen	III, 4
ZDZ	village	De Bilt	II, 428
ZDZ	village	De Rijp	

APPENDIX 1.1 *Overview of all places in the Dutch Deltas (cont.)*

Delta	Class.	Domicile	Ref. (AA)
ZDZ	village	De Streek	III, 498
ZDZ	village	Driehuizen	III, 516
ZDZ	village	Durgerdam	III, 575
ZDZ	village	Egmond	IV, 84
ZDZ	village	Gaast	IV, 405
ZDZ	village	Gaastmeer	IV, 409
ZDZ	village	Garyp	IV, 444
ZDZ	village	Graft	IV, 755
ZDZ	village	Graftdijk	IV, 755
ZDZ	village	Grootebroek	IV, 977
ZDZ	village	Heeg	V, 254
ZDZ	village	Heem	V, 410
ZDZ	village	Hemelum	V, 415
ZDZ	village	Holysloot	
ZDZ	village	Hommerts	V, 721
ZDZ	village	Hoogkarspel	V, 780
ZDZ	village	Hoogwoude	V, 790
ZDZ	village	Houwert	V, 873
ZDZ	village	Huisduinen	V, 881
ZDZ	village	Husum	V, 892
ZDZ	village	Ilpendam	VI, 113
ZDZ	village	Jisp	
ZDZ	village	Kalenburg	VI, 225
ZDZ	village	Katwoude	VI, 354
ZDZ	village	Kimswerd	VI, 448
ZDZ	village	Knollendam	VI, 501
ZDZ	village	Kolhoorn	VI, 545
ZDZ	village	Koudum	VI, 610
ZDZ	village	Krommenie	VI, 651
ZDZ	village	Krommeniedijk	VI, 653
ZDZ	village	Kuinre	VI, 686
ZDZ	village	Kwadijk	VI, 692
ZDZ	village	Lambertschagen	VII, 24
ZDZ	village	Landsmeer	VII, 56
ZDZ	village	Langweer	VII, 61
ZDZ	village	Limmen	VII, 334

Delta	Class.	Domicile	Ref. (AA)
ZDZ	village	Marken	VII, 689
ZDZ	village	Middeli	VII, 927
ZDZ	village	Minnertsga	VII, 979
ZDZ	village	Nieuwendam	VIII, 127
ZDZ	village	Nieuwendijk	VIII, 130
ZDZ	village	Nieuwendorp	VIII, 131
ZDZ	village	Obenhuizen	
ZDZ	village	Oldeboorn	VIII, 398
ZDZ	village	Oosterblokker	II, 495
ZDZ	village	Oosterleek	VIII, 529
ZDZ	village	Oosthuizen	VIII, 338
ZDZ	village	Oostwoude	VIII, 569
ZDZ	village	Oostzaan	VIII, 570
ZDZ	village	Opmeer	VIII, 590
ZDZ	village	Opperdoes	VIII, 592
ZDZ	village	Oudendijk	VIII, 670
ZDZ	village	Overleek	VIII, 755
ZDZ	village	Petten	IX, 141
ZDZ	village	Pingjum	IX, 187
ZDZ	village	Poppenhuzen	
ZDZ	village	Purmerland	IX, 308
ZDZ	village	Ransdorp	IX, 362
ZDZ	village	Rarop	
ZDZ	village	Ryptsjerk	IX, 788
ZDZ	village	Scharl	X, 114
ZDZ	village	Scharwoude	X, 118
ZDZ	village	Schellinkhout	X, 139
ZDZ	village	Schermer	X, 147
ZDZ	village	Schermerhorn	X, 149
ZDZ	village	Scherpenzeel	X, 156
ZDZ	village	Sloterdijk	X, 444
ZDZ	village	Spaarndam	X, 595
ZDZ	village	Spanbroek	X, 608
ZDZ	village	Surhuisterveen	X, 816
ZDZ	village	Terhorne	XI, 75
ZDZ	village	Terschelling	
ZDZ	village	Texel	XI, 103
ZDZ	village	Tjerkgash	XI, 199

APPENDIX 1.1 *Overview of all places in the Dutch Deltas (cont.)*

Delta	Class.	Domicile	Ref. (AA)
ZDZ	village	Tjummarum	XI, 272
ZDZ	village	Twisk	XI, 343
ZDZ	village	Uitdam	XI, 359
ZDZ	village	Uitgeest	XI, 363
ZDZ	village	Venhuizen	XI, 596
ZDZ	village	Vlieland	XI, 748
ZDZ	village	Warder	XII, 96
ZDZ	village	Warga	XII, 105
ZDZ	village	Warns	XII, 117
ZDZ	village	Wartena	XII, 121
ZDZ	village	Watergang	XII, 138
ZDZ	village	Wervershoof	XII, 280
ZDZ	village	Westerblokker	II, 495
ZDZ	village	Westzaan	XII, 357
ZDZ	village	Wieringen	XII, 385
ZDZ	village	Wijdenes	XII, 401
ZDZ	village	Wijk aan zee	XII, 410
ZDZ	village	Winkel	XII, 492
ZDZ	village	Witteveen	XII, 530
ZDZ	village	Woggelum	XII, 561
ZDZ	village	Wormer	XII, 611
ZDZ	village	Wormerveer	XII, 622
ZDZ	village	Woudsend	XII, 641
ZDZ	village	Zijbekarspel	
ZDZ	village	Zuiderwoude	XIII, 289
ZDZ	village	Zunderdorp	XIII, 329
ZDZ	village	Zuurwoude	
ZDZ	village	Zwaag	XIII, 379
ZDZ	village	Zwartsluis	XIII, 418
ZDZ	region	Friesland	
ZDZ	region	Holland	
ZDZ	region	Noord Holland	
ZDZ	region	Waterland	XII, 148
ZDZ	region	West-Friesland	
ZDZ	city	Alkmaar	I, 85–96

Delta	Class.	Domicile	Ref. (AA)
ZDZ	city	Amersfoort	I, 144
ZDZ	city	Amsterdam	I, 170–255
ZDZ	city	Arnhem	
ZDZ	city	Beverwijk	II, 393
ZDZ	city	Bolsward	II, 549
ZDZ	city	Deventer	
ZDZ	city	Doetinchem	III, 382
ZDZ	city	Driel	III, 519
ZDZ	city	Edam	
ZDZ	city	Elburg	IV, 126
ZDZ	city	Enkhuizen	IV, 203
ZDZ	city	Franeker	IV, 357
ZDZ	city	Haarlem	V, 30
ZDZ	city	Harderwijk	V, 156
ZDZ	city	Harlingen	V, 170
ZDZ	city	Hasselt	V, 208
ZDZ	city	Hindeloopen	V, 604
ZDZ	city	Hoorn	V, 801
ZDZ	city	Kampen	VI, 248
ZDZ	city	Leeuwarden	VII, 115
ZDZ	city	Medemblik	VII, 749
ZDZ	city	Monnikendam	VII, 1043
ZDZ	city	Muiden	VII, 1090
ZDZ	city	Naarden	VIII, 11
ZDZ	city	Nijmegen	VIII, 215
ZDZ	city	Purmerend	IX, 293
ZDZ	city	Sloten	X, 436
ZDZ	city	Sneek	X, 497
ZDZ	city	Stavoren	X, 677
ZDZ	city	Utrecht	XI, 425
ZDZ	city	Weesp	XII, 205
ZDZ	city	Workum	XII, 603
ZDZ	city	Ylst	XII, 671
ZDZ	city	Zaandam	XIII, 11
ZDZ	city	Zutphen	XIII, 334
ZDZ	city	Zwolle	XIII, 458
ZDZ	vlek	Blokzijl	II, 501

APPENDIX 1.1 *Overview of all places in the Dutch Deltas (cont.)*

Delta	Class.	Domicile	Ref. (AA)
ZDZ	vlek	Gorredijk	IV, 679
ZDZ	vlek	Grouw	IV, 991
ZDZ	vlek	Heerenveen	V, 295
ZDZ	vlek	Joure	VI, 170
ZDZ	vlek	Lemmer	VII, 181
ZDZ	vlek	Makkum	VII, 609
ZDZ	vlek	Molkwerum	VII, 1032
ZDZ	vlek	Schagen	X, 92

APPENDIX 1.2 *Location of maritime transport communities in the Low Countries before 1400, including number of registrations as found in the excerpted published primary sources*

Domicile	TOT
Lombartszijde	8
Yzer	18
Blankenberge	15
Duinkerken	11
Nieuwpoort	28
Oostende	10
Biervliet	9
Heist	13
Katwijk	7
Kortgene	12
Maasland	11
Meerland	5
Nieuwehaven	29
Noordwijk	12
Renesse	25
Scheveningen	10
Slepedamme	9
Westkapelle	23

Domicile	TOT
Wouterkeyne	5
Zwartewaal	5
Zeeland	5
Arnemuiden	22
Axel	7
Brielle	68
Brouwershaven	31
Dordrecht	67
Middelburg	33
Reimerswaal	6
Rotterdam	27
Schiedam	24
Sluis	43
Veere	119
Vlaardingen	50
Vlissingen	5
Zierikzee	47
Wieringen	5
Wijk aan zee	12
Amsterdam	8
Harderwijk	25
Hindeloopen	9
Hoorn	18
Kampen	65
Stavoren	44
Workum	26
Zutphen	17

The following places were mentioned less than 5 times: LDE: Oosterhuizen (4), Groningen (3); SMR: Lapschuere (2), Monnikereede (2), Nieuwvliet (3), Putten (2), Sint-Anna-ter-Muiden (3), Zoutelande (3), Antwerpen (4), Brugge (2), Gent (2), Steenbergen (2), Tholen (3); ZDZ: Kuinre (3), Friesland (3), Holland (3), Enkhuizen (4), Haarlem (4), Harlingen (3), Leeuwarden (2), Medemblik (4), Utrecht (4).

APPENDIX 1.3 *Location of maritime transport communities in the Low Countries 1400–1500, including number of registrations and location coefficient per area, as well as total number of registrations in published primary sources*

Unknown
Baltic: 567
North Sea – West: 898
Total: 1465
Aardenburg
Baltic: 1 (2,58)
Total: 1
Alkmaar
North Sea – West: 1 (1,63)
Total: 1
Amsterdam
Baltic: 75 (2,08)
North Sea – West: 18 (0,32)
Total: 93
Antwerpen
North Sea – West: 230 (1,63)
Total: 230
Arnemuiden
North Sea – West: 12 (1,63)
Total: 12
Bergen op Zoom
North Sea – West: 4 (1,63)
Total: 4
Biervliet
North Sea – West: 3 (1,63)
Total: 3
Brielle
North Sea – West: 17 (1,63)
Total: 17
Broek in Waterland
Baltic: 8 (2,58)
Total: 8
Brouwershaven
North Sea – West: 43 (1,63)
Total: 43
Deventer
Baltic: 1 (2,58)
Total: 1
Dordrecht
Baltic: 9 (0,73)
North Sea – West: 23 (1,17)
Total: 32
Duinkerken
North Sea – West: 2 (1,63)
Total: 2
Edam
Baltic: 24 (2,29)
North Sea – West: 3 (0,18)
Total: 27
Egmond
North Sea – West: 7 (1,63)
Total: 7
Enkhuizen
Baltic: 101 (2,56)
North Sea – West: 1 (0,02)
Total: 102
Goes
North Sea – West: 24 (1,63)
Total: 24
Gouda
North Sea – West: 48 (1,63)
Total: 48
Grootebroek
Baltic: 1 (2,58)
Total: 1
Haarlem
North Sea – West: 3 (1,63)
Total: 3
Harderwijk
Baltic: 3 (1,29)

North Sea – West: 3 (0,82)
Total: 6

Hasselt
Baltic: 2 (2,58)
Total: 2

Haukynswaard
North Sea – West: 1 (1,63)
Total: 1

Hindeloopen
Baltic: 4 (2,58)
Total: 4

Hoorn
Baltic: 25 (1,61)
North Sea – West: 15 (0,61)
Total: 40

Kampen
Baltic: 69 (2,00)
North Sea – West: 20 (0,37)
Total: 89

Katwijk
North Sea – West: 1 (1,63)
Total: 1

Katwoude
Baltic: 3 (2,58)
Total: 3

Kortgene
North Sea – West: 1 (1,63)
Total: 1

Landsmeer
Baltic: 6 (2,58)
Total: 6

Medemblik
Baltic: 8 (2,58)
Total: 8

Middelburg
North Sea – West: 37 (1,63)
Total: 37

Monnikendam
Baltic: 48 (2,58)
Total: 48

Nieuwehaven
North Sea – West: 2 (1,63)
Total: 2

Nieuwpoort
North Sea – West: 23 (1,63)
Total: 23

Noordwijk
North Sea – West: 1 (1,63)
Total: 1

Oostende
North Sea – West: 2 (1,63)
Total: 2

Oudewater
North Sea – West: 3 (1,63)
Total: 3

Purmerend
Baltic: 6 (2,58)
Total: 6

Ransdorp
Baltic: 22 (2,58)
Total: 22

Renesse
North Sea – West: 9 (1,63)
Total: 9

Rotterdam
North Sea – West: 8 (1,63)
Total: 8

Schellinkhout
Baltic: 7 (2,58)
Total: 7

Scheveningen
North Sea – West: 10 (1,63)
Total: 10

Schiedam
Baltic: 3 (0,08)
North Sea – West: 96 (1,58)
Total: 99

's-Gravenzande
Baltic: 2 (2,58)
Total: 2

APPENDIX 1.3 *Location of maritime transport communities (cont.)*

Sluis
North Sea – West: 20 (1,63)
Total: 20
Stavoren
Baltic: 21 (2,58)
Total: 21
Terschelling
Baltic: 36 (2,58)
Total: 36
Texel
North Sea – West: 1 (1,63)
Total: 1
Veere
Baltic: 4 (0,10)
North Sea – West: 97 (1,57)
Total: 101
Vlaardingen
North Sea – West: 14 (1,63)
Total: 14
Vlissingen
North Sea – West: 1 (1,63)
Total: 1
Waterland
Baltic: 58 (2,58)
Total: 58
Westenschouwen
North Sea – West: 30 (1,63)
Total: 30
Westkapelle
North Sea – West: 7 (1,63)
Total: 7
Wijk aan zee
North Sea – West: 5 (1,63)
Total: 5
Wormer
Baltic: 2 (2,58)
Total: 2
Zeeland
North Sea – West: 3 (1,63)
Total: 3
Zierikzee
Baltic: 7 (0,34)
North Sea – West: 46 (1,42)
Total: 53
Zoutelande
North Sea – West: 1 (1,63)
Total: 1
Zuiderwoude
Baltic: 7 (2,58)
Total: 7
Zunderdorp
Baltic: 4 (2,58)
Total: 4
Zwartewaal
North Sea – West: 2 (1,63)
Total: 2

APPENDIX 1.4 *Location of maritime transport communities in the Low Countries in 1500–1600, including number of registrations and location coefficient per area, as well as total number of registrations in published primary sources*[2]

Aartswoude
Atlantic North: 10 (6,60)
Baltic: 80 (1,10)
North Sea – Central: 80 (0,91)
North Sea – East: 1 (0,42)
Total: 171
Abbekerk
Baltic: 98 (1,21)
North Sea – Central: 92 (0,95)
Total: 190
Akersloot
Arctic: 1 (2,20)
Atlantic North: 1 (0,10)
Atlantic South: 2 (0,07)
Baltic: 560 (1,15)
North Sea – Central: 549 (0,94)
North Sea – West: 32 (2,32)
Total: 1145
Alkmaar
Atlantic North: 1 (0,26)
Baltic: 217 (1,17)
North Sea – Central: 212 (0,95)
North Sea – West: 5 (0,96)
Total: 435
Ameland
Baltic: 210 (1,18)
North Sea – Central: 206 (0,96)
North Sea – East: 1 (0,17)
Total: 417
Amersfoort
Baltic: 13 (2,03)
North Sea – Central: 2 (0,26)
Total: 15
Amsterdam
Arctic: 6 (1,12)
Atlantic North: 40 (0,33)
Atlantic South: 435 (1,24)
Baltic: 5408 (0,94)
North Sea – Central: 7507 (1,08)
North Sea – East: 43 (0,23)
North Sea – South: 1 (0,36)
North Sea – West: 99 (0,61)
Total: 13540
Ansloo
Atlantic North: 12 (104,23)
Baltic: 1 (0,18)
Total: 13
Antwerpen
Baltic: 130 (1,10)
North Sea – Central: 87 (0,61)
North Sea – West: 60 (18,02)
Total: 277
Appingedam
Baltic: 120 (1,32)
North Sea – Central: 94 (0,86)
Total: 214
Arnemuiden
Baltic: 11 (0,23)

2 For an overview of the sources that were excerpted to compile this list, see Introduction.

APPENDIX 1.4 *Location of maritime transport communities (cont.)*

North Sea – Central: 2 (0,04)
North Sea – West: 97 (73,34)
Total: 110
Arnhem
Baltic: 18 (2,35)
Total: 18
Assendelft
Baltic: 10 (1,02)
North Sea – Central: 13 (1,10)
Total: 23
Barsingerhorn
Baltic: 2 (0,78)
North Sea – Central: 4 (1,30)
Total: 6
Beets
Baltic: 81 (1,06)
North Sea – Central: 98 (1,07)
Total: 179
Bennebroek
Baltic: 1 (1,17)
North Sea – Central: 1 (0,98)
Total: 2
Bergen op Zoom
Baltic: 4 (0,34)
North Sea – Central: 1 (0,07)
North Sea – West: 23 (68,32)
Total: 28
Beverwijk
Baltic: 26 (1,20)
North Sea – Central: 25 (0,96)
Total: 51
Biervliet
North Sea – West: 4 (83,17)
Total: 4
Binnewijzend
Baltic: 1 (1,17)
North Sea – Central: 1 (0,98)
Total: 2
Blokker
Baltic: 16 (1,14)
North Sea – Central: 17 (1,01)
Total: 33
Bolsward
Atlantic North: 3 (0,30)
Baltic: 578 (1,22)
North Sea – Central: 528 (0,93)
North Sea – East: 4 (0,26)
Total: 1113
Bovenkarspel
Atlantic North: 2 (2,86)
Baltic: 40 (1,19)
North Sea – Central: 30 (0,74)
North Sea – West: 7 (7,37)
Total: 79
Brielle
Baltic: 160 (1,16)
North Sea – Central: 156 (0,94)
North Sea – West: 9 (2,30)
Total: 325
Broek in Waterland
Atlantic South: 4 (0,14)
Baltic: 784 (1,71)
North Sea – Central: 285 (0,52)
Total: 1073
Brouwershaven
Baltic: 46 (1,06)
North Sea – Central: 41 (0,78)
North Sea – West: 15 (12,23)
Total: 102
Brugge
Atlantic South: 1 (0,36)
Baltic: 62 (1,37)

North Sea – Central: 39 (0,72)
North Sea – West: 4 (3,14)
Total: 106
Buiksloot
Baltic: 230 (1,58)
North Sea – Central: 112 (0,64)
Total: 342
Callantsoog
Baltic: 2 (1,17)
North Sea – Central: 2 (0,98)
Total: 4
De Bilt
Atlantic North: 1 (4,91)
Baltic: 10 (1,02)
North Sea – Central: 11 (0,93)
North Sea – East: 1 (3,16)
Total: 23
De Rijp
Baltic: 295 (1,17)
North Sea – Central: 290 (0,95)
North Sea – East: 5 (0,61)
North Sea – South: 2 (16,39)
North Sea – West: 1 (0,14)
Total: 594
Delft
Atlantic South: 1 (1,29)
Baltic: 15 (1,17)
North Sea – Central: 7 (0,46)
North Sea – West: 7 (19,41)
Total: 30
Delftshaven
Baltic: 539 (1,19)
North Sea – Central: 525 (0,96)
North Sea – East: 1 (0,07)
Total: 1065
Delfzijl
Baltic: 5 (0,98)
North Sea – Central: 7 (1,14)
Total: 12
Deventer
Baltic: 66 (1,96)
North Sea – Central: 13 (0,32)
Total: 79
Doetinchem
Baltic: 1 (2,35)
Total: 1
Dokkum
Baltic: 158 (1,24)
North Sea – Central: 140 (0,91)
North Sea – East: 2 (0,48)
Total: 300
Domburg
Baltic: 4 (1,17)
North Sea – Central: 4 (0,98)
Total: 8
Dordrecht
Baltic: 249 (1,01)
North Sea – Central: 147 (0,49)
North Sea – West: 185 (26,48)
Total: 581
Duinkerken
Baltic: 18 (0,49)
North Sea – Central: 10 (0,22)
North Sea – West: 59 (56,40)
Total: 87
Durgerdam
Arctic: 1 (13,48)
Atlantic South: 4 (0,83)
Baltic: 91 (1,14)
North Sea – Central: 90 (0,94)
North Sea – West: 1 (0,44)
Total: 187
Edam
Arctic: 3 (4,12)

APPENDIX 1.4 *Location of maritime transport communities (cont.)*

Atlantic South: 2 (0,04)
Baltic: 1061 (1,36)
North Sea – Central: 767 (0,82)
North Sea – West: 2 (0,09)
Total: 1835
Egmond
Baltic: 18 (0,67)
North Sea – Central: 18 (0,56)
North Sea – West: 27 (35,64)
Total: 63
Elburg
Baltic: 10 (2,35)
Total: 10
Enkhuizen
Arctic: 16 (2,46)
Atlantic North: 3 (0,01)
Atlantic South: 750 (1,76)
Baltic: 6046 (0,86)
North Sea – Central: 9535 (1,14)
North Sea – East: 15 (0,03)
North Sea – West: 60 (0,29)
Total: 16425
Esens
North Sea – East: 1 (72,64)
Total: 1
Farmsum
Baltic: 14 (1,43)
North Sea – Central: 9 (0,76)
Total: 23
Franeker
Baltic: 228 (1,26)
North Sea – Central: 196 (0,90)
North Sea – East: 1 (0,17)
Total: 425
Goeree
North Sea – West: 5 (83,17)
Total: 5
Goes
North Sea – West: 5 (83,17)
Total: 5
Gorcum
Baltic: 1 (0,78)
North Sea – West: 2 (55,45)
Total: 3
Gorinchem
Baltic: 19 (1,17)
North Sea – Central: 19 (0,98)
Total: 38
Gouda
Baltic: 157 (0,96)
North Sea – Central: 131 (0,67)
North Sea – West: 94 (20,47)
Total: 382
Graft
Baltic: 340 (1,17)
North Sea – Central: 340 (0,97)
North Sea – South: 1 (7,14)
North Sea – West: 1 (0,12)
Total: 682
Grijpswoude
Baltic: 1 (2,35)
Total: 1
Groningen
Baltic: 200 (1,37)
North Sea – Central: 137 (0,78)
North Sea – West: 5 (1,22)
Total: 342
Grootebroek
Atlantic North: 1 (0,04)
Baltic: 1331 (1,23)
North Sea – Central: 1190 (0,92)
North Sea – East: 1 (0,03)
North Sea – West: 12 (0,39)
Total: 2535

Haarlem
Atlantic North: 2 (0,24)
Atlantic South: 3 (0,12)
Baltic: 500 (1,24)
North Sea – Central: 413 (0,86)
North Sea – East: 5 (0,39)
North Sea – West: 20 (1,76)
Total: 943
Harderwijk
Baltic: 96 (2,35)
Total: 96
Harlingen
Arctic: 1 (0,33)
Atlantic North: 2 (0,03)
Atlantic South: 58 (0,29)
Baltic: 2608 (0,80)
North Sea – Central: 4999 (1,27)
North Sea – East: 7 (0,07)
North Sea – West: 3 (0,03)
Total: 7678
Hasselt
Baltic: 3 (1,76)
North Sea – Central: 1 (0,49)
Total: 4
Heem
Baltic: 200 (1,17)
North Sea – Central: 201 (0,98)
Total: 401
Hindeloopen
Atlantic North: 214 (14,83)
Atlantic South: 2 (0,05)
Baltic: 751 (1,08)
North Sea – Central: 659 (0,79)
North Sea – East: 1 (0,04)
Total: 1629
Holland
North Sea – Central: 2 (0,49)
North Sea – West: 6 (62,38)
Total: 8
Holysloot
Baltic: 4 (1,04)
North Sea – Central: 5 (1,08)
Total: 9
Hoogkarspel
Baltic: 19 (1,31)
North Sea – Central: 15 (0,86)
Total: 34
Hoogwoude
Atlantic North: 3 (3,99)
Baltic: 40 (1,10)
North Sea – Central: 42 (0,96)
Total: 85
Hoorn
Arctic: 5 (1,06)
Atlantic North: 2 (0,02)
Atlantic South: 783 (2,54)
Baltic: 4332 (0,86)
Mediterranean: 1 (11,89)
North Sea – Central: 6736 (1,11)
North Sea – East: 6 (0,04)
North Sea – South: 1 (0,41)
North Sea – West: 7 (0,05)
Total: 11873
Horssens
North Sea – East: 1 (72,64)
Total: 1
Huisduinen
Baltic: 114 (1,12)
North Sea – Central: 121 (0,99)
North Sea – East: 1 (0,31)
North Sea – West: 2 (0,70)
Total: 238
Husum
Atlantic North: 21 (71,85)
North Sea – East: 12 (26,42)
Total: 33
Ilpendam
Baltic: 223 (1,66)

APPENDIX 1.4 *Location of maritime transport communities (cont.)*

North Sea – Central: 92 (0,57)
Total: 315
Jisp
Arctic: 1 (30,74)
Baltic: 42 (1,20)
North Sea – Central: 39 (0,93)
Total: 82
Joure
North Sea – Central: 1 (1,95)
Total: 1
Kalenburg
North Sea – East: 1 (72,64)
Total: 1
Kampen
Atlantic North: 1 (0,07)
Baltic: 1050 (1,49)
North Sea – Central: 597 (0,70)
North Sea – East: 3 (0,13)
North Sea – South: 3 (8,83)
Total: 1654
Kamperduin
Baltic: 2 (2,35)
Total: 2
Katwijk
Baltic: 38 (1,19)
North Sea – Central: 35 (0,91)
North Sea – West: 2 (2,22)
Total: 75
Katwoude
Baltic: 14 (1,17)
North Sea – Central: 14 (0,98)
Total: 28
Kolhoorn
Baltic: 1 (1,17)
North Sea – Central: 1 (0,98)
Total: 2
Koudum
Atlantic North: 37 (18,65)
Baltic: 92 (0,96)
North Sea – Central: 94 (0,82)
North Sea – East: 1 (0,32)
Total: 224
Krommeniedijk
Baltic: 81 (1,18)
North Sea – Central: 80 (0,97)
Total: 161
Kuinre
Baltic: 96 (1,17)
North Sea – Central: 97 (0,98)
Total: 193
Landsmeer
Arctic: 1 (2,87)
Atlantic North: 3 (0,39)
Atlantic South: 3 (0,13)
Baltic: 549 (1,47)
North Sea – Central: 321 (0,71)
North Sea – East: 1 (0,08)
Total: 878
Langendijk
Baltic: 1 (2,35)
Total: 1
Leek
Baltic: 54 (1,17)
North Sea – Central: 54 (0,98)
Total: 108
Leeuwarden
Baltic: 100 (1,22)
North Sea – Central: 92 (0,94)
Total: 192
Leiden
Baltic: 26 (1,25)
North Sea – Central: 23 (0,92)
Total: 49
Lemmer
Atlantic North: 1 (22,58)
Baltic: 2 (0,94)

North Sea – Central: 2 (0,78)
Total: 5
Lutjegast
Baltic: 48 (1,14)
North Sea – Central: 51 (1,01)
Total: 99
Maasland
Baltic: 1 (2,35)
Total: 1
Makkum
Baltic: 14 (1,03)
North Sea – Central: 18 (1,10)
Total: 32
Marken
Baltic: 16 (2,09)
North Sea – Central: 2 (0,22)
Total: 18
Mechelen
North Sea – West: 4 (83,17)
Total: 4
Medemblik
Arctic: 3 (0,67)
Atlantic North: 5 (0,05)
Atlantic South: 760 (2,61)
Atlantic West: 1 (0,18)
Baltic: 3829 (0,80)
North Sea – Central: 6607 (1,15)
North Sea – East: 2 (0,01)
North Sea – West: 6 (0,04)
Total: 11214
Middelburg
Baltic: 92 (0,92)
North Sea – Central: 89 (0,74)
North Sea – West: 53 (18,84)
Total: 234
Middeli
Baltic: 1 (2,35)
Total: 1
Molkwerum
Arctic: 2 (2,41)
Atlantic North: 58 (3,13)
Atlantic South: 1 (0,02)
Baltic: 1042 (1,17)
North Sea – Central: 987 (0,92)
North Sea – East: 1 (0,03)
Total: 2091
Monnikendam
Baltic: 1511 (1,48)
North Sea – Central: 881 (0,72)
North Sea – West: 1 (0,03)
Total: 2393
Muiden
Baltic: 35 (1,17)
North Sea – Central: 34 (0,95)
North Sea – West: 1 (1,19)
Total: 70
Naarden
Baltic: 4 (2,35)
Total: 4
Nieuwendam
Baltic: 45 (1,70)
North Sea – Central: 17 (0,54)
Total: 62
Nieuwendorp
Baltic: 41 (1,25)
North Sea – Central: 36 (0,91)
Total: 77
Nieuwpoort
Baltic: 16 (0,77)
North Sea – Central: 14 (0,56)
North Sea – West: 19 (32,25)
Total: 49
Nijmegen
Baltic: 4 (2,35)
Total: 4
Noordwijk
Baltic: 6 (0,74)
North Sea – Central: 6 (0,62)
North Sea – West: 7 (30,64)
Total: 19

APPENDIX 1.4 *Location of maritime transport communities (cont.)*

Oostende
Baltic: 227 (1,01)
North Sea – Central: 220 (0,81)
North Sea – West: 81 (12,76)
Total: 528
Oosterblokker
Baltic: 23 (1,15)
North Sea – Central: 24 (1,00)
Total: 47
Oosthuizen
Baltic: 385 (1,35)
North Sea – Central: 282 (0,83)
Total: 667
Oostwoude
Baltic: 1 (1,17)
North Sea – Central: 1 (0,98)
Total: 2
Oostzaan
Arctic: 2 (8,10)
Baltic: 366 (1,38)
North Sea – Central: 254 (0,80)
Total: 622
Oudendijk
Baltic: 1 (0,59)
North Sea – Central: 3 (1,46)
Total: 4
Overleek
Baltic: 1 (2,35)
Total: 1
Petten
Baltic: 3 (0,88)
North Sea – Central: 2 (0,49)
North Sea – West: 3 (31,19)
Total: 8
Purmerend
Arctic: 2 (5,43)
Atlantic North: 1 (0,12)
Baltic: 597 (1,51)
North Sea – Central: 324 (0,68)
North Sea – East: 1 (0,08)
North Sea – West: 3 (0,27)
Total: 928
Purmerland
Baltic: 2 (1,17)
North Sea – Central: 2 (0,98)
Total: 4
Ransdorp
Baltic: 1914 (1,44)
North Sea – Central: 1216 (0,76)
Total: 3130
Rarop
Atlantic North: 4 (15,57)
Atlantic South: 4 (5,32)
Baltic: 21 (1,70)
Total: 29
Reimerswaal
Baltic: 1 (1,17)
North Sea – Central: 1 (0,98)
Total: 2
Rinchoove
Atlantic North: 1 (112,91)
Total: 1
Roosendaal
Baltic: 38 (0,49)

North Sea – Central: 23 (0,25)
North Sea – West: 122 (55,45)
Total: 183

Rotterdam
Baltic: 1264 (1,15)
North Sea – Central: 1266 (0,96)
North Sea – East: 5 (0,14)
North Sea – West: 45 (1,45)
Total: 2580

Schagen
Baltic: 9 (1,24)
North Sea – Central: 8 (0,92)
Total: 17

Scharl
Atlantic North: 3 (112,91)
Total: 3

Scharwoude
Atlantic North: 1 (112,91)
Total: 1

Schellinkhout
Atlantic South: 1 (0,02)
Baltic: 1234 (1,46)
North Sea – Central: 753 (0,74)
Total: 1988

Schermer
Baltic: 4 (1,04)
North Sea – Central: 3 (0,65)
North Sea – West: 2 (18,48)
Total: 9

Schermerhorn
Baltic: 5 (1,07)
North Sea – Central: 6 (1,07)
Total: 11

Scherpenzeel
Baltic: 1 (2,35)
Total: 1

Scheveningen
Baltic: 4 (1,04)
North Sea – Central: 4 (0,87)
North Sea – West: 1 (9,24)
Total: 9

Schiedam
Baltic: 657 (1,18)
North Sea – Central: 640 (0,96)
North Sea – West: 7 (0,45)
Total: 1304

Sloten
Baltic: 16 (1,25)
North Sea – Central: 14 (0,91)
Total: 30

Sloterdijk
Baltic: 12 (1,22)
North Sea – Central: 11 (0,93)
Total: 23

Sloterzijl
Baltic: 1 (1,17)
North Sea – Central: 1 (0,98)
Total: 2

Sluis
Baltic: 6 (1,28)
North Sea – Central: 4 (0,71)
North Sea – West: 1 (7,56)
Total: 11

Sneek
Baltic: 45 (1,65)
North Sea – Central: 19 (0,58)
Total: 64

Spaarndam
Baltic: 1 (2,35)
Total: 1

Stavoren
Arctic: 4 (1,30)

APPENDIX 1.4 *Location of maritime transport communities (cont.)*

Atlantic North: 29 (0,42)
Atlantic South: 463 (2,30)
Baltic: 2980 (0,90)
North Sea – Central: 4284 (1,08)
North Sea – South: 1 (0,63)
North Sea – West: 2 (0,02)
Total: 7763

Ter Hagen
Baltic: 4 (2,35)
Total: 4

Ter Heide
Baltic: 3 (1,01)
North Sea – Central: 4 (1,12)
Total: 7

Terschelling
Arctic: 1 (0,25)
Atlantic North: 4 (0,05)
Atlantic South: 114 (0,44)
Baltic: 3544 (0,84)
North Sea – Central: 6260 (1,23)
North Sea – East: 10 (0,07)
North Sea – West: 1 (0,08)
Total: 9935

Texel
Baltic: 271 (1,19)
North Sea – Central: 249 (0,91)
North Sea – East: 6 (0,81)
North Sea – West: 10 (1,55)
Total: 536

Tiel
Baltic: 3 (2,35)
Total: 3

Twisk
Baltic: 158 (1,16)
North Sea – Central: 161 (0,99)
Total: 319

Uitdam
Baltic: 47 (2,25)
North Sea – Central: 2 (0,08)
Total: 49

Uitgeest
Atlantic North: 1 (0,80)
Baltic: 74 (1,23)
North Sea – Central: 65 (0,90)
North Sea – South: 1 (34,52)
Total: 141

Utrecht
Baltic: 1 (2,35)
Total: 1

Veere
Baltic: 172 (0,94)
North Sea – Central: 143 (0,65)
North Sea – West: 114 (22,10)
Total: 429

Venhuizen
Arctic: 1 (1,24)
Atlantic South: 1 (0,02)
Baltic: 1049 (1,21)
North Sea – Central: 988 (0,95)
Total: 2039

Vlaanderen
North Sea – West: 4 (83,17)
Total: 4

Vlaardingen
Baltic: 26 (1,30)
North Sea – Central: 21 (0,87)
Total: 47

Vlieland
Atlantic North: 1 (0,01)
Atlantic South: 150 (0,67)
Baltic: 2975 (0,81)
North Sea – Central: 5504 (1,24)

North Sea – East: 2 (0,02)
North Sea – West: 1 (0,10)
Total: 8633
Vlissingen
Baltic: 436 (1,01)
North Sea – Central: 434 (0,83)
North Sea – West: 146 (11,95)
Total: 1016
Warder
Baltic: 2 (0,67)
North Sea – Central: 5 (1,39)
Total: 7
Warns
Arctic: 5 (77,31)
Atlantic North: 6 (4,16)
Baltic: 76 (1,09)
North Sea – Central: 76 (0,91)
Total: 163
Watergang
Baltic: 1 (1,17)
North Sea – Central: 1 (0,98)
Total: 2
Waterland
Baltic: 201 (2,34)
North Sea – West: 1 (0,41)
Total: 202
Weesp
Baltic: 37 (1,19)
North Sea – Central: 36 (0,96)
Total: 73
Westenschouwen
Baltic: 2 (2,35)
Total: 2
Westerblokker
Baltic: 14 (0,94)
North Sea – Central: 21 (1,17)
Total: 35
Westerwoude
Baltic: 23 (1,29)
North Sea – Central: 19 (0,88)
Total: 42
Westkapelle
Baltic: 26 (1,09)
North Sea – Central: 27 (0,94)
North Sea – West: 3 (4,46)
Total: 56
Westzaan
Baltic: 70 (1,24)
North Sea – Central: 60 (0,89)
North Sea – West: 2 (1,26)
Total: 132
Wieringen
Atlantic North: 12 (2,15)
Atlantic South: 1 (0,06)
Baltic: 311 (1,16)
North Sea – Central: 306 (0,95)
North Sea – East: 1 (0,12)
Total: 631
Wijdenes
Baltic: 30 (1,26)
North Sea – Central: 26 (0,91)
Total: 56
Wijk aan zee
Baltic: 2 (0,43)
North Sea – Central: 3 (0,53)
North Sea – West: 6 (45,37)
Total: 11
Winkel
Arctic: 1 (13,27)
Atlantic North: 3 (1,78)
Baltic: 93 (1,15)
North Sea – Central: 93 (0,96)
Total: 190
Workum
Atlantic North: 29 (4,34)
Baltic: 375 (1,17)
North Sea – Central: 349 (0,90)
North Sea – East: 1 (0,10)

APPENDIX 1.4 *Location of maritime transport communities (cont.)*

North Sea – West: 1 (0,11)
Total: 755
Wormer
Baltic: 100 (1,51)
North Sea – Central: 54 (0,68)
North Sea – South: 1 (31,40)
Total: 155
Zaandam
Baltic: 135 (1,35)
North Sea – Central: 98 (0,81)
North Sea – East: 1 (0,31)
North Sea – West: 1 (0,35)
Total: 235
Zierikzee
Baltic: 52 (1,03)
North Sea – Central: 17 (0,28)
North Sea – West: 49 (34,54)
Total: 118
Zijbekarspel
Baltic: 24 (1,15)
North Sea – Central: 25 (1,00)
Total: 49
Zoutelande
Baltic: 36 (1,17)
North Sea – Central: 36 (0,98)
Total: 72
Zuiderdam
Baltic: 1 (2,35)
Total: 1
Zuiderwoude
Baltic: 429 (1,54)
North Sea – Central: 225 (0,67)
Total: 654
Zunderdorp
Baltic: 466 (1,37)
North Sea – Central: 332 (0,81)
Total: 798
Zutphen
Baltic: 29 (2,35)
Total: 29
Zuurwoude
Atlantic North: 1 (18,82)
Atlantic South: 3 (19,29)
Baltic: 2 (0,78)
Total: 6
Zwartewaal
Baltic: 29 (1,17)
North Sea – Central: 29 (0,98)
Total: 58
Zwartsluis
Baltic: 1 (2,35)
Total: 1
Zwolle
Atlantic South: 1 (3,22)
Baltic: 11 (2,15)
Total: 12
Grand Total
Arctic: 56
Atlantic North: 1250
Atlantic South: 3658
Atlantic West: 70
Baltic: 60140
Mediterranean: 1
North Sea – Central: 72287
North Sea – East: 1943
North Sea – South: 29
North Sea – West: 1697
Total: 141142

APPENDIX 1.5 *Location of maritime transport communities in the Low Countries in 1600–1700, including number of registrations and location coefficient per area, as well as total number of registrations in published primary sources*[3]

Aalsmeer
North Sea – East: 1 (79,62)
Total: 2 (incl. 1 without spec. region)
Aartswoude
Arctic: 1 (0,19)
Atlantic North: 1 (0,65)
Atlantic South: 33 (1,89)
Baltic: 205 (0,99)
Mediterranean: 1 (0,56)
North Sea – Central: 166 (0,92)
North Sea – East: 1 (0,36)
North Sea – West: 1 (0,96)
Total: 439 (incl. 30 without spec. region)
Abbekerk
Atlantic South: 14 (5,95)
Baltic: 30 (1,08)
Mediterranean: 1 (4,18)
North Sea – Central: 14 (0,57)
Total: 59
Akersloot
Arctic: 114 (10,69)
Atlantic: 1 (1,84)
Atlantic South: 81 (2,30)
Baltic: 368 (0,89)
Mediterranean: 8 (2,23)
North Sea – Central: 276 (0,76)
North Sea – East: 1 (0,18)
North Sea – West: 2 (0,95)
Total: 883 (incl. 32 without spec. region)
Alkmaar
Arctic: 7 (18,70)
Atlantic South: 3 (2,43)
Baltic: 10 (0,69)
Mediterranean: 2 (15,91)
North Sea – Central: 7 (0,55)
Total: 31 (incl. 2 without spec. region)
Ameland
Arctic: 14 (0,17)
Atlantic: 1 (0,24)
Atlantic North: 22 (0,94)
Atlantic South: 120 (0,45)
Baltic: 3331 (1,06)
North Sea – Central: 3039 (1,10)
North Sea – East: 111 (2,64)
North Sea – West: 10 (0,63)
Total: 6704 (incl. 56 without spec. region)
Amsterdam
Arctic: 144 (1,45)
Atlantic: 9 (1,78)
Atlantic North: 59 (2,07)
Atlantic South: 523 (1,60)
Baltic: 3772 (0,98)
Mediterranean: 93 (2,80)
North Sea – Central: 3087 (0,91)
North Sea – East: 101 (1,96)
North Sea – West: 48 (2,46)
Total: 8201 (incl. 365 without spec. region)
Assendelft
Mediterranean: 1 (123,28)
North Sea – Central: 1 (1,21)
Total: 2
Avenhorn
Mediterranean: 2 (246,57)
Total: 2

3 For an overview of the sources that were excerpted to compile this list, see Introduction.

APPENDIX 1.5 *Location of maritime transport communities in 1600–1700 (cont.)*

Axwijk
Atlantic South: 54 (3,97)
Baltic: 173 (1,08)
North Sea – Central: 99 (0,70)
North Sea – West: 1 (1,23)
Total: 341 (incl. 14 without spec. region)
Beets
Arctic: 5 (1,49)
Atlantic South: 27 (2,44)
Baltic: 108 (0,83)
Mediterranean: 3 (2,67)
North Sea – Central: 68 (0,59)
North Sea – West: 3 (4,55)
Total: 277 (incl. 63 without spec. region)
Berkhout
Mediterranean: 1 (246,57)
Total: 1
Blokker
Atlantic South: 1 (1,39)
Baltic: 6 (0,71)
North Sea – Central: 5 (0,67)
Total: 18 (incl. 6 without spec. region)
Blokzijl
North Sea – Central: 2 (1,61)
Total: 3 (incl. 1 without spec. region)
Bolsward
Atlantic South: 2 (0,63)
Baltic: 28 (0,75)
North Sea – Central: 18 (0,55)
North Sea – East: 5 (10,08)
Total: 79 (incl. 26 without spec. region)
Brielle
Baltic: 22 (1,06)
Mediterranean: 1 (5,60)
North Sea – Central: 18 (0,99)
Total: 44 (incl. 3 without spec. region)
Broek in Waterland
Atlantic South: 60 (1,45)
Baltic: 522 (1,07)
Mediterranean: 5 (1,19)
North Sea – Central: 402 (0,94)
North Sea – East: 7 (1,07)
North Sea – West: 1 (0,40)
Total: 1037 (incl. 40 without spec. region)
Buiksloot
Arctic: 22 (2,17)
Atlantic South: 53 (1,59)
Baltic: 370 (0,94)
North Sea – Central: 327 (0,94)
North Sea – West: 1 (0,50)
Total: 838 (incl. 65 without spec. region)
Cadoelen
Atlantic South: 2 (1,32)
Baltic: 19 (1,06)
North Sea – Central: 17 (1,08)
Total: 38
Callantsoog
Arctic: 3 (82,82)
Total: 3
De Rijp
Arctic: 5 (0,38)
Atlantic South: 91 (2,11)
Baltic: 505 (0,99)
Mediterranean: 5 (1,14)
North Sea – Central: 371 (0,83)
North Sea – East: 2 (0,29)
North Sea – West: 3 (1,16)
Total: 1082 (incl. 100 without spec. region)
De Streek
Atlantic South: 2 (1,36)

Baltic: 16 (0,92)
North Sea – Central: 15 (0,98)
Total: 37 (incl. 4 without spec. region)

Delft
Atlantic South: 4 (5,57)
Baltic: 7 (0,83)
Mediterranean: 2 (27,40)
North Sea – Central: 5 (0,67)
Total: 18

Delftshaven
Atlantic South: 10 (0,96)
Baltic: 129 (1,05)
Mediterranean: 6 (5,69)
North Sea – Central: 109 (1,01)
North Sea – East: 1 (0,61)
North Sea – West: 2 (3,23)
Total: 260 (incl. 3 without spec. region)

Delfzijl
Atlantic South: 2 (1,43)
Baltic: 17 (1,03)
North Sea – Central: 9 (0,62)
North Sea – East: 7 (31,85)
Total: 35

Den Helder
Atlantic South: 5 (1,84)
Baltic: 33 (1,03)
North Sea – Central: 29 (1,03)
Total: 68 (incl. 1 without spec. region)

d'Hoef
Atlantic South: 4 (0,81)
Baltic: 62 (1,06)
North Sea – Central: 56 (1,09)
Total: 124 (incl. 2 without spec. region)

Dokkum
Atlantic South: 1 (0,37)
Baltic: 34 (1,08)
North Sea – Central: 29 (1,05)
North Sea – East: 2 (4,75)
North Sea – West: 1 (6,27)
Total: 67

Dordrecht
Atlantic South: 13 (1,36)
Baltic: 118 (1,05)
North Sea – Central: 102 (1,03)
North Sea – East: 1 (0,67)
North Sea – West: 3 (5,27)
Total: 239 (incl. 2 without spec. region)

Driehuizen
Baltic: 2 (1,06)
North Sea – Central: 2 (1,21)
Total: 4

Driel
Atlantic South: 1 (0,49)
Baltic: 11 (0,46)
North Sea – Central: 26 (1,23)
North Sea – West: 1 (8,23)
Total: 51 (incl. 12 without spec. region)

Duinkerken
Mediterranean: 1 (246,57)
Total: 1

Durgerdam
Arctic: 32 (1,87)
Atlantic North: 1 (0,20)
Atlantic South: 52 (0,92)
Baltic: 636 (0,95)
Mediterranean: 15 (2,61)
North Sea – Central: 570 (0,97)
North Sea – East: 3 (0,34)
North Sea – West: 4 (1,18)
Total: 1418 (incl. 105 without spec. region)

Edam
Arctic: 33 (1,67)
Atlantic: 2 (1,98)
Atlantic North: 9 (1,57)

APPENDIX 1.5 *Location of maritime transport communities in 1600–1700 (cont.)*

Atlantic South: 219 (3,35)
Baltic: 767 (0,99)
Mediterranean: 8 (1,20)
North Sea – Central: 491 (0,72)
North Sea – East: 6 (0,58)
North Sea – West: 5 (1,28)
Total: 1641 (incl. 101 without spec. region)

Egmond
Atlantic South: 16 (2,43)
Baltic: 81 (1,04)
North Sea – Central: 67 (0,98)
Total: 165 (incl. 1 without spec. region)

Elburg
Baltic: 1 (2,12)
Total: 1

Enkhuizen
Arctic: 97 (1,53)
Atlantic: 3 (0,93)
Atlantic North: 10 (0,55)
Atlantic South: 305 (1,45)
Baltic: 2476 (1,00)
Mediterranean: 82 (3,84)
North Sea – Central: 1898 (0,87)
North Sea – East: 30 (0,91)
North Sea – West: 13 (1,04)
Total: 5264 (incl. 350 without spec. region)

Franeker
Baltic: 5 (0,89)
North Sea – Central: 7 (1,41)
Total: 12

Friesland
Arctic: 1 (6,90)
Baltic: 5 (0,89)
North Sea – Central: 5 (1,01)
North Sea – East: 1 (13,27)
Total: 12

Gaast
Baltic: 1 (1,06)
North Sea – Central: 1 (1,21)
Total: 2

Gorinchem
Baltic: 2 (0,85)
North Sea – Central: 3 (1,45)
Total: 5

Gouda
Baltic: 4 (1,06)
North Sea – Central: 4 (1,21)
Total: 8

Graft
Atlantic South: 9 (2,15)
Baltic: 45 (0,91)
Mediterranean: 3 (7,04)
North Sea – Central: 30 (0,69)
North Sea – West: 1 (4,00)
Total: 105 (incl. 17 without spec. region)

Graftdijk
Arctic: 2 (2,72)
Atlantic South: 2 (0,82)
Atlantic West: 1 (1145,94)
Baltic: 27 (0,94)
North Sea – Central: 27 (1,07)
North Sea – West: 1 (6,88)
Total: 61 (incl. 1 without spec. region)

Groningen
Atlantic: 1 (7,63)
Atlantic South: 2 (0,24)
Baltic: 109 (1,09)
North Sea – Central: 97 (1,10)
North Sea – East: 4 (2,99)
Total: 213

Grootebroek
Arctic: 2 (0,82)
Atlantic North: 1 (1,41)
Atlantic South: 15 (1,85)
Baltic: 97 (1,02)

North Sea – Central: 58 (0,69)
North Sea – West: 2 (4,14)
Total: 203 (incl. 28 without spec. region)

Haarlem
Mediterranean: 1 (246,57)
Total: 1

Harlingen
Arctic: 33 (0,26)
Atlantic: 17 (2,59)
Atlantic North: 54 (1,45)
Atlantic South: 190 (0,45)
Baltic: 5149 (1,02)
Mediterranean: 2 (0,05)
North Sea – Central: 4654 (1,05)
North Sea – East: 131 (1,95)
North Sea – West: 15 (0,59)
Total: 10689 (incl. 444 without spec. region)

Heem
Atlantic South: 7 (2,02)
Baltic: 43 (1,05)
Mediterranean: 1 (2,83)
North Sea – Central: 35 (0,97)
North Sea – East: 1 (1,83)
Total: 87

Hindeloopen
Arctic: 115 (1,22)
Atlantic: 5 (1,04)
Atlantic North: 51 (1,87)
Atlantic South: 75 (0,24)
Baltic: 3529 (0,96)
North Sea – Central: 3717 (1,15)
North Sea – East: 21 (0,43)
North Sea – West: 7 (0,38)
Total: 7809 (incl. 289 without spec. region)

Holland
Atlantic North: 2 (10,83)
Atlantic South: 1 (0,47)
Baltic: 23 (0,92)
North Sea – Central: 12 (0,55)
North Sea – West: 13 (102,98)
Total: 53 (incl. 2 without spec. region)

Holysloot
Arctic: 1 (16,56)
Baltic: 1 (0,42)
North Sea – Central: 2 (0,97)
North Sea – West: 1 (83,97)
Total: 5

Hoogkarspel
Arctic: 2 (27,61)
Baltic: 2 (0,71)
Total: 6 (incl. 2 without spec. region)

Hoogwoude
Atlantic South: 7 (1,81)
Baltic: 46 (1,01)
Mediterranean: 2 (5,08)
North Sea – Central: 24 (0,60)
Total: 97 (incl. 18 without spec. region)

Hoorn
Arctic: 64 (0,59)
Atlantic: 4 (0,73)
Atlantic North: 26 (0,83)
Atlantic South: 828 (2,32)
Baltic: 4291 (1,02)
Mediterranean: 86 (2,36)
North Sea – Central: 3290 (0,89)
North Sea – East: 33 (0,59)
North Sea – West: 14 (0,66)
Total: 8967 (incl. 331 without spec. region)

Houwert
Atlantic South: 10 (2,32)
Baltic: 51 (1,00)
North Sea – Central: 45 (1,01)
Total: 108 (incl. 2 without spec. region)

Huisduinen
Arctic: 20 (5,65)
Atlantic North: 2 (1,96)
Atlantic South: 20 (1,71)
Baltic: 129 (0,94)

APPENDIX 1.5 *Location of maritime transport communities in 1600–1700 (cont.)*

North Sea – Central: 112 (0,93)
Total: 293 (incl. 10 without spec. region)
Ilpendam
Arctic: 8 (22,09)
Atlantic South: 2 (1,67)
Baltic: 9 (0,64)
Mediterranean: 4 (32,88)
North Sea – Central: 7 (0,56)
Total: 30
Jisp
Arctic: 2 (0,97)
Baltic: 83 (1,03)
Mediterranean: 1 (1,44)
North Sea – Central: 85 (1,20)
Total: 171
Joure
Atlantic South: 1 (0,30)
Baltic: 42 (1,06)
North Sea – Central: 41 (1,18)
Total: 84
Katwijk
Baltic: 1 (1,06)
North Sea – Central: 1 (1,21)
Total: 2
Katwoude
Baltic: 5 (0,97)
North Sea – Central: 4 (0,88)
Total: 11 (incl. 2 without spec. region)
Knollendam
Atlantic South: 1 (3,13)
Baltic: 5 (1,33)
North Sea – Central: 2 (0,61)
Total: 8
Koudum
Arctic: 27 (1,52)
Atlantic North: 8 (1,56)
Atlantic South: 25 (0,43)
Baltic: 667 (0,96)
North Sea – Central: 633 (1,04)
North Sea – East: 3 (0,33)
North Sea – West: 6 (1,71)
Total: 1469 (incl. 100 without spec. region)
Krommenie
Atlantic South: 4 (3,71)
Baltic: 13 (1,02)
North Sea – Central: 10 (0,90)
Total: 27
Krommeniedijk
Arctic: 10 (6,73)
Atlantic North: 1 (2,33)
Atlantic South: 5 (1,02)
Baltic: 31 (0,54)
Mediterranean: 4 (8,02)
North Sea – Central: 42 (0,83)
Total: 123 (incl. 30 without spec. region)
Kuinre
Arctic: 3 (22,59)
Atlantic South: 1 (2,28)
Baltic: 1 (0,19)
Mediterranean: 3 (67,25)
North Sea – Central: 3 (0,66)
Total: 11
Kwadijk
Arctic: 7 (4,79)
Atlantic South: 12 (2,49)
Baltic: 52 (0,91)
North Sea – Central: 39 (0,78)
North Sea – East: 1 (1,32)
North Sea – West: 1 (3,47)
Total: 121 (incl. 9 without spec. region)
Lambertschagen
Atlantic South: 2 (1,73)
Baltic: 17 (1,25)

North Sea – Central: 9 (0,75)
North Sea – East: 1 (5,49)
Total: 29

Landsmeer
Arctic: 16 (1,47)
Atlantic: 1 (1,80)
Atlantic North: 1 (0,32)
Atlantic South: 63 (1,75)
Baltic: 432 (1,02)
Mediterranean: 3 (0,82)
North Sea – Central: 369 (0,99)
North Sea – West: 2 (0,93)
Total: 904 (incl. 17 without spec. region)

Langendijk
Baltic: 4 (1,06)
North Sea – Central: 3 (0,91)
Total: 8 (incl. 1 without spec. region)

Langweer
Baltic: 2 (1,06)
North Sea – Central: 2 (1,21)
Total: 4

Leek
Baltic: 1 (0,71)
Mediterranean: 2 (164,38)
Total: 3

Leerdam
Baltic: 2 (1,42)
North Sea – Central: 1 (0,81)
Total: 3

Leeuwarden
Baltic: 12 (1,06)
North Sea – Central: 12 (1,21)
Total: 24

Lemmer
North Sea – Central: 1 (2,42)
Total: 1

Limmen
Mediterranean: 2 (246,57)
Total: 2

Maas
Atlantic North: 1 (95,69)
Baltic: 2 (1,42)
Total: 3

Maassluis
Atlantic South: 6 (2,15)
Baltic: 34 (1,03)
North Sea – Central: 29 (1,00)
Total: 70 (incl. 1 without spec. region)

Makkum
Arctic: 1 (0,20)
Atlantic North: 3 (2,11)
Atlantic South: 6 (0,37)
Baltic: 208 (1,08)
North Sea – Central: 180 (1,07)
North Sea – East: 5 (1,95)
North Sea – West: 1 (1,03)
Total: 408 (incl. 4 without spec. region)

Medemblik
Arctic: 70 (3,01)
Atlantic: 2 1,69)
Atlantic North: 10 (1,49)
Atlantic South: 179 (2,33)
Baltic: 917 (1,01)
Mediterranean: 37 (4,73)
North Sea – Central: 658 (0,83)
North Sea – East: 13 (1,07)
North Sea – West: 3 (0,65)
Total: 1929 (incl. 40 without spec. region)

Middelburg
Atlantic South: 33 (2,22)
Baltic: 180 (1,03)
Mediterranean: 8 (5,30)
North Sea – Central: 125 (0,81)
North Sea – East: 14 (5,99)
North Sea – West: 3 (3,39)
Total: 372 (incl. 9 without spec. region)

Middeli
Atlantic North: 1 (0,78)

APPENDIX 1.5 *Location of maritime transport communities in 1600–1700 (cont.)*

Atlantic South: 49 (3,36)
Baltic: 171 (0,99)
Mediterranean: 2 (1,35)
North Sea – Central: 108 (0,71)
Total: 366 (incl. 35 without spec. region)

Molkwerum
Arctic: 43 (0,67)
Atlantic North: 35 (1,88)
Atlantic South: 66 (0,31)
Baltic: 2346 (0,93)
North Sea – Central: 2330 (1,06)
North Sea – East: 18 (0,54)
North Sea – West: 8 (0,63)
Total: 5332 (incl. 486 without spec. region)

Monnikendam
Arctic: 9 (3,64)
Atlantic South: 17 (2,08)
Baltic: 83 (0,86)
Mediterranean: 12 (14,43)
North Sea – Central: 71 (0,84)
North Sea – East: 2 (1,55)
Total: 205 (incl. 11 without spec. region)

Muiden
North Sea – Central: 1 (2,42)
Total: 1

Naarden
Mediterranean: 1 (246,57)
Total: 1

Nederveen
Mediterranean: 1 (246,57)
Total: 1

Nieuwendam
Arctic: 8 (1,93)
Atlantic: 1 (4,74)
Atlantic North: 1 (0,84)
Atlantic South: 29 (2,12)
Baltic: 160 (0,99)
Mediterranean: 2 (1,44)
North Sea – Central: 128 (0,90)
Total: 343 (incl. 14 without spec. region)

Nieuwendijk
Baltic: 1 (1,06)
North Sea – Central: 1 (1,21)
Total: 2

Nieuweschans
North Sea – Central: 1 (2,42)
Total: 1

Nieuwpoort
Mediterranean: 2 (246,57)
Total: 2

Oldendam
Baltic: 1 (1,06)
Total: 2 (incl. 1 without spec. region)

Oosterbeek
Baltic: 11 (1,06)
North Sea – Central: 10 (1,10)
Total: 22 (incl. 1 without spec. region)

Oosterblokker
Arctic: 1 (1,34)
Atlantic South: 5 (2,02)
Baltic: 29 (0,99)
North Sea – Central: 24 (0,94)
Total: 62 (incl. 3 without spec. region)

Oosthuizen
Arctic: 6 (4,97)
Atlantic South: 15 (3,76)
Baltic: 36 (0,76)
Mediterranean: 4 (9,86)
North Sea – Central: 22 (0,53)
Total: 100 (incl. 17 without spec. region)

Oostwoude
Atlantic South: 3 (2,15)
Baltic: 16 (0,97)

North Sea – Central: 15 (1,04)
Total: 35 (incl. 1 without spec. region)

Oostzaan
Arctic: 12 (2,48)
Atlantic South: 20 (1,25)
Baltic: 240 (1,27)
Mediterranean: 3 (1,85)
North Sea – Central: 67 (0,41)
North Sea – West: 1 (1,05)
Total: 400 (incl. 57 without spec. region)

Opmeer
Arctic: 3 (2,46)
Atlantic South: 7 (1,74)
Baltic: 41 (0,86)
North Sea – Central: 34 (0,81)
North Sea – West: 2 (8,31)
Total: 101 (incl. 14 without spec. region)

Opperdoes
Atlantic South: 12 (2,21)
Baltic: 66 (1,03)
Mediterranean: 1 (1,81)
North Sea – Central: 52 (0,93)
Total: 136 (incl. 5 without spec. region)

Oudega
North Sea – Central: 1 (2,42)
Total: 1

Oudendijk
Arctic: 2 (3,76)
Atlantic South: 8 (4,56)
Baltic: 18 (0,87)
Mediterranean: 1 (5,60)
North Sea – Central: 7 (0,39)
Total: 44 (incl. 8 without spec. region)

Petten
Atlantic South: 1 (0,90)
Baltic: 14 (1,06)
North Sea – Central: 13 (1,12)
Total: 28

Purmerend
Arctic: 15 (4,85)
Atlantic South: 29 (2,84)
Baltic: 110 (0,91)
Mediterranean: 3 (2,89)
North Sea – Central: 89 (0,84)
Total: 256 (incl. 10 without spec. region)

Ransdorp
Arctic: 9 (31,06)
Atlantic South: 2 (2,09)
Baltic: 8 (0,71)
North Sea – Central: 4 (0,40)
North Sea – West: 1 (17,49)
Total: 24

Rarop
Atlantic South: 8 (2,54)
Baltic: 35 (0,94)
North Sea – Central: 13 (0,40)
Total: 79 (incl. 23 without spec. region)

Rotterdam
Atlantic North: 11 (1,91)
Atlantic South: 127 (1,92)
Baltic: 819 (1,05)
Mediterranean: 12 (1,79)
North Sea – Central: 640 (0,94)
North Sea – East: 6 (0,58)
North Sea – West: 12 (3,04)
Total: 1655 (incl. 28 without spec. region)

Schagen
Arctic: 3 (35,50)
Baltic: 2 (0,61)
Mediterranean: 1 (35,22)
North Sea – Central: 1 (0,35)
Total: 7

Schellinkhout
Arctic: 4 (0,27)
Atlantic North: 2 (0,47)
Atlantic South: 88 (1,80)

APPENDIX 1.5 *Location of maritime transport communities in 1600–1700 (cont.)*

Baltic: 597 (1,04)
Mediterranean: 4 (0,81)
North Sea – Central: 497 (0,98)
North Sea – East: 2 (0,26)
North Sea – West: 1 (0,34)
Total: 1223 (incl. 28 without spec. region)
Schermer
Arctic: 2 (82,82)
Total: 2
Schermerhorn
Arctic: 4 (0,34)
Atlantic North: 1 (0,29)
Atlantic South: 52 (1,33)
Baltic: 465 (1,01)
North Sea – Central: 378 (0,93)
North Sea – East: 5 (0,81)
North Sea – West: 1 (0,43)
Total: 979 (incl. 73 without spec. region)
Schiedam
Atlantic: 1 (1,25)
Atlantic North: 1 (0,22)
Atlantic South: 83 (1,60)
Baltic: 643 (1,05)
Mediterranean: 7 (1,33)
North Sea – Central: 513 (0,96)
North Sea – East: 5 (0,61)
North Sea – West: 1 (0,32)
Total: 1299 (incl. 45 without spec. region)
Schiermonnikoog
North Sea – West: 1 (419,83)
Total: 1
Sluis
North Sea – Central: 1 (2,42)
Total: 1
Sneek
North Sea – Central: 2 (1,61)
Total: 3 (incl. 1 without spec. region)
Spanbroek
Baltic: 3 (1,06)
North Sea – Central: 2 (0,81)
Total: 6 (incl. 1 without spec. region)
Stavoren
Arctic: 100 (1,73)
Atlantic: 3 (1,02)
Atlantic North: 35 (2,10)
Atlantic South: 161 (0,84)
Baltic: 2125 (0,94)
Mediterranean: 16 (0,82)
North Sea – Central: 1880 (0,95)
North Sea – East: 14 (0,47)
North Sea – West: 9 (0,79)
Total: 4785 (incl. 442 without spec. region)
Terhorne
Baltic: 2 (0,85)
North Sea – Central: 3 (1,45)
Total: 5
Terschelling
Arctic: 95 (0,39)
Atlantic: 8 (0,65)
Atlantic North: 53 (0,76)
Atlantic South: 391 (0,49)
Atlantic West: 1 (3,49)
Baltic: 9500 (1,01)
Mediterranean: 2 (0,02)
North Sea – Central: 8842 (1,07)
North Sea – East: 123 (0,98)
North Sea – West: 43 (0,90)
Total: 20016 (incl. 958 without spec. region)
Texel
Arctic: 34 (1,78)
Atlantic: 1 (1,03)
Atlantic North: 2 (0,36)
Atlantic South: 63 (1,00)
Baltic: 777 (1,04)

Mediterranean: 4 (0,62)
North Sea – Central: 656 (1,00)
North Sea – East: 16 (1,61)
North Sea – West: 4 (1,06)
Total: 1581 (incl. 24 without spec. region)

Twisk
Atlantic South: 29 (2,29)
Baltic: 149 (1,00)
Mediterranean: 1 (0,78)
North Sea – Central: 103 (0,78)
Total: 318 (incl. 36 without spec. region)

Uitdam
Mediterranean: 2 (246,57)
Total: 2

Uitgeest
Arctic: 4 (66,26)
Atlantic South: 1 (5,01)
Total: 5

Veendam
Baltic: 6 (1,16)
North Sea – Central: 5 (1,10)
Total: 11

Veere
Atlantic South: 7 (5,48)
Baltic: 17 (1,13)
Mediterranean: 2 (15,41)
North Sea – Central: 5 (0,38)
North Sea – West: 1 (13,12)
Total: 32

Venhuizen
Arctic: 25 (0,58)
Atlantic North: 5 (0,40)
Atlantic South: 245 (1,73)
Baltic: 1730 (1,03)
Mediterranean: 5 (0,35)
North Sea – Central: 1390 (0,95)
North Sea – East: 16 (0,72)
North Sea – West: 6 (0,71)
Total: 3559 (incl. 137 without spec. region)

Vlaanderen
Mediterranean: 5 (246,57)
Total: 5

Vlaardingen
Arctic: 1 (1,66)
Baltic: 29 (1,23)
North Sea – Central: 19 (0,92)
North Sea – West: 1 (8,40)
Total: 50

Vlieland
Arctic: 187 (0,68)
Atlantic: 21 (1,51)
Atlantic North: 34 (0,43)
Atlantic South: 381 (0,42)
Baltic: 10805 (1,01)
Mediterranean: 3 (0,03)
North Sea – Central: 10251 (1,10)
North Sea – East: 109 (0,77)
North Sea – West: 36 (0,67)
Total: 22643 (incl. 816 without spec. region)

Vlissingen
Atlantic: 1 (1,88)
Atlantic North: 2 (0,66)
Atlantic South: 38 (1,10)
Baltic: 416 (1,02)
Mediterranean: 11 (3,14)
North Sea – Central: 325 (0,91)
North Sea – East: 7 (1,29)
North Sea – West: 3 (1,46)
Total: 864 (incl. 61 without spec. region)

Walls
Atlantic North: 1 (143,54)
Baltic: 1 (1,06)
Total: 2

Warder
Arctic: 14 (5,32)
Atlantic South: 14 (1,61)
Baltic: 101 (0,98)
North Sea – Central: 82 (0,91)
Total: 218 (incl. 7 without spec. region)

APPENDIX 1.5 *Location of maritime transport communities in 1600–1700 (cont.)*

Warga
Baltic: 1 (1,06)
North Sea – Central: 1 (1,21)
Total: 2
Warns
Arctic: 74 (2,10)
Atlantic North: 7 (0,69)
Atlantic South: 48 (0,41)
Baltic: 1311 (0,95)
North Sea – Central: 1277 (1,06)
North Sea – East: 3 (0,16)
North Sea – West: 5 (0,72)
Total: 2918 (incl. 193 without spec. region)
Watergang
Arctic: 2 (1,95)
Atlantic South: 5 (1,47)
Baltic: 41 (1,02)
North Sea – Central: 33 (0,94)
Total: 85 (incl. 4 without spec. region)
Waterland
Atlantic South: 2 (1,93)
Baltic: 12 (0,98)
Mediterranean: 2 (18,97)
North Sea – Central: 8 (0,74)
Total: 26 (incl. 2 without spec. region)
Weer
Mediterranean: 1 (246,57)
Total: 1
Wervershoof
Atlantic South: 1 (6,27)
Baltic: 2 (1,06)
North Sea – Central: 1 (0,61)
Total: 4
Westerblokker
Atlantic South: 2 (1,62)
Baltic: 14 (0,96)
North Sea – Central: 12 (0,94)
Total: 31 (incl. 3 without spec. region)
Westzaan
Arctic: 7 (15,67)
Atlantic South: 3 (2,03)
Baltic: 14 (0,80)
North Sea – Central: 11 (0,72)
North Sea – West: 1 (11,35)
Total: 37 (incl. 1 without spec. region)
Wieringen
Baltic: 5 (0,89)
North Sea – Central: 6 (1,21)
Total: 12 (incl. 1 without spec. region)
Wijdenes
Baltic: 13 (0,99)
Mediterranean: 1 (8,81)
North Sea – Central: 14 (1,21)
Total: 28
Wijk aan zee
Arctic: 2 (15,06)
Baltic: 8 (1,55)
North Sea – Central: 1 (0,22)
Total: 11
Willemstad
Baltic: 1 (1,06)
North Sea – Central: 1 (1,21)
Total: 2
Winkel
Arctic: 6 (1,06)
Atlantic South: 38 (2,04)
Baltic: 217 (0,99)
North Sea – Central: 154 (0,80)
North Sea – East: 1 (0,34)
North Sea – West: 8 (7,18)
Total: 468 (incl. 44 without spec. region)
Woggelum
Atlantic South: 4 (3,58)
Baltic: 11 (0,83)
Mediterranean: 1 (8,81)
North Sea – Central: 9 (0,78)
Total: 28 (incl. 3 without spec. region)

Workum
Arctic: 27 (1,24)
Atlantic: 1 (0,90)
Atlantic North: 23 (3,67)
Atlantic South: 29 (0,40)
Baltic: 851 (1,01)
North Sea – Central: 761 (1,02)
North Sea – East: 19 (1,68)
North Sea – West: 11 (2,57)
Total: 1798 (incl. 76 without spec. region)
Wormer
Atlantic North: 1 (6,24)
Atlantic South: 3 (1,64)
Baltic: 24 (1,11)
North Sea – Central: 14 (0,74)
Total: 46 (incl. 4 without spec. region)
Woudsend
Baltic: 3 (2,12)
Total: 3
Ylst
Atlantic South: 1 (0,22)
Baltic: 45 (0,85)
North Sea – Central: 43 (0,92)
North Sea – East: 2 (2,82)
Total: 113 (incl. 22 without spec. region)
Zaandam
Arctic: 107 (4,24)
Atlantic: 1 (0,78)
Atlantic North: 3 (0,41)
Atlantic South: 202 (2,42)
Baltic: 920 (0,94)
Mediterranean: 8 (0,94)
North Sea – Central: 664 (0,77)
North Sea – East: 15 (1,14)
North Sea – West: 6 (1,21)
Total: 2089 (incl. 163 without spec. region)
Zeeland
Mediterranean: 1 (246,57)
Total: 1
Zierikzee
Atlantic North: 6 (6,09)
Atlantic South: 6 (0,53)
Baltic: 147 (1,10)
North Sea – Central: 111 (0,95)
North Sea – East: 1 (0,56)
Total: 283 (incl. 12 without spec. region)
Zijbekarspel
Atlantic South: 3 (1,93)
Baltic: 17 (0,93)
North Sea – Central: 14 (0,87)
Total: 39 (incl. 5 without spec. region)
Zuiderdorp
Mediterranean: 1 (246,57)
Total: 1
Zuiderwoude
Arctic: 15 (3,06)
Atlantic: 2 (8,01)
Atlantic North: 1 (0,71)
Atlantic South: 40 (2,47)
Baltic: 195 (1,02)
Mediterranean: 1 (0,61)
North Sea – Central: 145 (0,86)
North Sea – East: 2 (0,78)
Total: 406 (incl. 5 without spec. region)
Zunderdorp
Arctic: 6 (1,16)
Atlantic South: 26 (1,52)
Baltic: 213 (1,06)
North Sea – Central: 175 (0,99)
North Sea – East: 5 (1,86)
North Sea – West: 2 (1,96)
Total: 428 (incl. 1 without spec. region)
Zuurwoude
Mediterranean: 9 (246,57)
Total: 9
Zwaag
Atlantic South: 8 (3,93)
Baltic: 24 (1,00)
North Sea – Central: 14 (0,66)
Total: 51 (incl. 5 without spec. region)

APPENDIX 1.5 *Location of maritime transport communities in 1600–1700 (cont.)*

Grand Total
Arctic: 1688
Atlantic: 86
Atlantic North: 487
Atlantic South: 5576
Atlantic West: 2
Baltic: 65800
Mediterranean: 567
North Sea – Central: 57759
North Sea – East: 878
North Sea – West: 333
Total: 139805 (incl. 6629 without spec. region)

APPENDIX 1.6 *Location of maritime transport communities in the Low Countries in 1700–1800, including number of registrations and location coefficient per area, as well as total number of registrations in published primary sources*[4]

Aartswoude
Arctic: 8 (15,74)
Atlantic South: 1 (0,20)
Baltic: 32 (0,91)
North Sea – Central: 31 (1,14)
Total: 72
Abbekerk
Baltic: 1 (1,03)
North Sea – Central: 1 (1,32)
Total: 2
Abt
Baltic: 3 (1,03)
North Sea – Central: 3 (1,32)
Total: 6
Adorp
Atlantic South: 4 (2,21)
Baltic: 13 (1,03)
North Sea – Central: 9 (0,91)
Total: 26
Akersloot
Arctic: 9 (26,02)
Atlantic South: 3 (0,88)
Baltic: 20 (0,84)
North Sea – Central: 16 (0,86)
North Sea – East: 1 (0,51)
Total: 49
Akkrum
Atlantic South: 1 (0,28)
Baltic: 26 (1,03)
North Sea – Central: 25 (1,27)
Total: 52
Alkmaar
Arctic: 1 (12,88)
Atlantic North: 3 (76,60)
Atlantic South: 2 (2,61)
Baltic: 5 (0,93)
Total: 11
Ameland
Arctic: 249 (2,42)
Atlantic North: 16 (0,31)
Atlantic South: 526 (0,52)
Baltic: 7171 (1,01)

4 For an overview of the sources that were excerpted to compile this list, see Introduction.

Mediterranean: 100 (0,68)
North Sea – Central: 6354 (1,15)
North Sea – East: 134 (0,23)
North Sea – South: 20 (0,60)
Total: 14571

Amerongen
Baltic: 1 (1,03)
North Sea – East: 1 (12,47)
Total: 2

Amsterdam
Africa: 2 (0,93)
Arctic: 338 (0,58)
Atlantic: 4 (1,87)
Atlantic North: 278 (0,94)
Atlantic South: 6827 (1,18)
Atlantic West: 11 (2,77)
Baltic: 39778 (0,98)
Mediterranean: 1753 (2,09)
North Sea – Central: 29617 (0,94)
North Sea – East: 4424 (1,33)
North Sea – South: 130 (0,68)
Total: 83170

Anjum
Atlantic South: 4 (0,87)
Baltic: 33 (1,03)
North Sea – Central: 28 (1,12)
North Sea – East: 1 (0,38)
Total: 66

Appingedam
Atlantic North: 1 (0,46)
Atlantic South: 11 (0,26)
Baltic: 304 (1,03)
Mediterranean: 1 (0,16)
North Sea – Central: 212 (0,92)
North Sea – East: 72 (2,96)
North Sea – South: 6 (4,33)
Total: 607

Arum
Baltic: 1 (1,03)
North Sea – South: 1 (218,91)
Total: 2

Avenhorn
Arctic: 3 (141,68)
Total: 3

Axwijk
Arctic: 4 (8,33)
Atlantic South: 6 (1,27)
Baltic: 32 (0,97)
North Sea – Central: 26 (1,01)
Total: 68

Bakhuizen
Atlantic South: 9 (2,53)
Baltic: 27 (1,09)
Mediterranean: 1 (1,95)
North Sea – Central: 13 (0,67)
North Sea – East: 1 (0,49)
Total: 51

Balk
Atlantic South: 11 (1,97)
Baltic: 40 (1,03)
North Sea – Central: 27 (0,89)
North Sea – East: 2 (0,62)
Total: 80

Beets
Arctic: 1 (20,24)
Atlantic South: 1 (2,05)
Baltic: 4 (1,17)
North Sea – Central: 1 (0,38)
Total: 7

Bergen (NH)
Arctic: 1 (14,17)
Mediterranean: 9 (89,34)
Total: 10

Bergum
Atlantic South: 2 (3,59)
Baltic: 4 (1,03)
North Sea – Central: 2 (0,66)
Total: 8

APPENDIX 1.6 *Location of maritime transport communities in 1700–1800 (cont.)*

Berkhout
Arctic: 1 (141,68)
Total: 1
Berltsum
Atlantic South: 3 (2,39)
Baltic: 9 (1,03)
North Sea – Central: 5 (0,73)
North Sea – South: 1 (24,32)
Total: 18
Bildtziel
Atlantic South: 1 (0,55)
Baltic: 13 (1,03)
North Sea – Central: 11 (1,12)
North Sea – South: 1 (16,84)
Total: 26
Black
Atlantic South: 1 (1,79)
Baltic: 4 (1,03)
North Sea – Central: 3 (0,99)
Total: 8
Bolsward
Atlantic North: 1 (0,68)
Atlantic South: 80 (2,77)
Baltic: 208 (1,03)
Mediterranean: 2 (0,48)
North Sea – Central: 121 (0,77)
North Sea – East: 2 (0,12)
Total: 414
Bovenkarspel
Baltic: 8 (1,03)
North Sea – Central: 8 (1,32)
Total: 16
Brielle
Atlantic South: 1 (1,03)
Baltic: 7 (1,03)
North Sea – Central: 6 (1,13)
Total: 14
Broek in Waterland
Arctic: 6 (2,96)
Atlantic South: 7 (0,35)
Baltic: 140 (1,00)
Mediterranean: 1 (0,35)
North Sea – Central: 131 (1,20)
North Sea – South: 1 (1,53)
Total: 287
Brugge
Mediterranean: 1 (99,26)
Total: 1
Buiksloot
Arctic: 9 (2,63)
Atlantic North: 4 (2,32)
Atlantic South: 31 (0,92)
Baltic: 237 (1,00)
Mediterranean: 10 (2,05)
North Sea – Central: 172 (0,93)
North Sea – East: 21 (1,08)
North Sea – South: 1 (0,90)
Total: 485
Callantsoog
Arctic: 10 (70,84)
Atlantic South: 1 (0,72)
Baltic: 2 (0,21)
Mediterranean: 6 (29,78)
North Sea – Central: 1 (0,13)
Total: 20
De Lier
Baltic: 13 (1,03)
North Sea – Central: 5 (0,51)
North Sea – East: 8 (7,68)
Total: 26

De Rijp
Atlantic South: 2 (1,03)
Baltic: 14 (1,03)
North Sea – Central: 12 (1,13)
Total: 28
Delft
Atlantic South: 1 (1,03)
Baltic: 3 (0,44)
North Sea – Central: 2 (0,38)
North Sea – East: 8 (14,26)
Total: 14
Delftshaven
Arctic: 2 (6,75)
Atlantic North: 1 (6,69)
Atlantic South: 2 (0,68)
Baltic: 20 (0,98)
North Sea – Central: 17 (1,07)
Total: 42
Delfzijl
Atlantic North: 6 (0,34)
Atlantic South: 77 (0,22)
Baltic: 2500 (1,03)
Mediterranean: 2 (0,04)
North Sea – Central: 1901 (1,00)
North Sea – East: 489 (2,45)
North Sea – South: 13 (1,14)
Total: 4988
Den Helder
Arctic: 5 (5,67)
Atlantic South: 8 (0,92)
Baltic: 55 (0,90)
Mediterranean: 15 (11,91)
North Sea – Central: 42 (0,89)
Total: 125
d'Hoef
Atlantic South: 2 (1,59)
Baltic: 9 (1,03)
North Sea – Central: 7 (1,03)
Total: 18
Dokkum
Atlantic North: 22 (0,90)
Atlantic South: 397 (0,83)
Baltic: 3424 (1,02)
Mediterranean: 16 (0,23)
North Sea – Central: 2606 (1,00)
North Sea – East: 407 (1,47)
North Sea – South: 17 (1,08)
Total: 6889
Dordrecht
Arctic: 3 (0,15)
Atlantic North: 9 (0,88)
Atlantic South: 213 (1,07)
Baltic: 1422 (1,02)
Indian: 1 (94,73)
Mediterranean: 57 (1,97)
North Sea – Central: 1085 (1,00)
North Sea – East: 70 (0,61)
North Sea – South: 10 (1,53)
Total: 2870
Driel
Baltic: 2 (1,03)
North Sea – Central: 1 (0,66)
North Sea – South: 1 (109,45)
Total: 4
Durgerdam
Arctic: 28 (40,07)
Atlantic South: 6 (0,87)
Baltic: 35 (0,73)
Mediterranean: 2 (2,01)
North Sea – Central: 28 (0,75)
Total: 99
Edam
Arctic: 5 (0,35)
Atlantic North: 5 (0,70)
Atlantic South: 103 (0,73)
Baltic: 1005 (1,02)
Mediterranean: 15 (0,74)
North Sea – Central: 782 (1,02)

APPENDIX 1.6 *Location of maritime transport communities in 1700–1800 (cont.)*

North Sea – East: 96 (1,19)
North Sea – South: 6 (1,30)
Total: 2017
Egmond
Baltic: 5 (1,03)
North Sea – Central: 5 (1,32)
Total: 10
Enkhuizen
Arctic: 2 (0,23)
Atlantic North: 2 (0,46)
Atlantic South: 63 (0,75)
Baltic: 498 (0,84)
Mediterranean: 9 (0,74)
North Sea – Central: 404 (0,88)
North Sea – East: 234 (4,82)
Total: 1212
Ezumazijl
North Sea – East: 1 (24,95)
Total: 1
Farmsum
Arctic: 2 (3,46)
Atlantic North: 1 (3,43)
Baltic: 40 (1,00)
North Sea – Central: 39 (1,25)
Total: 82
Franeker
Atlantic North: 1 (2,44)
Atlantic South: 13 (1,62)
Baltic: 57 (1,02)
Mediterranean: 2 (1,73)
North Sea – Central: 42 (0,96)
Total: 115
Friesland
Africa: 1 (5,83)
Atlantic: 1 (5,83)
Atlantic North: 13 (0,55)
Atlantic South: 921 (1,98)
Baltic: 3405 (1,05)
Mediterranean: 20 (0,30)
North Sea – Central: 1954 (0,77)
North Sea – East: 303 (1,13)
North Sea – South: 46 (3,02)
Total: 6664
Gaast
Atlantic South: 4 (1,30)
Baltic: 22 (1,03)
North Sea – Central: 18 (1,08)
Total: 44
Gaastmeer
Atlantic South: 1 (0,80)
Baltic: 9 (1,03)
North Sea – Central: 8 (1,17)
Total: 18
Garyp
Baltic: 5 (1,03)
North Sea – Central: 5 (1,32)
Total: 10
Gent
Mediterranean: 4 (79,41)
North Sea – East: 1 (4,99)
Total: 5
Gorinchem
Baltic: 2 (1,03)
North Sea – Central: 2 (1,32)
Total: 4
Gorredijk
Atlantic South: 21 (1,96)
Baltic: 77 (1,03)
Mediterranean: 1 (0,64)
North Sea – Central: 50 (0,86)
North Sea – East: 5 (0,81)
Total: 154

Gouda
North Sea – East: 88 (24,95)
Total: 88
Graft
Baltic: 3 (1,03)
North Sea – Central: 3 (1,32)
Total: 6
Groningen
Atlantic North: 31 (0,88)
Atlantic South: 411 (0,60)
Baltic: 5000 (1,04)
Mediterranean: 16 (0,16)
North Sea – Central: 3861 (1,03)
North Sea – East: 479 (1,21)
North Sea – South: 99 (4,38)
Total: 9898
Grouw
Atlantic North: 4 (1,15)
Atlantic South: 120 (1,76)
Baltic: 492 (1,03)
North Sea – Central: 342 (0,92)
North Sea – East: 19 (0,48)
North Sea – South: 3 (1,34)
Total: 980
Haarlem
Atlantic South: 1 (7,18)
Baltic: 1 (1,03)
Total: 2
Harlingen
Arctic: 23 (0,37)
Atlantic: 1 (4,46)
Atlantic North: 78 (2,51)
Atlantic South: 553 (0,91)
Baltic: 4309 (1,02)
Mediterranean: 55 (0,63)
North Sea – Central: 3366 (1,02)
North Sea – East: 300 (0,86)
North Sea – South: 26 (1,31)
Total: 8716
Heeg
Atlantic North: 1 (0,71)
Atlantic South: 47 (1,70)
Baltic: 197 (1,02)
Mediterranean: 1 (0,25)
North Sea – Central: 138 (0,92)
North Sea – East: 12 (0,75)
North Sea – South: 1 (1,10)
Total: 397
Heem
Arctic: 2 (14,17)
Atlantic South: 3 (2,15)
Baltic: 9 (0,92)
North Sea – Central: 5 (0,66)
North Sea – South: 1 (21,89)
Total: 20
Heerenveen
Arctic: 1 (0,03)
Atlantic North: 35 (1,98)
Atlantic South: 440 (1,27)
Baltic: 2483 (1,03)
Mediterranean: 5 (0,10)
North Sea – Central: 1884 (1,00)
North Sea – East: 98 (0,49)
North Sea – South: 8 (0,71)
Total: 4954
Hellevoetsluis
Baltic: 10 (1,03)
North Sea – Central: 10 (1,32)
Total: 20
Hemelum
Africa: 1 (19420,21)
Baltic: 1 (1,03)
Total: 2
Hindeloopen
Arctic: 171 (1,41)
Atlantic North: 25 (0,41)
Atlantic South: 108 (0,09)
Baltic: 8473 (1,01)

APPENDIX 1.6 *Location of maritime transport communities in 1700–1800 (cont.)*

Mediterranean: 9 (0,05)
North Sea – Central: 8327 (1,28)
North Sea – East: 44 (0,06)
North Sea – South: 2 (0,05)
Total: 17181

Holland
Atlantic North: 1 (3,65)
Atlantic South: 2 (0,37)
Baltic: 63 (1,68)
North Sea – Central: 5 (0,17)
North Sea – East: 6 (1,94)
Total: 77

Holwerd
Baltic: 1 (1,03)
North Sea – Central: 1 (1,32)
Total: 2

Holysloot
Atlantic South: 1 (2,39)
Baltic: 3 (1,03)
North Sea – Central: 2 (0,88)
Total: 6

Hommerts
Atlantic South: 2 (0,84)
Baltic: 17 (1,03)
North Sea – Central: 14 (1,09)
North Sea – East: 1 (0,73)
Total: 34

Hooge
Arctic: 3 (47,23)
Baltic: 3 (0,68)
North Sea – Central: 2 (0,59)
North Sea – East: 1 (2,77)
Total: 9

Hoogezand
Atlantic South: 1 (0,51)
Baltic: 14 (1,03)
North Sea – Central: 13 (1,22)
Total: 28

Hoogwoude
Atlantic North: 4 (140,44)
Baltic: 4 (1,03)
Total: 8

Hoorn
Arctic: 21 (0,40)
Atlantic North: 7 (0,26)
Atlantic South: 760 (1,47)
Baltic: 3694 (1,02)
Mediterranean: 22 (0,29)
North Sea – Central: 2841 (1,01)
North Sea – East: 83 (0,28)
North Sea – South: 4 (0,24)
Total: 7433

Houwert
Baltic: 1 (1,03)
North Sea – Central: 1 (1,32)
Total: 2

Huisduinen
Arctic: 5 (15,74)
Baltic: 20 (0,91)
North Sea – Central: 20 (1,17)
Total: 45

Ilpendam
Baltic: 2 (1,03)
North Sea – Central: 2 (1,32)
Total: 4

Jisp
Arctic: 1 (4,57)
Atlantic South: 1 (0,46)
Baltic: 15 (0,99)
North Sea – Central: 13 (1,11)
North Sea – East: 1 (0,80)
Total: 31

Joure
Arctic: 6 (0,28)
Atlantic North: 13 (1,19)
Atlantic South: 278 (1,30)

Baltic: 1555 (1,04)
Mediterranean: 4 (0,13)
North Sea – Central: 1160 (0,99)
North Sea – East: 57 (0,46)
North Sea – South: 8 (1,14)
Total: 3081

Kampen
North Sea – East: 1 (24,95)
Total: 1

Katwijk
Mediterranean: 3 (99,26)
Total: 3

Kimswerd
Baltic: 1 (1,03)
North Sea – Central: 1 (1,32)
Total: 2

Kinderdijk
North Sea – East: 4 (24,95)
Total: 4

Kollum
Baltic: 10 (1,03)
North Sea – Central: 8 (1,05)
North Sea – East: 2 (2,49)
Total: 20

Koudum
Arctic: 46 (5,46)
Atlantic North: 1 (0,24)
Atlantic South: 24 (0,29)
Baltic: 568 (0,98)
Mediterranean: 11 (0,92)
North Sea – Central: 533 (1,18)
North Sea – East: 8 (0,17)
Total: 1193

Krimpen
North Sea – East: 3 (24,95)
Total: 3

Kuinre
Atlantic South: 2 (1,69)
Baltic: 8 (0,97)
Mediterranean: 1 (5,84)
North Sea – Central: 6 (0,93)
Total: 17

Kwadijk
Atlantic South: 1 (3,59)
Baltic: 2 (1,03)
North Sea – Central: 1 (0,66)
Total: 4

Lambertschagen
Atlantic South: 1 (7,18)
Baltic: 1 (1,03)
Total: 2

Landsmeer
Atlantic South: 1 (0,19)
Baltic: 37 (1,03)
North Sea – Central: 34 (1,21)
North Sea – East: 2 (0,67)
Total: 74

Lange Akkerschans
Baltic: 1 (1,03)
North Sea – Central: 1 (1,32)
Total: 2

Langweer
Baltic: 1 (1,03)
North Sea – Central: 1 (1,32)
Total: 2

Leek
Atlantic South: 3 (4,31)
Baltic: 5 (1,03)
North Sea – Central: 1 (0,26)
North Sea – East: 1 (2,49)
Total: 10

Leerdam
Baltic: 2 (1,03)
North Sea – Central: 2 (1,32)
Total: 4

Leeuwarden
Arctic: 1 (0,11)
Atlantic North: 3 (0,65)
Atlantic South: 230 (2,56)
Baltic: 643 (1,02)

APPENDIX 1.6 *Location of maritime transport communities in 1700–1800 (cont.)*

Mediterranean: 9 (0,69)
North Sea – Central: 342 (0,70)
North Sea – East: 58 (1,12)
North Sea – South: 3 (1,02)
Total: 1289
Leiden
Baltic: 1 (1,03)
North Sea – East: 1 (12,47)
Total: 2
Lek (De)
North Sea – East: 2 (24,95)
Total: 2
Lemmer
Atlantic North: 32 (1,20)
Atlantic South: 842 (1,61)
Baltic: 3776 (1,03)
Mediterranean: 13 (0,17)
North Sea – Central: 2689 (0,94)
North Sea – East: 146 (0,48)
North Sea – South: 13 (0,76)
Total: 7511
Maas
Baltic: 2 (1,03)
North Sea – Central: 2 (1,32)
Total: 4
Maasland
Atlantic South: 1 (7,18)
Baltic: 1 (1,03)
Total: 2
Maassluis
Atlantic South: 23 (1,34)
Baltic: 121 (1,00)
Mediterranean: 1 (0,40)
North Sea – Central: 94 (1,00)
North Sea – East: 8 (0,81)
Total: 247
Makkum
Atlantic North: 35 (1,62)
Atlantic South: 530 (1,25)
Baltic: 2945 (1,00)
Mediterranean: 20 (0,33)
North Sea – Central: 2142 (0,93)
North Sea – East: 379 (1,56)
North Sea – South: 16 (1,15)
Total: 6067
Medemblik
Arctic: 1 (0,23)
Atlantic North: 6 (2,70)
Atlantic South: 45 (1,03)
Baltic: 289 (0,95)
Mediterranean: 7 (1,11)
North Sea – Central: 218 (0,92)
North Sea – East: 55 (2,20)
North Sea – South: 2 (1,40)
Total: 625
Meeden
Baltic: 2 (1,03)
North Sea – Central: 1 (0,66)
North Sea – East: 1 (6,24)
Total: 4
Middelburg
Arctic: 3 (0,25)
Atlantic North: 2 (0,33)
Atlantic South: 54 (0,46)
Baltic: 841 (1,02)
Mediterranean: 54 (3,17)
North Sea – Central: 699 (1,09)
North Sea – East: 35 (0,52)
North Sea – South: 3 (0,78)
Total: 1691
Middeli
Arctic: 6 (17,71)
Atlantic South: 6 (1,79)
Baltic: 21 (0,90)
North Sea – Central: 15 (0,82)
Total: 48

Minnertsga
Atlantic South: 1 (2,39)
Baltic: 3 (1,03)
North Sea – Central: 2 (0,88)
Total: 6

Molkwerum
Arctic: 16 (1,83)
Atlantic North: 4 (0,90)
Atlantic South: 19 (0,22)
Baltic: 610 (1,01)
North Sea – Central: 586 (1,24)
North Sea – East: 3 (0,06)
North Sea – South: 1 (0,35)
Total: 1242

Monnikendam
Arctic: 1 (1,21)
Atlantic North: 1 (2,40)
Atlantic South: 6 (0,74)
Baltic: 48 (0,84)
North Sea – Central: 33 (0,74)
North Sea – East: 28 (5,97)
Total: 117

Muiden
Baltic: 3 (0,56)
North Sea – Central: 1 (0,24)
North Sea – East: 7 (15,88)
Total: 11

Muntendam
Baltic: 1 (1,03)
North Sea – Central: 1 (1,32)
Total: 2

Munterziel
Baltic: 6 (1,03)
North Sea – Central: 6 (1,32)
Total: 12

Nieuwe Zijlen
Baltic: 3 (0,25)
North Sea – Central: 3 (0,32)
North Sea – East: 19 (18,96)
Total: 25

Nieuwendam
Africa: 1 (20,10)
Arctic: 2 (0,15)
Atlantic North: 13 (1,89)
Atlantic South: 61 (0,45)
Baltic: 972 (1,03)
Mediterranean: 9 (0,46)
North Sea – Central: 761 (1,04)
North Sea – East: 107 (1,38)
North Sea – South: 6 (1,36)
Total: 1932

Nieuwendorp
Baltic: 1 (1,03)
North Sea – Central: 1 (1,32)
Total: 2

Nieuweschans
Atlantic South: 1 (0,11)
Baltic: 46 (0,71)
North Sea – Central: 35 (0,70)
North Sea – East: 49 (9,26)
North Sea – South: 1 (3,32)
Total: 132

Nieuwolda
Baltic: 1 (1,03)
North Sea – Central: 1 (1,32)
Total: 2

Nieuwpoort
North Sea – East: 4 (24,95)
Total: 4

Noord Holland
Baltic: 1 (1,03)
North Sea – Central: 1 (1,32)
Total: 2

Obenhuizen
Baltic: 2 (1,03)
North Sea – Central: 2 (1,32)
Total: 4

Oldeboorn
Atlantic North: 15 (5,46)
Atlantic South: 82 (1,52)

APPENDIX 1.6 *Location of maritime transport communities in 1700–1800 (cont.)*

Baltic: 389 (1,03)
Mediterranean: 2 (0,26)
North Sea – Central: 268 (0,92)
North Sea – East: 14 (0,45)
North Sea – South: 2 (1,13)
Total: 772

Oostende
Arctic: 1 (1,06)
Mediterranean: 1 (0,74)
North Sea – East: 132 (24,58)
Total: 134

Oosterbeek
Atlantic South: 1 (3,59)
Baltic: 2 (1,03)
North Sea – Central: 1 (0,66)
Total: 4

Oosthuizen
Atlantic South: 1 (0,62)
Baltic: 11 (0,98)
Mediterranean: 2 (8,63)
North Sea – Central: 9 (1,03)
Total: 23

Oostwoude
Atlantic South: 1 (0,40)
Baltic: 18 (1,03)
North Sea – Central: 17 (1,24)
Total: 36

Oostzaan
Arctic: 1 (10,90)
Baltic: 6 (0,95)
North Sea – Central: 6 (1,22)
Total: 13

Opmeer
Atlantic South: 2 (0,68)
Baltic: 21 (1,03)
North Sea – Central: 19 (1,19)
Total: 42

Oudebom
Atlantic South: 3 (1,13)
Baltic: 20 (1,08)
North Sea – Central: 14 (0,97)
North Sea – East: 1 (0,66)
Total: 38

Oudega
Atlantic South: 3 (1,03)
Baltic: 21 (1,03)
North Sea – Central: 17 (1,07)
North Sea – East: 1 (0,59)
Total: 42

Peesens
Atlantic South: 1 (0,51)
Baltic: 14 (1,03)
Mediterranean: 1 (3,55)
North Sea – Central: 12 (1,13)
Total: 28

Pekela
Atlantic North: 12 (0,57)
Atlantic South: 136 (0,33)
Baltic: 2900 (1,01)
North Sea – Central: 2473 (1,11)
North Sea – East: 330 (1,40)
North Sea – South: 38 (2,82)
Total: 5890

Petten
Atlantic South: 1 (0,51)
Baltic: 14 (1,03)
North Sea – Central: 12 (1,13)
North Sea – South: 1 (15,64)
Total: 28

Pingjum
North Sea – East: 1 (24,95)
Total: 1

Poppenhuzen
Baltic: 4 (1,03)
North Sea – Central: 4 (1,32)
Total: 8

Purmerend
Atlantic South: 2 (7,18)
Baltic: 2 (1,03)
Total: 4

Ransdorp
Baltic: 1 (1,03)
North Sea – South: 1 (218,91)
Total: 2
Rotterdam
Africa: 2 (12,02)
Arctic: 1 (0,02)
Atlantic North: 9 (0,39)
Atlantic South: 610 (1,35)
Atlantic West: 2 (6,47)
Baltic: 2998 (0,95)
Mediterranean: 188 (2,89)
North Sea – Central: 2062 (0,84)
North Sea – East: 551 (2,13)
North Sea – South: 42 (2,84)
Total: 6465
Ryptsjerk
Baltic: 13 (1,03)
North Sea – Central: 13 (1,32)
Total: 26
Sappemeer
Atlantic North: 2 (1,26)
Atlantic South: 23 (0,74)
Baltic: 231 (1,06)
Mediterranean: 1 (0,22)
North Sea – Central: 159 (0,94)
North Sea – East: 28 (1,56)
North Sea – South: 3 (2,94)
Total: 447
Sas van Gent
Baltic: 1 (1,03)
North Sea – Central: 1 (1,32)
Total: 2
Schagen
Atlantic North: 6 (8,69)
Baltic: 97 (1,03)
North Sea – Central: 91 (1,24)
Total: 194
Scharl
Arctic: 1 (141,68)
Total: 1
Schellinkhout
Arctic: 2 (0,18)
Atlantic North: 2 (0,35)
Atlantic South: 33 (0,30)
Baltic: 812 (1,04)
Mediterranean: 2 (0,12)
North Sea – Central: 737 (1,21)
North Sea – East: 12 (0,19)
North Sea – South: 2 (0,55)
Total: 1602
Schermer
Arctic: 1 (141,68)
Total: 1
Schermerhorn
Atlantic South: 2 (0,65)
Baltic: 22 (1,03)
North Sea – Central: 19 (1,14)
North Sea – East: 1 (0,57)
Total: 44
Schiedam
Atlantic North: 1 (0,53)
Atlantic South: 35 (0,95)
Baltic: 263 (1,02)
Mediterranean: 3 (0,57)
North Sea – Central: 218 (1,09)
North Sea – East: 7 (0,33)
Total: 527
Schiermonnikoog
Atlantic North: 1 (2,34)
Atlantic South: 3 (0,36)
Baltic: 61 (1,04)
North Sea – Central: 51 (1,12)
North Sea – East: 3 (0,62)
North Sea – South: 1 (3,65)
Total: 120
Schoonhoven
North Sea – East: 2 (24,95)
Total: 2
Sloten
Atlantic North: 3 (1,28)
Atlantic South: 84 (1,84)

APPENDIX 1.6 *Location of maritime transport communities in 1700–1800 (cont.)*

Baltic: 330 (1,03)
North Sea – Central: 229 (0,92)
North Sea – East: 9 (0,34)
North Sea – South: 1 (0,67)
Total: 656

Sloterdijk
Baltic: 1 (1,03)
North Sea – Central: 1 (1,32)
Total: 2

Sluis
Baltic: 1 (1,03)
North Sea – Central: 1 (1,32)
Total: 2

Sneek
Atlantic North: 8 (1,75)
Atlantic South: 164 (1,84)
Baltic: 641 (1,03)
Mediterranean: 1 (0,08)
North Sea – Central: 439 (0,90)
North Sea – East: 28 (0,54)
North Sea – South: 1 (0,34)
Total: 1282

Statenzijl
North Sea – East: 21 (24,95)
Total: 21

Stavoren
Arctic: 46 (1,25)
Atlantic North: 32 (1,73)
Atlantic South: 531 (1,46)
Baltic: 2589 (1,02)
Mediterranean: 13 (0,25)
North Sea – Central: 1878 (0,95)
North Sea – East: 112 (0,54)
North Sea – South: 4 (0,34)
Total: 5206

Surhuisterveen
Baltic: 1 (1,03)
North Sea – Central: 1 (1,32)
Total: 2

Temse
North Sea – East: 95 (24,95)
Total: 95

Terhorne
Atlantic North: 13 (1,16)
Atlantic South: 294 (1,34)
Baltic: 1580 (1,03)
Mediterranean: 2 (0,06)
North Sea – Central: 1210 (1,01)
North Sea – East: 49 (0,39)
North Sea – South: 3 (0,42)
Total: 3151

Termunten
Baltic: 4 (1,03)
North Sea – Central: 4 (1,32)
Total: 8

Termunterzijl
Baltic: 9 (0,97)
North Sea – Central: 7 (0,97)
North Sea – East: 3 (3,94)
Total: 19

Terschelling
Arctic: 93 (1,11)
Atlantic North: 41 (0,97)
Atlantic South: 309 (0,37)
Baltic: 5882 (1,02)
Mediterranean: 66 (0,55)
North Sea – Central: 5410 (1,20)
North Sea – East: 35 (0,07)
North Sea – South: 4 (0,15)
Total: 11843

Texel
Arctic: 24 (2,79)
Atlantic North: 1 (0,23)
Atlantic South: 57 (0,67)
Baltic: 597 (1,00)
Mediterranean: 16 (1,30)
North Sea – Central: 512 (1,11)
North Sea – East: 9 (0,18)

North Sea – South: 2 (0,72)
Total: 1220
Tjerkgash
Baltic: 12 (1,03)
North Sea – Central: 12 (1,32)
Total: 24
Tjummarum
Atlantic South: 1 (2,39)
Baltic: 3 (1,03)
North Sea – Central: 2 (0,88)
Total: 6
Twisk
Arctic: 1 (12,88)
Baltic: 4 (0,75)
Mediterranean: 2 (18,05)
North Sea – Central: 4 (0,96)
Total: 11
Uitgeest
Arctic: 1 (141,68)
Total: 1
Uithuizen
Mediterranean: 1 (99,26)
Total: 1
Utrecht
North Sea – East: 8 (24,95)
Total: 8
Veendam
Atlantic North: 2 (0,69)
Atlantic South: 17 (0,30)
Baltic: 420 (1,06)
North Sea – Central: 345 (1,11)
North Sea – East: 21 (0,64)
North Sea – South: 11 (5,90)
Total: 816
Veere
Atlantic North: 1 (1,72)
Atlantic South: 14 (1,23)
Baltic: 81 (1,02)
Mediterranean: 2 (1,22)
North Sea – Central: 12 (0,19)
North Sea – East: 52 (7,96)
North Sea – South: 1 (2,69)
Total: 163
Venhuizen
Arctic: 6 (3,28)
Atlantic South: 7 (0,39)
Baltic: 126 (1,00)
Mediterranean: 1 (0,38)
North Sea – Central: 119 (1,21)
Total: 259
Vianen
North Sea – East: 2 (24,95)
Total: 2
Vlaardingen
Atlantic South: 1 (0,13)
Baltic: 31 (0,56)
North Sea – Central: 28 (0,65)
North Sea – East: 53 (11,60)
North Sea – South: 1 (3,84)
Total: 114
Vlieland
Arctic: 90 (2,56)
Atlantic North: 23 (1,30)
Atlantic South: 170 (0,49)
Baltic: 2440 (1,01)
Mediterranean: 20 (0,40)
North Sea – Central: 2205 (1,17)
North Sea – East: 20 (0,10)
North Sea – South: 2 (0,18)
Total: 4976
Vlissingen
Atlantic South: 12 (0,51)
Baltic: 164 (1,00)
Mediterranean: 4 (1,18)
North Sea – Central: 130 (1,02)
North Sea – East: 26 (1,93)
Total: 336
Walls
Atlantic South: 1 (1,20)
Baltic: 6 (1,03)

APPENDIX 1.6 *Location of maritime transport communities in 1700–1800 (cont.)*

North Sea – Central: 4 (0,88)
North Sea – South: 1 (36,48)
Total: 12
Warder
Arctic: 170 (33,45)
Atlantic South: 17 (0,34)
Baltic: 275 (0,78)
Mediterranean: 14 (1,93)
North Sea – Central: 234 (0,86)
North Sea – East: 10 (0,35)
Total: 720
Warffum
Baltic: 6 (1,03)
North Sea – Central: 6 (1,32)
Total: 12
Warga
Atlantic North: 3 (1,23)
Atlantic South: 78 (1,63)
Baltic: 345 (1,03)
Mediterranean: 1 (0,14)
North Sea – Central: 242 (0,93)
North Sea – East: 17 (0,62)
Total: 686
Warns
Arctic: 98 (10,28)
Atlantic North: 2 (0,42)
Atlantic South: 55 (0,58)
Baltic: 621 (0,94)
Mediterranean: 5 (0,37)
North Sea – Central: 557 (1,09)
North Sea – East: 9 (0,17)
North Sea – South: 2 (0,65)
Total: 1351
Wartena
Arctic: 2 (0,43)
Atlantic North: 1 (0,42)
Atlantic South: 47 (1,02)
Baltic: 332 (1,03)
Mediterranean: 2 (0,30)
North Sea – Central: 264 (1,05)
North Sea – East: 14 (0,53)
Total: 662
Watergang
Arctic: 1 (141,68)
Total: 1
Waterland
Baltic: 1 (1,03)
North Sea – Central: 1 (1,32)
Total: 2
Westerblokker
Baltic: 5 (1,03)
Mediterranean: 1 (9,93)
North Sea – Central: 3 (0,79)
North Sea – South: 1 (43,78)
Total: 10
West-Friesland
Atlantic North: 7 (20,92)
Atlantic South: 3 (0,46)
Baltic: 55 (1,20)
Mediterranean: 1 (1,06)
North Sea – Central: 18 (0,50)
North Sea – East: 8 (2,12)
North Sea – South: 2 (9,32)
Total: 94
Westzaan
Atlantic South: 1 (2,39)
Baltic: 3 (1,03)
North Sea – Central: 2 (0,88)
Total: 6
Wieringen
Arctic: 1 (47,23)
Baltic: 1 (0,68)
Mediterranean: 1 (33,09)
Total: 3
Wijdenes
Atlantic South: 2 (0,72)

Baltic: 20 (1,03)
North Sea – Central: 18 (1,19)
Total: 40

Wijk aan zee
Baltic: 2 (0,68)
Mediterranean: 2 (33,09)
North Sea – Central: 2 (0,88)
Total: 6

Wildervank
North Sea – East: 1 (24,95)
Total: 1

Winkel
Arctic: 4 (18,89)
Baltic: 13 (0,89)
North Sea – Central: 13 (1,14)
Total: 30

Winneweer
Baltic: 3 (1,54)
North Sea – Central: 1 (0,66)
Total: 4

Winschoten
Baltic: 6 (0,88)
North Sea – Central: 3 (0,56)
North Sea – East: 5 (8,91)
Total: 14

Witteveen
Atlantic South: 1 (1,79)
Baltic: 4 (1,03)
North Sea – Central: 3 (0,99)
Total: 8

Workum
Arctic: 15 (0,24)
Atlantic North: 46 (1,48)
Atlantic South: 930 (1,53)
Baltic: 4384 (1,03)
Mediterranean: 26 (0,30)
North Sea – Central: 3131 (0,95)
North Sea – East: 181 (0,52)
North Sea – South: 19 (0,95)
Total: 8732

Wormerveer
Atlantic South: 1 (1,20)
Baltic: 7 (1,20)
North Sea – Central: 3 (0,66)
North Sea – East: 1 (2,08)
Total: 12

Woudsend
Arctic: 2 (0,04)
Atlantic North: 56 (2,08)
Atlantic South: 951 (1,80)
Baltic: 3805 (1,03)
Mediterranean: 13 (0,17)
North Sea – Central: 2558 (0,89)
North Sea – East: 181 (0,60)
North Sea – South: 11 (0,64)
Total: 7578

Ylst
Atlantic North: 4 (1,03)
Atlantic South: 96 (1,27)
Baltic: 546 (1,03)
North Sea – Central: 412 (1,00)
North Sea – East: 28 (0,64)
North Sea – South: 2 (0,80)
Total: 1088

Zaandam
Arctic: 33 (2,27)
Atlantic North: 4 (0,54)
Atlantic South: 147 (1,02)
Baltic: 841 (0,84)
Mediterranean: 68 (3,27)
North Sea – Central: 588 (0,75)
North Sea – East: 381 (4,61)
North Sea – South: 1 (0,21)
Total: 2063

Zeeland
Atlantic South: 4 (1,59)
Baltic: 18 (1,03)
Mediterranean: 2 (5,51)
North Sea – Central: 8 (0,59)
North Sea – East: 4 (2,77)
Total: 36

Zierikzee
Arctic: 25 (5,03)
Atlantic: 1 (55,17)

APPENDIX 1.6 *Location of maritime transport communities in 1700–1800 (cont.)*

Atlantic North: 14 (5,59)
Atlantic South: 60 (1,22)
Baltic: 336 (0,98)
Mediterranean: 8 (1,13)
North Sea – Central: 218 (0,82)
North Sea – East: 31 (1,10)
North Sea – South: 11 (6,84)
Total: 704

Zoutkamp
Baltic: 2 (1,03)
North Sea – Central: 2 (1,32)
Total: 4

Zuiderwoude
Baltic: 1 (1,03)
North Sea – Central: 1 (1,32)
Total: 2

Zunderdorp
Arctic: 5 (101,20)
Baltic: 1 (0,29)
North Sea – Central: 1 (0,38)
Total: 7

Zwartsluis
North Sea – East: 6 (24,95)
Total: 6

Zwolle
Atlantic South: 2 (1,20)
Baltic: 12 (1,03)
North Sea – Central: 10 (1,10)
Total: 24

Grand Total
Africa: 7
Arctic: 1919
Atlantic: 7
Atlantic North: 968
Atlantic South: 18942
Atlantic West: 13
Baltic: 132579
Indian: 1
Mediterranean: 2739
North Sea – Central: 103127
North Sea – East: 10898
North Sea – South: 621
Total: 271883

Appendix to Chapter 2

APPENDIX 2.1 *ZDZ*

	entrepot				gateway							hub				marina					TOTAL
	TOT	HMO		SO	TOT	HMO			SO			TOT	HMO	SO		TOT	HMO	SO			
		−1	L	−1		−1	L	S	−1	L	S		−1	−1	S		−1	−1	L	S	
1670	1			1	2		1				1	5		5		38	1	26	4	7	46
1671					2		1				1	11	1	10		35		22	4	9	48
1672					2		1		1			7		7		23		15	2	6	32
1673					3		1		1		1	4		4		28	2	13	2	11	35

	entrepot			gateway				hub			marina					TOTAL
	TOT	HMO	SO	TOT	HMO	SO		TOT	HMO	SO	TOT	HMO	SO			
1674				2	1	1		5		5	39	4	21	3	11	46
1675				2	1	1		9		9	27		19	2	6	38
1676				3	1	2		9	1	8	26		14	2	10	38
1677				1	1			7		7	36		21	3	12	44
1678				2	1	1		6	3	3	33	1	15	4	13	41
1679				1	1			9	1	8	37		19	5	13	47
1680				1	1			8		8	35		19	4	12	44
1681				2	1	1		8		8	38	1	21	4	12	48
1682				1	1			6	2	4	36		19	5	12	43
1683				2	1	1		5		5	39		24	5	10	46
1684				1	1			5		5	39		25	4	10	45
1685				1	1			6	2	4	39	1	24	4	10	46
1686				2	1	1		8		8	36	1	22	5	8	46
1687				1	1			6	1	5	42	1	25	5	11	49
1688				1	1			4	1	3	45		31	5	9	50
1689				1	1			8		8	42		28	5	9	51
1690				1	1			7		7	35		22	5	8	43
1691				1	1			12		12	32		17	3	12	45
1692				1			1	14		14	18			5	13	33
1693				1	1			7		7	43		32	3	8	51
1694				1	1			8		8	37		24		13	46
1695				1	1			15		15	31		21	3	7	47
1697				1	1			6		6	25		17	2	6	32
1698				1	1			4		4	35		21	5	9	40
1699				1	1			4		4	33		20	4	9	38
1700				2	1	1		6		6	30		15	5	10	38
1701				1	1			7		7	29		17	4	8	37
1702				2	1	1		11		11	24		13	4	7	37
1703				2	1	1		7		7	23		16	1	6	32
1704				1	1			5		5	28		18	2	8	34
1705				1	1			1		1	26		19		7	28
1706				1	1			4		4	22		14	4	4	27
1707				1	1											1
1708				1	1			3		3	26		19	3	4	30
1709				1			1									1
1710				1	1			6		6	24		14	5	5	31

APPENDIX 2.1 *ZDZ* (*cont.*)

	entrepot			gateway				hub				marina					TOTAL
	TOT	HMO	SO	TOT	HMO		SO	TOT	HMO	SO		TOT	HMO	SO			
1711				1		1		5		5		20		10	5	5	26
1712				1		1		7	2	5		24		13	5	6	32
1713				1		1		6		6		33		24	5	4	40
1714				1		1		5		5		33		24	3	6	39
1715				1		1		7		7		22		15	2	5	30
1716				1		1		9		9		20		11	3	6	30
1717	1	1						4		4		21		11	1	9	26
1718				1		1		7		7		24		12	4	8	32
1719				1		1		12		12		29		20	5	4	42
1720				1		1		8		8		27		15	4	8	36
1721				1		1		5		5		32		20	5	7	38
1722				1		1		6		6		15		9	3	3	22
1723				3		1	2	2		2		30		20	3	7	35
1724				2		1	1	7		7		30		19	2	9	39
1725				2		1	1	6		6		33	1	20	3	9	41
1726				2		1	1	5		5		33		20	2	11	40
1727				3	1	1	1	8		8		26		14	1	11	37
1728				2		1	1	7		7		29		20	2	7	38
1729				3	1	1	1	6		6		33		20	1	12	42
1730				2		1	1	8	1	6	1	32		19	2	11	42
1731	1		1	3		1	2	4		4		30		17	1	12	38
1732				1		1		7		7		25		9	1	15	33
1733				2		1	1	7		7		27		16	1	10	36
1734				3		1	2	8	1	7		29		17	1	11	40
1735				2		1	1	4		4		31		21	1	9	37
1736				2		1	1	7	1	6		26		16	2	8	35
1737				1		1		7	1	6		28		17	2	9	36
1738				2		1	1	9		9		32		20	2	10	43
1739				2		1	1	5		5		31		16	2	13	38
1740				3		1	2	7		7		33		19	2	12	43
1741	1	1		3		1	2	6		6		31		19	1	11	41
1742	1	1		2		1	1	4		4		32		20		12	39
1743				2		1	1	6	1	5		30		17		13	38

	entrepot			gateway				hub			marina					TOTAL
	TOT	HMO	SO	TOT	HMO		SO	TOT	HMO	SO	TOT	HMO	SO			
1744				2		1	1	8		8	32		17	3	12	42
1745				3		1	2	4	1	3	33		22	1	10	40
1746	1	1		3		1	2	8		8	29		16		13	41
1747				3	1	1	1	6		6	28		17	1	10	37
1748	1	1		3		1	2	6		6	31		19	1	11	41
1749				2		1	1	7	1	6	32		20		12	41
1750	1	1		2		1	1	5	1	4	35		22		13	43
1751				2		1	1	7	1	6	40		28	1	11	49
1752				3		1	2	7		7	36		23	1	12	46
1753				3		1	2	4	1	3	37		24	1	12	44
1754				3		1	2	9		9	34		21	1	12	46
1755				3		1	2	9	1	8	34	1	21		12	46
1756				3		1	2	8	2	6	32		21		11	43
1757				2		1	1	8	1	7	36		25		11	46
1758				3		1	2	8	1	7	35		23		12	46
1759	1	1		3		1	2	6		6	35		23		12	45
1760				3		1	2	6	1	5	33		20		13	42
1761				3		1	2	8	1	7	38		26		12	49
1762				3		1	2	10		10	35		23		12	48
1763				2		1	1	8	1	7	34		22		12	44
1764				2		1	1	9		9	36		25	1	10	47
1765	2		2	4		1	3	7	1	6	34		24		10	47
1766	1	1		5		1	4	7	2	5	34	1	22		11	47
1767				4		1	3	7	1	6	35		23	1	11	46
1768	1	1		3		1	2	3	1	2	36		27	1	8	43
1769				2		1	1	10		10	36		23		13	48
1770				3		1	2	7		7	34		21	1	12	44
1771				2		1	1	7		7	36		22		14	45
1772				3		1	2	10		10	36		27		9	49
1773				4		1	3	10		10	36		30		6	50
1774				3		1	2	13		13	35	1	26		8	51
1775				2		1	1	14		14	38		23		15	54
1776				2		1	1	8		8	30		21		9	40
1777				2		1	1	2		2	35		28		7	39
1778				3		1	2	9		9	31		24		7	43

APPENDIX 2.1 *ZDZ* (*cont.*)

	entrepot			gateway					hub			marina				TOTAL
	TOT	HMO	SO	TOT	HMO	SO			TOT	HMO	SO	TOT	HMO	SO		
1779				3	1	2			5		5	31		25	6	39
1780				2	1	1			8		8	30		25	5	40
1781				2	1	1			4	1	3					6
1782				2	1	1			2		2	4		4		8
1783	1		1	4	1	3			6		6	16		16		27
1784				2	1	1			7		7	24		23	1	33
1785				3	1	2			3		3	23		21	2	29
1786	1		1	3	1	2			4		4	23		21	2	31
1787				3	1	2			8		8	21		20	1	32
1788				1	1				1		1	26		23	3	28
1789				3	1	2			3		3	29		26	3	35
1790				2	1	1			6	1	5	26		23	3	34
1791				3	1	2			5		5	28		26	2	36
1792	1		1	2	1	1			3		3	27		22	5	33
1793				1	1				4		4	21		20	1	26
1794				4	1	3			6		6	22		22		32
1795				2		1		1	1		1	2		2		5
1796				5		4	1		3		3					8
1797				5		4	1		2		2					7
1798				4		3		1	1		1					5
1799				1		1			2		2					3
1800				3		2		1	1		1	1		1		5
1801				3		2		1				1		1		4
1802				4		3	1		1		1					5
1803				3	1	2			6		6	4		4		13
1804				3		2		1								3
1805				2		1		1	3		3	1		1		6
1806				2		1		1	1		1					3
1807				1		1										1
1809				1				1								1
1810				1				1	4		4	1		1		6
1811				1		1						1		1		2
1812				1		1										1
1813				1		1										1
1814	1		1	1	1				6		6	9		9		17

	entrepot			gateway					hub			marina				TOTAL
	TOT	HMO	SO	TOT	HMO		SO		TOT	HMO	SO	TOT	HMO	SO		
1815				4		1	3		4		4	8		8		16
1816	1		1	1		1			1		1	13		13		16
1817	1		1	5		1	4		2		2	14		14		22
1818	1		1	3		1	2		2		2	11		11		17
1819	1		1	4		1	3		3		3	10		10		18
1820				2			1	1	2	1	1					4
1821	2		2	3		1	2		4		4	8		8		17
1822	1	1		1		1			2		2	7		7		11
1823	2		2	2			1	1	4	1	3	6		6		14
1824	1	1		2		1	1		3		3	5		5		11
1825	1		1	4		1	3					7		7		12
1826	1	1		4		1	3		1		1	7		7		13
1827				1			1									1
1828	1	1		4		1	3		2		2	5		5		12
1829				4		1	3		1		1	10		10		15
1830	1		1	6		1	5		1		1	7		7		15
1831	1		1	5		1	4		3		3	8		8		17
1832				6	2	1	3		2		2	8		7	1	16
1833	1	1		5		1	4		1		1	7		7		14
1834	1	1		6	1	1	4					7		7		14
1835	1	1		7	1	1	5		2		2	7		7		17
1836				6	1	1	4					6		6		12
1837				7	1	1	5		3		3	5		5		15
1838	2	1	1	5		1	4					7		7		14
1839				8		1	7		2		2	8		8		18
1840	1	1		6		1	5		2		2	7		7		16
1841				8	1	1	6		4		4	6		6		18
1842	1		1	6		1	5		1		1	8		8		16
1843	1		1	10			9	1				7		7		18
1844	1		1	7	1	1	5					6		6		14
1845	2		2	9		1	8		3		3	6		6		20
1846	1		1	6	3	1	2		2		2	6		6		15
1847	2	1	1	6	1	1	4		5		5	11		11		24
1848	1		1	6	1		4	1	3		3	10		10		20
1849	1		1	5	3	1	1		2		2	9		9		17
1850	1		1	5			4	1	4		4	11		11		21

APPENDIX 2.2 *SMR*

	entrepot					gateway					hub			marina				TOTAL
	TOT	HMO		SO		TOT	HMO		SO		TOT	HMO	SO	TOT	HMO	SO		
		−1	S	−1	S		−1	S	−1	S		−1	−1		−1	−1	S	
1670	2	2									1		1	2		1	1	5
1671	1			1		1			1		1		1	1		1		4
1672											2		2	3		2	1	5
1673											1		1	3		2	1	4
1674											4		4	3		1	2	7
1675						1			1					3		3		4
1676						1			1		1	1		2		2		4
1677	1			1							1		1	2		1	1	4
1678											1	1		2		1	1	3
1679											5	1	4	3		2	1	8
1680						1			1		3		3	4		3	1	8
1681	1	1									2		2	4		3	1	7
1682						1			1		3	1	2	6		5	1	10
1683	2			2							3	1	2	4	1	3		9
1684	1			1							4		4	4	1	3		9
1685	2	1		1										5		5		7
1686	1			1							4	1	3	1		1		6
1687	2	1		1							2	1	1	5		5		9
1688						2			2		4	1	3	2		2		8
1689	2			2							1	1		3		3		6
1690														5		4	1	5
1691						1			1					3		2	1	4
1692											2		2	1			1	3
1693						1			1					4		4		5
1694	1			1							1		1	4		4		6
1695											1		1	3		3		4
1697											1		1	2		2		3
1698	2	2									2	1	1	3	1	2		7
1699	2	1		1							2		2	4		4		8
1700						1			1		1		1	6		5	1	8
1701	1			1							2	1	1	2		2		5
1702											2		2					2
1703						1			1		1		1	3		3		5

	entrepot				gateway			hub			marina				TOTAL
	TOT	HMO	SO		TOT	HMO	SO	TOT	HMO	SO	TOT	HMO	SO		
1704								2		2	5		5		7
1705								1		1	2		2		3
1706								1		1	1		1		2
1708											2		2		2
1710								1		1	2			2	3
1711								1		1	2		1	1	3
1712								4		4	1		1		5
1713											6		4	2	6
1714								3		3	4		4		7
1715								2		2	3		2	1	5
1716								2		2	2		1	1	4
1717								1	1		2			2	3
1718	1		1					2		2	2		1	1	5
1719											5		5		5
1720								1		1	5		5		6
1721	2	1	1					3		3	2		2		7
1722					1	1		1		1	3		3		5
1723	1	1						3		3	3		3		7
1724					2		2	4		4	1		1		7
1725								4	1	3	2		2		6
1726								4	2	2	4		4		8
1727	1	1			2		2	4	1	3	1		1		8
1728	1	1						4	1	3	1		1		6
1729	1	1			1		1	3		3	1		1		6
1730	1		1		1		1	2	1	1	1			1	5
1731	1	1			2		2	3		3	1		1		7
1732	1	1						2		2	1		1		4
1733	1	1			1	1		2		2	2		2		6
1734					2	1	1	1		1	2	1	1		5
1735	1		1		1	1		1		1	4		4		7
1736					1	1		4		4	3		3		8
1737	1		1		1	1		5		5	1		1		8
1738	1		1		2	1	1	1		1	5	1	4		9
1739	1		1		1		1	4	1	3	2		2		8
1740	2		1	1	1	1		2		2	5		4	1	10
1741					1		1	2		2	5		5		8

APPENDIX 2.2 *SMR* (*cont.*)

	entrepot					gateway					hub			marina				TOTAL
	TOT	HMO		SO		TOT	HMO		SO		TOT	HMO	SO	TOT	HMO	SO		
1742	1	1									4		4	2		2		7
1743	1	1				2			2		2		2	4		4		9
1744	1			1		1	1				1	1		2		2		5
1745						2	1		1		1		1	1		1		4
1746	1			1		2	1		1		1	1						4
1747						4	2		2		1		1	2		2		7
1748						3	1		2		2		2	1		1		6
1749	1	1									2		2	6		6		9
1750	1			1							3	1	2	4		4		8
1751	2	1			1						4	1	3	4		3	1	10
1752	1				1						4	2	2	5		4	1	10
1753	2	1	1			4			4		2		2	3		2	1	11
1754	2	1		1		1			1					5		5		8
1755	3	1		2		1			1		1		1	1		1		6
1756	3	1		2							1		1	1		1		5
1757	3	2		1							1		1	4		4		8
1758	2	1		1										3		3		5
1759	1	1									3	2	1	1		1		5
1760	1	1				1			1		3		3	3		3		8
1761	1			1							2	1	1	2		2		5
1762	2			2		1			1		5	1	4	1			1	9
1763											4	2	2	4		3	1	8
1764	1	1				1			1		4	2	2	2		2		8
1765	1	1									3	1	2	2		2		6
1766	1	1									3	2	1	1		1		5
1767	3	2		1							2	1	1	2		2		7
1768	1		1			2			2		3	2	1	3		3		9
1769	1	1									5	2	3	1		1		7
1770						2			1	1	1	1		5		4	1	8
1771	1		1			1			1		4	1	3	2		2		8
1772	1		1								2	2		5		4	1	8
1773	1			1		3		1	2		4	2	2					8
1774						2		1	1		3	1	2	4		4		9
1775	1		1			2			2		4	2	2	1			1	8
1776						2		1	1		3	1	2	2		2		7

	entrepot				gateway					hub			marina				TOTAL
	TOT	HMO	SO		TOT	HMO		SO		TOT	HMO	SO	TOT	HMO	SO		
1777	1	1			2		1	1		7	2	5	2		1	1	12
1778	1	1			4	1		2	1	1	1		4		3	1	10
1779	1	1			2			1	1	2		2	3		3		8
1780	1			1	2	2				1		1	4		4		8
1781					2			2		1		1					3
1782					5			5									5
1783	1		1		6	1		5					1		1		8
1784	1	1			3	1	1	1		1		1	3		3		8
1785	2		2		2		1	1		2		2	1		1		7
1786	1		1		2			2		2		2	1		1		6
1787					5	1		4		1		1	3		3		9
1788					2	1		1		1	1		5		5		8
1789					3			2	1	5	1	4	1		1		9
1790	1		1		2		1	1		1		1	3		3		7
1791	1		1		2			1	1	2		2	3	1	2		8
1792					5	1		3	1	4	1	3	3		3		12
1793					5			4	1	3		3					8
1794					5	2	1	2		3		3	2		2		10
1795					4			4									4
1796					4			3	1								4
1797					5			3	2								5
1798					3			3									3
1800					3			3									3
1801					3			3									3
1802					4			3	1								4
1803					3	1		2									3
1804					3			3									3
1805					5			5									5
1806					3			3									3
1809					1				1								1
1810					1			1		2		2					3
1811					1			1									1
1812					1			1		1		1					2
1813					1			1									1
1814					3	1		2		1		1	1		1		5
1815					4	1	1	2		1	1						5
1816					5	1		2	2				2		2		7

APPENDIX 2.2 *SMR* (*cont.*)

	entrepot			gateway					hub			marina			TOTAL
	TOT	HMO	SO	TOT	HMO		SO		TOT	HMO	SO	TOT	HMO	SO	
1817				5	1	1	3		1	1		2		2	8
1818				4	2		2		2		2	1		1	7
1819				3	1	1	1		2		2	2	1	1	7
1820				4			4								4
1821	1		1	4		1	2	1				1	1		6
1822				3			1	2	1		1	1		1	5
1823	1		1	4		1	2	1				1		1	6
1824				2			2					2		2	4
1825	1		1	3			3		1	1					5
1826	1		1	4			2	2	1	1					6
1827				2			2								2
1828	2		2	3			2	1	3		3				8
1829	1		1	7		1	6					1		1	9
1830				5		1	3	1				1		1	6
1831	1		1	6	1		4	1				1		1	8
1832	2	1	1	6		1	3	2				2		2	10
1833				6	1		5					1		1	7
1834	1	1		6	1		5								7
1835	1	1		3			3		1	1		1		1	6
1836				4	1		3								4
1837	1		1	6	3		3								7
1838	1	1		3	1		2		1		1				5
1839	1		1	4			3	1	2	1	1	1		1	8
1840	1	1		4	2		2		1		1	1		1	7
1841				5			5					1		1	6
1842	1	1		5	1		4								6
1843				5	1		4								5
1844				5	1		4								5
1845				4	1		3					1		1	5
1846	3	1	2	4	1		3		1		1				8
1847	1		1	4	2		2		2	2		2		2	9
1848	2	1	1	4	1		3		2	2					8
1849	1	1		4	1		3		3	2	1	1		1	9
1850	1		1	3	1		2		2		2				6

APPENDIX 2.3 *LDE*

	entrepot			gateway			hub			marina					TOTAL
	TOT	HMO	SO	TOT	HMO	SO	TOT	HMO	SO	TOT	HMO	SO			
		−1	−1		−1	−1		−1	−1		−1	−1	L	S	
1670							3		3	1			1		4
1671										1			1		1
1672							1		1	1				1	2
1673										2		1		1	2
1674							1	1		1				1	2
1675				1		1	2		2	1				1	4
1676										2		1		1	2
1677										3		2		1	3
1678				1		1	1		1	1			1		3
1679							1		1	2		1	1		3
1680	1		1	1		1				3		2		1	5
1681				1		1	1		1	2		1	1		4
1682										3		2	1		3
1683										3		2	1		3
1684										1			1		1
1685										2		1	1		2
1686										2		1	1		2
1687				1		1				2	1		1		3
1688										2		1	1		2
1689										2		1	1		2
1690										2		1	1		2
1691										2		1		1	2
1692										1			1		1
1693							1		1	2		1		1	3
1694										2		1		1	2
1695										2		1		1	2
1697										2		1		1	2
1698							1	1		2		1	1		3
1699							2	1	1	1			1		3
1700							1		1	3		2	1		4
1701										3		2	1		3
1702										3	1	1	1		3
1703							1		1	1			1		2
1704										2		1	1		2

APPENDIX 2.3 *LDE* (*cont.*)

	entrepot			gateway			hub			marina					TOTAL
	TOT	HMO	SO	TOT	HMO	SO	TOT	HMO	SO	TOT	HMO	SO			
1705										2		1		1	2
1706							1		1	1			1		2
1708										1			1		1
1710										1			1		1
1711										1			1		1
1712							1		1	2		1	1		3
1713										2		1	1		2
1714							2		2	2		1	1		4
1715										2		1	1		2
1716										1			1		1
1717										2		1		1	2
1718							1		1	1			1		2
1719				1		1	1		1	3		1	1	1	5
1720										4		3	1		4
1721										4		3	1		4
1722							1		1	2		1	1		3
1723							1	1		2			1	1	3
1724							1	1		2		1	1		3
1725							1	1		4		2	1	1	5
1726							2	2		3		2	1		5
1727										4		2	1	1	4
1728							2	1	1	3		2	1		5
1729							1		1	4		2	1	1	5
1730							1	1		3		1	1	1	4
1731	1	1								3		1	1	1	4
1732							2	1	1	2			1	1	4
1733							2	1	1	3		1	1	1	5
1734							1	1		3		1	1	1	4
1735							1	1		3		1	1	1	4
1736	1	1								3		1	1	1	4
1737							2		2	4		2	1	1	6
1738							2		2	4		2	1	1	6
1739							2		2	4		1	1	2	6
1740				1		1	3		3	5		2	1	2	9

	entrepot			gateway			hub			marina					TOTAL
	TOT	HMO	SO	TOT	HMO	SO	TOT	HMO	SO	TOT	HMO	SO			
1741										4		1	1	2	4
1742				1		1	1		1	4		1	1	2	6
1743										4				4	4
1744							1		1	4		1	1	2	5
1745										4		1		3	4
1746										4		2		2	4
1747										4		2		2	4
1748										4		3		1	4
1749										4		2		2	4
1750							1		1	4		2	1	1	5
1751							1		1	3			1	2	4
1752				1		1	2		2	3		1		2	6
1753							1		1	3		1		2	4
1754				1		1				4		2		2	5
1755	1		1	1		1	1		1	3		1		2	6
1756							1	1		4		1		3	5
1757							2		2	4		1		3	6
1758	1	1								4		1		3	5
1759				1		1	1	1		5		2	1	2	7
1760				1		1	1		1	6		3		3	8
1761	2		2				2		2	7		4	2	1	11
1762				3	1	2	1		1	7		4	2	1	11
1763				1		1	3	1	2	7		4		3	11
1764				1		1	1		1	6		3		3	8
1765							1		1	6		2		4	7
1766							2	1	1	6		3		3	8
1767							1		1	7		3		4	8
1768				1		1				8		4		4	9
1769										8	1	3		4	8
1770										10	1	4		5	10
1771							1		1	9		5		4	10
1772				1		1	2		2	7		2		5	10
1773				1		1	5		5	6	1	1		4	12
1774				1		1	4		4	8		3		5	13
1775	1	1					3		3	9	1	4	1	3	13
1776							4	1	3	4				4	8

APPENDIX 2.3 *LDE* (*cont.*)

	entrepot			gateway			hub			marina					TOTAL
	TOT	HMO	SO	TOT	HMO	SO	TOT	HMO	SO	TOT	HMO	SO			
1777				2		2	2		2	8		4		4	12
1778				1		1	4		4	8		4		4	13
1779				4		4	3		3	8		4		4	15
1780				3		3	3		3	9		5		4	15
1781				1		1				2		2			3
1782				6		6									6
1783	1		1	3		3	2		2	4		4			10
1784				5		5	1		1	7		5		2	13
1785				2	1	1	3		3	7		5		2	12
1786				4		4	3		3	9		7		2	16
1787				3		3	2		2	9		7		2	14
1788				3		3	3		3	9		7		2	15
1789				3		3				10		7	1	2	13
1790				2		2	4		4	9		7		2	15
1791				3		3	2		2	10		8		2	15
1792				3		3	2		2	13		10		3	18
1793				1		1	3		3	6		4		2	10
1794				2		2	3		3	9	1	6		2	14
1795				2		2	1		1	1		1			4
1796				4	1	3	2		2	1		1			7
1797				3		3	1		1						4
1798				3		3	1		1						4
1799				1		1									1
1800				2		2									2
1801				1		1				1		1			2
1802				3		3				1		1			4
1803				2		2	2		2	5	1	4			9
1804				3		3				1		1			4
1805				3		3				1		1			4
1806				1		1									1
1810							3		3	1		1			4
1814				2	1	1				8		7		1	10
1815	1		1	1		1				10		7		3	12

	entrepot			gateway			hub			marina					TOTAL
	TOT	HMO	SO	TOT	HMO	SO	TOT	HMO	SO	TOT	HMO	SO			
1816				2	1	1	1		1	9		6		3	12
1817	1		1	2		2				9		6		3	12
1818	1		1				2		2	7		4		3	10
1819	1		1	1		1				8		5		3	10
1820										1		1			1
1821				1		1	1		1	8		5		3	10
1822				1		1	1		1	7		6		1	9
1823							2		2	6		4		2	8
1824				1		1				8		6		2	9
1825				1		1	1		1	9		7		2	11
1826				1		1	2		2	6		4		2	9
1828				1		1	1		1	9		7		2	11
1829	1		1	1		1	2		2	7		5		2	11
1830				1		1	1		1	10		7	1	2	12
1831				2		2	1		1	8		6		2	11
1832				1		1	2		2	9		5	1	3	12
1833	1		1							7		5		2	8
1834	1		1	1		1	1		1	7		4		3	10
1835							1		1	6		4		2	7
1836				1		1	1		1	6		4		2	8
1837	1		1	1		1	1		1	5		3		2	8
1838	1		1	1		1				7		4		3	9
1839	1		1	1		1	1		1	7		4		3	10
1840	1		1				1		1	8		5		3	10
1841				2		2	1		1	6		4		2	9
1842	1		1							9		6		3	10
1843	1		1	1		1				9		6		3	11
1844				2		2	1		1	9		6		3	12
1845	1		1							9		7		2	10
1846	1		1	1		1				10		7		3	12
1847	1		1				1		1	9		7		2	11
1848	1		1	1		1				10		7		3	12
1849	1		1	1		1				10		7		3	12
1850	1		1				1		1	10		7		3	12

Appendix to Chapter 3

Abbreviations: Gtrad = traditional grain routes (Danzig-Amsterdam, Königsberg-Amsterda
Gadd = additional grain routes; Gem = emerging grain routes; Ttrad = traditional timber rou
Tadd = additional timber routes; Tem = emerging timber routes.

All values are expressed in tonkm.

APPENDIX 3.1 *Estimates of the production of maritime transport services by shipmasters of group 1. Based on STRO*

year	GRAIN	Gtrad	Gadd	Gem
1670	258.972.294	193.166.208	17.339.526	230.92
1671	170.180.888	109.667.678	8.773.602	131.17
1672	50.242.951	42.220.937	3.455.008	
1673	70.460.635	49.150.835	2.389.721	
1674	158.070.258	110.562.240	10.041.336	
1675	130.151.619	99.091.783	10.938.023	
1676	88.489.207	58.529.209	9.140.336	
1677	153.690.622	107.384.748	20.388.780	
1678	157.072.736	128.917.024	4.395.452	
1679	235.838.730	189.321.216	14.603.736	
1680	257.716.390	181.662.698	28.480.835	
1681	286.702.548	210.367.029	37.452.640	
1682	299.320.050	202.417.986	38.604.480	
1683	289.919.679	218.690.579	32.584.505	
1684	327.613.882	259.099.457	23.370.965	
1685	252.473.984	194.663.395	19.627.198	
1686	248.116.905	194.958.912	25.236.472	
1687	200.078.066	146.928.669	20.648.976	532.12
1688	243.111.092	173.954.625	22.335.765	
1689	92.028.780	63.471.507	11.179.367	
1690	61.722.834	38.226.818	8.463.301	
1691	43.574.988	27.574.925	4.206.225	
1692	1.085.399	608.401	476.998	
1693	209.256.536	108.762.468	48.415.863	
1694	165.665.066	76.454.849	63.642.845	
1695	125.761.298	62.773.640	35.025.859	
1696				

MBER	Ttrad	Tadd	Tem	OTHER
.718.568		793.789		30.084.847
.536.600	166.995	819.329		46.651.426
14.164		4.491		9.369.823
18.490		4.865		16.085.875
.198.260		16.898		33.623.918
719.177		678.145		35.506.145
		50.524		40.097.488
.641.415	25.824	4.615.591		32.523.175
60.454	17.216	6.737		23.774.922
.725.108	28.438.226	25.823		39.454.895
666.176		44.910		51.020.741
868.626		24.701		33.685.271
139.890		77.245		44.505.426
97.996	1.139	65.120		41.288.007
399.699		121.258		50.697.456
627.779	1.656	499.464		65.118.243
367.347	136.867	84.625		56.396.485
.774.329	28.858	32.681		58.709.490
873.583	43.116	76.539		35.186.407
15.187		13.949		27.763.187
338.956	56.015	24.077		25.280.976
361.234	123.238	12.524		23.235.692
				292.495
708.362		685.691		56.207.812
366.323		49.140		36.036.699
29.420		6.341		50.020.256

APPENDIX 3.1 *Estimates of the production of maritime transport services by shipmasters of group 1 (co*

year	GRAIN	Gtrad	Gadd	Gem
1697	42.926.902	38.639.058	928.200	
1698	125.326.794	98.350.621	15.862.999	
1699	235.974.473	195.573.992	29.500.670	
1700	105.853.702	82.010.724	8.930.839	
1701	111.577.193	82.251.625	15.415.489	
1702	121.501.777	103.519.748	5.246.556	
1703	54.730.246	49.645.317	682.942	
1704	106.130.508	100.744.733	842.889	
1705	100.087.501	90.265.617	5.268.113	
1706	4.647.615	4.314.950	238.499	
1707				
1708	6.139.162	5.431.519	89.941	
1709				
1710	29.286.155	28.099.380	674.170	
1711	123.701.012	115.045.646	3.767.440	
1712	122.288.953	107.498.523	6.455.326	
1713	83.594.315	70.129.342	5.895.840	
1714	58.775.644	48.213.071	4.055.137	
1715	13.924.689	6.746.469	3.859.999	
1716	24.152.786	17.125.071	3.913.062	
1717	10.037.249	8.140.248	876.484	
1718	41.102.321	39.891.673		
1719	66.569.992	63.203.306	22.485	
1720	44.973.128	38.816.112	766.502	
1721	79.110.846	76.096.521		
1722	35.437.001	32.842.752	732.920	
1723	110.380.380	99.909.066	788.140	
1724	147.765.440	130.411.628	4.022.136	
1725	128.929.590	116.742.785	3.047.821	
1726	72.371.167	64.794.497	1.750.515	
1727	34.984.150	33.063.154	1.326.322	
1728	99.246.373	92.289.740	2.923.100	
1729	181.338.008	161.164.512	8.569.663	
1730	66.213.687	53.800.719	2.140.532	
1731	86.885.894	62.919.491	6.055.718	
1732	29.603.807	28.554.909	190.760	

BER	Ttrad	Tadd	Tem	OTHER
50.682		50.682		39.421.594
761.688		154.047		58.526.383
046.120	243.302	8.266		81.222.866
050.047	167.526	371.445		35.913.343
246.644		42.282		40.140.073
73.657		16.784		34.128.919
14.773		3.894		22.925.855
278.004		27.910		29.445.579
457.701	75.961	36.976		20.264.155
34.220		20.719		12.371.048
77.347		40.370		12.330.811
322.100		217.054		21.022.886
131.539		125.745		15.592.909
245.620		135.870		19.442.490
6.526		1.354		29.813.762
115.919		115.763		31.102.945
52.140	43.059	9.081		28.922.260
13.305		1.083		25.272.397
718.585		714.456		6.511.396
068.660	107.649	938.116		22.360.869
102.965		832.218		22.728.549
523.826		376.213		26.733.998
095.195	16.561	132.965		26.210.532
121.930		121.819		10.155.967
661.516	157.096	744.388		35.484.847
268.738	1.003.553	878.078		25.536.311
148.587	48.871.749	164.313		50.094.611
070.641	106.103	1.720.948		32.697.922
.367.729	746.907	359.282	496.619	25.863.311
.862.090	3.635.673	1.000.863		21.222.455
722.828	300.164	305.305		27.767.175
.251.465	1.026.802	153.163		6.750.496
.549.758	1.313.463	568.031		18.034.558
.106.386	328.687	461.736		10.131.204

APPENDIX 3.1 *Estimates of the production of maritime transport services by shipmasters of group 1 (c*

year	GRAIN	Gtrad	Gadd	Gem
1733	46.354.386	42.112.683	3.972.063	
1734	55.931.818	44.430.058	5.594.128	
1735	49.818.459	43.812.750	2.361.673	321.2
1736	11.093.784	4.476.885	4.469.249	301.2
1737	4.112.911	1.694.367	985.241	
1738	37.398.065	17.712.247	15.275.419	
1739	59.397.080	33.333.745	13.161.099	
1740	136.849.703	77.066.714	30.716.965	458.1
1741	64.553.360	36.538.351	9.258.103	
1742	29.667.014	24.395.000	978.112	281.0
1743	26.085.475	22.881.459	538.500	
1744	42.349.380	39.491.486	1.737.928	
1745	10.476.690	7.866.357	1.353.048	
1746	10.096.825	8.411.649		
1747	21.483.224	19.986.141	471.722	
1748	16.987.236	15.899.861	926.555	
1749	39.417.493	38.006.305	873.480	
1750	62.336.395	56.258.034	2.883.790	356.4
1751	109.283.194	92.494.378	4.140.952	399.1
1752	108.018.740	74.378.102	1.420.659	
1753	44.431.296	28.867.170	818.260	1.668.8
1754	50.879.415	42.448.730	1.609.554	
1755	27.469.561	24.215.040	1.721.415	
1756	6.631.023	6.631.023		
1757	27.853.622	26.037.531	715.497	
1758	12.076.097	11.554.314		
1759	33.190.698	30.362.573	963.840	
1760	52.527.239	38.487.953	2.112.869	
1761	54.995.219	36.457.533	7.941.892	1.395.2
1762	27.027.224	11.344.715	8.758.565	
1763	28.990.487	21.058.041	2.978.888	661.8
1764	55.639.627	36.206.613	2.062.619	1.240.1
1765	38.502.491	29.797.982	357.749	
1766	34.846.766	29.583.998	1.120.539	
1767	36.895.757	29.116.817	2.219.181	
1768	37.284.347	25.347.762	1.318.930	512.7

MBER	Ttrad	Tadd	Tem	OTHER
.199.673	760	502.871		6.568.251
.620.537	858.429	1.851.008		12.640.237
.184.878	1.514.921	348.354	5.520	12.335.432
.254.187	3.657.318	2.030.405		19.545.386
.103.781	272.924	466.385		18.078.193
.815.894	3.558.646	505.178		26.820.026
.486.451	211.304	1.045.034		40.241.123
.908.334	2.529.301	420.757		23.036.410
.421.220	2.716.271	951.412		31.219.459
.327.194	647.652	1.162.450		40.533.408
.288.313	40.266	1.376.896		13.494.631
.125.925	949.681	890.971	2.794	27.235.474
.239.022	237.019	252.758		23.088.760
.990.755		3.980.389		36.485.284
.365.224	1.850.010	375.782		49.696.461
.790.173		10.409.820		33.794.838
.367.732	2.887.694	554.441		31.079.078
.554.673	1.077.199	555.399		31.353.930
.465.077	395.248	250.889		23.196.434
.423.195	527.517	1.044.615		21.292.679
.255.660	2.010.545	954.735		18.104.897
.190.140		1.501.561		27.639.518
.844.036	1.646.341	1.126.455		18.206.330
.581.943	633.968	2.143.929		27.565.496
.914.756		11.130.978		45.463.592
3.567.197	5.064	1.634.070		14.471.017
.352.484		382.462		14.479.877
		1.201.847		18.554.732
2.864.417		692.784		14.398.117
.233.716		18.060.960	8.022	28.135.771
.115.662	1.289.796	599.521	1.610	21.595.974
3.137.500	2.532	652.382		16.354.552
3.197.236	2.172.846	10.249.287	1.863	17.950.375
3.875.500	1.149.356	313.068	5.487	20.753.663
3.537.078		133.167		8.915.630
.620.627		1.403.814		13.837.227

APPENDIX 3.1 *Estimates of the production of maritime transport services by shipmasters of group 1 (co*

year	GRAIN	Gtrad	Gadd	Gem
1769	42.077.064	26.232.725	6.237.618	
1770	227.178.581	39.485.069	4.517.713	1.042.66
1771	31.297.959	19.373.967	6.427.216	1.909.60
1772	36.230.541	11.988.741	8.972.825	5.356.22
1773	23.025.835	11.696.535	5.726.718	2.732.81
1774	29.713.725	11.730.718	7.523.474	4.135.39
1775	28.335.241	9.379.190	8.443.764	4.074.37
1776	15.833.222	9.643.880	787.008	3.207.81
1777	33.577.357	19.199.423	2.200.525	4.908.83
1778	36.163.279	22.613.646	5.361.321	2.407.72
1779	23.166.372	12.752.940	1.068.839	1.043.43
1780	23.149.751	11.985.162	4.211.519	500.86
1781				
1782				
1783	7.263.780	5.268.490		953.99
1784	9.573.875	5.069.858	450.792	476.99
1785	14.421.678	4.112.102	547.227	
1786	2.590.155	1.251.967	1.303.689	
1787	801.555	801.555		
1788	2.956.225	1.644.876		
1789	12.764.954	4.318.048	2.499.030	2.588.62
1790	21.680.134	3.028.065	3.966.997	3.979.46
1791	13.071.795	3.687.259		4.198.09
1792	17.428.100	3.030.820	2.154.609	5.430.44
1793	2.503.035	749.209		560.47
1794	7.434.603			
1795				
1796				
1797	29.214			
1798	27.865			
1799	25.006			
1800				
1801				
1802				
1803				
1804				

MBER	Ttrad	Tadd	Tem	OTHER
.246.019		5.577.890	281.391	14.277.984
.613.399		2.141.473		17.048.137
.430.546	55.040	2.590.483	823.057	12.169.813
.879.987		6.948.483	585.760	16.931.315
.902.397		2.276.554	2.054.266	10.492.845
.672.350		2.001.896	1.214.762	18.290.071
.056.778		2.013.910	1.000.959	25.258.578
.676.955		345.905		6.315.494
.284.846	774.718	2.844.132	633.307	15.503.063
.480.744		12.237.428		10.037.994
.528.974	1.291.197	4.926.456	813.799	13.816.794
.309.742		484.451	800.112	13.579.957
.002.356		1.002.356		241.982
.964.423		23.997		1.986.941
.981.847	759.502	1.157.390		8.292.220
.625.299	1.996.528	3.238.572	769.477	42.536.625
.612.639	1.853.956	2.427.471		3.982.282
.871.407	1.884.138	1.185.817		9.506.297
.624.596	1.939.662	1.344.804		12.708.338
.031.259	1.091.631	1.313.315		10.748.594
		238.846.867		5.921.330
.674.343	1.778.792	1.874.336	637.792	4.459.044
.856.387	1.857.867	1.701.398		5.248.845
.536.755	3.559.658	852.879		380.262
.701.645	2.841.055	827.374		3.880.069
16.040				1.604
				705
				237
				3

219

APPENDIX 3.1 *Estimates of the production of maritime transport services by shipmasters of group 1 (co*

year	GRAIN	Gtrad	Gadd	Gem
1805				
1806				
1807				
1808				
1809				
1810				
1811				
1812				
1813				
1814				
1815				
1816	3.013.647	1.848.553		
1817	9.980.527	128.747	2.961.214	1.470.37
1818	3.130.546	200.800	1.051.816	222.59
1819	971.960	374.907		
1820				
1821	719.719	198.073	225.900	
1822	209.828			
1823	318.301			
1824	297.109	297.109		
1825	282.477	282.477		
1826	168.362	168.362		
1828	2.323.020	549.648	399.486	500.29
1829	9.329.970	858.943	476.998	2.071.68
1830	8.212.155	3.230.260	959.959	1.999.94
1831	7.710.936	905.482	846.672	1.121.09
1832	9.991.819	1.302.992	888.409	3.518.06
1833	391.254			
1834	422.660	284.759		
1835				
1836	1.376.136			
1837	1.803.354	158.732	433.276	953.99
1838	4.588.030	370.978		924.18
1839	3.482.128	1.609.389		596.24
1840	3.005.625	1.118.092		517.16
1841	431.134	184.198		

BER	Ttrad	Tadd	Tem	OTHER
124.912		40.278		559.149
465.676	1.278.740			1.679.295
981.278	1.276.386	24.647		1.781.258
437.050	1.285.914			936.568
454.291	1.144.785	76.540		1.067.644
516.579	4.937.021	1.316		3.541.612
				324.128
197.760	3.427.165			2.473.781
055.653	3.220.787	91.954		797.816
817.609	2.042.650	219.680		1.707.277
377.013	1.399.666	159.153		1.975.573
137.509		149.121		4.785.124
663.709	820.378	55.182		3.730.595
879.160	213.672	59.824		755.005
635.655	412.516	185.411		7.017.633
767.375	467.974	41.586		3.748.319
455.924	774.412			2.382.495
240.420	227.664	194.944		2.936.594
950.965	243.779			1.254.962
460.575	710.973	49.586		2.673.289
566.003	959.703	25.014		1.626.110
911.891				2.450.380
909.542	281.523			3.325.311
310.194	178.466			2.626.408
609.458		23.765		5.112.926
646.942				1.762.923
457.565	162.613			1.390.298

APPENDIX 3.1 *Estimates of the production of maritime transport services by shipmasters of group 1 (c*

year	GRAIN	Gtrad	Gadd	Gem
1842	2.905.466	1.004.050		1.901.4
1843	6.765.884	5.020.713		324.3
1844	6.984.332	4.526.144		
1845	1.136.705	813.231		
1846	981.152	709.117		
1847	2.675.015	811.777		652.0
1848	3.122.056	2.655.486		466.5
1849	1.414.635	438.854		

APPENDIX 3.2 *Estimates of the production of maritime transport services by shipmasters of group 2. Based on* STRO

year	GRAIN	Gtrad	Gadd	Gem
1670	16.881.073	10.444.154	1.489.670	431.
1671	21.549.355	17.082.948	1.503.321	
1672	12.397.340	10.465.989	1.206.182	
1673	3.906.627	2.893.821		
1674	14.378.609	8.826.939	1.481.570	
1675	14.780.597	12.474.727	804.934	
1676	10.316.008	5.541.457	1.353.482	
1677	26.520.097	9.296.220	9.165.904	
1678	27.975.618	24.578.914	1.401.056	
1679	23.635.489	15.265.777	3.036.459	
1680	46.246.768	24.246.694	7.831.957	
1681	37.558.839	23.775.258	7.479.238	
1682	42.148.482	19.008.426	10.483.999	
1683	35.292.101	19.649.835	11.334.369	
1684	33.826.012	16.144.969	3.022.640	
1685	15.200.458	9.925.844	1.721.635	
1686	17.592.294	10.067.499	3.272.480	
1687	15.691.564	5.928.772	5.459.650	
1688	20.239.683	6.167.925	6.379.047	
1689	13.204.825	5.727.621	1.020.790	

MBER	Ttrad	Tadd	Tem	OTHER
.786.912	410.217	77.732		3.053.260
.391.960	605.616			3.200.695
762.093	212.187	61.360		1.837.814
.904.943	441.547	970		1.503.967
.533.384		233		900.103
267.738		2.456		2.699.910
.026.012	105.440	6.055		1.947.085
105.676		468		17.342.450

MBER	Ttrad	Tadd	Tem	OTHER
1.459.617		3.917		1.469.050
959.819	1.055	79.716		4.245.917
181.063		2.246		1.587.393
2.420		2.246		1.089.610
61.212		5.115		8.863.731
730.712		673.654		4.432.265
19.154				3.722.593
106.147.517				10.560.310
1.494				15.676.433
15.742		5.614		15.554.187
131.071	34.229	26.644		102.841.125
32.240	439	31.437		8.791.574
21.054	9.228			16.701.124
92.032	80.674			13.714.955
364.201	51	2.794		15.614.937
19.886	6.393	11.552		11.480.495
2.899.703	2.850.656	21.479		13.757.813
252.303	126.396	24.423		22.519.346
3.037.312	236.150	39		11.533.293
199.198	47.158	20.003		8.699.704

APPENDIX 3.2 *Estimates of the production of maritime transport services by shipmasters of group 2 (co*

year	GRAIN	Gtrad	Gadd	Gem
1690	7.422.097	4.023.654	1.496.585	
1691	7.252.012	2.873.830	996.479	
1692				
1693	36.481.958	8.704.783	11.984.182	
1694	45.166.606	7.362.532	31.270.076	
1695	19.157.275	9.484.622	6.216.227	
1696				
1697	6.155.137	5.347.073	562.186	
1698	30.581.730	24.730.997	1.457.708	
1699	29.774.010	17.994.847	8.927.296	
1700	14.505.590	7.530.697	1.293.378	
1701	13.767.761	8.460.896	3.710.079	
1702	28.129.121	26.374.910	416.660	
1703	9.085.669	8.594.806		
1704	18.026.182	16.445.152	860.064	
1705	20.743.397	19.300.495		
1706	119.065	114.045		
1707				
1708	2.252.719	2.016.266		
1709				
1710	13.427.903	12.285.901	838.340	
1711	61.625.909	51.344.890	4.540.211	
1712	69.368.574	59.065.206	7.071.728	
1713	56.249.150	39.987.407	5.407.544	
1714	28.954.846	15.749.987	5.851.729	
1715	13.367.985	1.342.626	1.490.619	
1716	25.818.676	16.188.343	3.344.949	
1717	12.140.040	11.524.149	250.424	
1718	18.755.127	15.745.341	2.147.414	
1719	27.991.293	22.341.865	1.647.446	
1720	13.225.646	10.284.198	251.000	
1721	28.954.563	23.619.152	44.971	
1722	12.226.624	10.073.528	447.186	
1723	64.291.565	51.717.628	1.220.552	
1724	60.415.849	46.533.485	646.453	
1725	58.935.626	50.294.810	1.488.274	

MBER	Ttrad	Tadd	Tem	OTHER
329.660	292.335	1.828		4.581.733
1.053.838	771.506	25.483		3.214.079
70.612.951	184.059	70.093.162		7.502.979
1.019.716	27.913	64.485		10.860.492
132.090	96.118	81		4.848.218
2.289.103	1.575.260	4.869		7.795.651
3.871.219	2.437.387	87.152		9.658.701
7.672.501	5.159.989	35.928		14.795.131
6.252.820	4.913.307	40.204		15.355.913
2.573.300		392.992		4.856.817
28.583		25.854		7.420.634
279		252		8.754.470
42.410		17.136		14.696.580
557.307	22.082	89.578		4.874.114
315.889	19.874			2.650.639
1.239				276.570
140.172		2.266		203.261.058
2.178.532		137		19.513.824
33.854		14.279		22.335.244
241.165		19.073		24.230.728
21.891		1.388		30.079.932
260.438		193.302		23.793.118
19.228		16.930		32.284.429
72.978		27.431		23.741.576
795.011		49.339		20.844.669
345.448	86.671	234.825		27.226.762
1.811.211	1.101.328	497.387		34.090.049
13.099.776	1.161.857	199.218		29.535.526
1.226.768		153.057		25.351.274
449.679	416.009			10.164.733
3.989.472	2.821.568	355.926		32.239.064
446.104	5.064	237.772		23.241.129
28.549.155	23.130.050	1.832.323		31.373.220

APPENDIX 3.2 *Estimates of the production of maritime transport services by shipmasters of group 2 (co*

year	GRAIN	Gtrad	Gadd	Gem
1726	34.708.100	22.736.743	489.292	
1727	14.946.258	14.206.122	379.899	
1728	37.570.635	21.465.353	7.964.613	
1729	102.659.510	71.924.249	13.779.974	
1730	34.598.774	21.438.942	3.159.915	
1731	59.046.051	31.094.702	12.976.587	
1732	16.299.058	10.424.014	2.611.889	
1733	19.913.998	14.929.731	1.513.375	
1734	52.797.720	17.919.038	19.954.150	
1735	37.882.309	20.407.767	10.072.414	
1736	10.412.709	3.373.487	2.878.080	
1737	4.297.112	1.056.389	2.290.563	
1738	33.502.817	10.975.054	10.551.429	
1739	68.522.709	20.690.991	22.575.676	
1740	148.370.919	49.549.253	47.275.319	753.2
1741	110.751.467	28.528.505	34.079.046	1.047.9
1742	35.326.396	22.662.124	2.074.272	371.0
1743	25.387.091	19.790.076	912.024	309.1
1744	19.332.702	16.270.184	775.744	
1745	10.011.951	5.217.187	1.450.357	
1746	9.355.085	6.583.053		
1747	21.794.365	15.400.503	153.110	
1748	21.758.538	12.773.446	3.888.152	
1749	36.029.509	26.944.837	3.167.109	291.1
1750	44.317.967	28.908.599	6.035.478	
1751	72.029.551	45.206.278	7.623.402	688.6
1752	96.817.305	50.747.119	12.660.112	2.068.6
1753	58.419.395	35.041.658	8.582.559	736.3
1754	46.683.659	30.745.001	5.644.315	868.4
1755	25.260.083	17.440.202	2.018.919	465.0
1756	3.032.330	2.535.560		
1757	16.568.082	14.166.571	918.437	
1758	13.821.334	10.437.581	456.820	
1759	13.849.613	11.373.035	673.354	
1760	45.284.051	24.475.402	3.402.901	269.8
1761	64.180.076	33.394.315	14.808.426	971.8

BER	Ttrad	Tadd	Tem	OTHER
3.432.083	2.135.423	113.511		35.491.158
19.530.203	16.005.815	1.595.176		55.466.869
28.472.049	25.959.984	282.126		56.665.188
14.988.084	13.309.971	98.940		62.046.156
3.415.397	11.759	3.115.379		32.084.870
21.972.054	18.089.340	1.199.057		59.976.619
1.516.386	1.078.970	79.167		23.094.514
1.094.759	96.056	590.086		23.865.036
2.931.574	1.617.259	204.027		62.389.387
5.447.153	2.864.371	922.610		57.324.390
19.163.167	16.266.531	1.560.821		69.453.348
1.566.124	430.662	41.191		39.199.874
12.155.215	7.226.148	1.206.044	121.726	68.315.316
23.788.961	322.061.827	1.058.853		65.596.136
6.012.234	4.061.630	336.862		70.511.410
8.456.812	4.087.685	1.130.729		90.765.781
56.882.530	4.232.280	612.200	116	90.696.332
12.057.844	106.205.001	1.144.526		94.528.885
44.431.134	340.362.924	704.342		88.900.949
15.438.038	529.628.338	215.396		83.689.615
22.358.469	9.126.598	114.323		64.827.316
84.482.259	376.409.215	942.953		54.008.721
33.441.423	16.219.775	1.272.704		45.912.508
24.758.638	15.824.373	381.212	127.853	60.835.804
32.786.252	115.071.516	538.045		28.400.503
57.965.383	490.026.600	240.727	30	33.995.311
40.759.593	12.582.743	1.078.571	1.093.710	59.510.494
41.101.704	24.468.240	2.798.556		53.608.283
40.857.640	310.632.105	3.513.458		59.234.345
34.944.152	12.750.098	1.617.089		43.943.837
63.419.971	413.006.385	2.195.909		45.292.645
61.940.383	30.735.279	7.650.537		62.896.112
58.648.447	2.843.456	19.157.971		35.212.904
88.345.507	45.944.767	92.700.095		51.031.447
06.889.856	685.408.549	2.578.671		44.027.633
49.492.082	15.118.257	1.421.801		30.314.095

APPENDIX 3.2 *Estimates of the production of maritime transport services by shipmasters of group 2* (c

year	GRAIN	Gtrad	Gadd	Gem
1762	40.456.379	9.012.639	7.345.886	449.
1763	38.565.558	22.338.827	2.733.601	909.
1764	45.292.571	28.849.686	2.236.514	4.063.
1765	36.506.702	17.728.082	1.372.526	
1766	42.208.377	29.648.378	3.397.308	441.
1767	53.927.104	37.021.771	4.583.412	1.714.
1768	54.680.251	20.750.549	8.660.459	
1769	87.072.764	29.143.560	7.077.931	524.
1770	119.086.665	53.329.322	11.078.179	2.050.
1771	71.378.641	12.354.408	18.771.130	1.031.
1772	61.240.269	19.146.363	12.019.510	3.879.
1773	57.877.746	17.964.313	15.594.292	5.675.
1774	97.444.492	13.841.677	35.181.453	3.380.
1775	100.543.123	15.585.657	31.641.174	10.972.
1776	35.929.380	7.426.204	6.835.939	4.307.
1777	78.743.370	12.844.466	10.436.264	8.409.
1778	71.415.390	7.435.652	22.137.335	5.835.
1779	57.818.240	4.471.918	11.987.534	7.533.
1780	48.351.133	5.956.196	11.784.061	3.648.
1781	1.964.264	1.584.644		
1782				
1783	19.290.186	1.826.655	4.986.550	2.460.
1784	35.654.658	4.297.308	4.123.841	6.833.
1785	35.516.169	7.546.373	2.819.550	5.348.
1786	9.012.900	2.360.152	1.968.171	292.
1787	3.094.847	301.200	890.586	331.
1788	11.237.323	2.176.194	993.516	
1789	27.385.958	3.374.717	11.721.638	2.007.
1790	31.312.604	4.146.901	8.320.583	9.624.
1791	34.347.443	6.587.586	5.619.204	8.565.
1792	45.375.756	7.688.953	15.340.737	10.131.
1793	19.481.169	2.529.426	3.718.643	5.497.
1794	24.651.949	1.539.501	3.297.452	3.949.
1795	314.795		314.795	
1796	675			
1797				

MBER	Ttrad	Tadd	Tem	OTHER
102.894.542	21.526.036	40.910.235		77.205.176
89.416.157	25.457.347	888.115		44.114.180
45.447.616	36.169.405	1.055.541		34.104.244
256.688.197	194.999.313	2.351.716	7.555.668	37.258.867
66.747.951	15.866.932	1.151.914	20.839	48.584.936
38.953.282	6.583.991	3.405.420		64.743.108
101.659.624	44.637.142	5.841.905	2.123.855	49.805.054
235.983.012	165.766.633	10.924.273	1.747.214	111.669.396
80.871.189	27.952.234	5.057.637	1.862.758	88.389.542
79.124.138	20.236.769	7.529.950	1.527.371	44.339.234
141.598.074	36.258.585	16.636.115	2.639.074	60.690.119
360.258.724	41.263.844	12.844.084	508.221	60.974.371
144.351.334	42.788.045	24.485.682	3.376.819	64.877.147
103.278.678	36.871.260	13.821.988	3.655.383	75.246.250
143.474.316	26.994.465	65.909.111	1.968.185	14.607.961
811.254.161	179.642.950	25.043.148	9.822.906	74.888.636
140.669.452	45.061.886	20.793.224	7.093.864	74.476.116
129.673.886	36.886.823	21.429.873	6.094.864	44.086.602
71.193.208	17.631.926	30.259.709	1.912.099	46.369.431
10.391.719	9.889.761	501.958		190.765
1.648.121	772.566	875.555		699
91.821.794	18.965.313	12.450.641	1.495.819	11.615.088
65.565.866	24.372.578	1.440.281	2.054.598	40.692.355
96.126.943	51.606.045	9.553.773	1.733.110	75.603.295
99.892.705	53.621.494	17.974.679	2.155.092	55.590.862
71.398.592	41.943.173	7.012.382		31.843.881
87.607.854	28.872.109	2.842.403	2.398.411	48.924.290
219.513.314	46.847.062	104.102.748		63.400.097
87.597.336	36.663.302	1.647.795	2.112.575	55.067.433
72.594.683	48.456.092	2.350.947	658.561	34.792.845
84.835.181	43.615.712	4.495.885		35.172.831
32.149.164	22.519.012	3.135.239		17.942.166
65.744.391	48.358.533	4.313.563	968.246	17.924.489

APPENDIX 3.2 *Estimates of the production of maritime transport services by shipmasters of group 2 (co*

year	GRAIN	Gtrad	Gadd	Gem
1798				
1799				
1800				
1801				
1802				
1803	1.824.450	138.050	1.686.400	
1804				
1805				
1806				
1807				
1808				
1809				
1810				
1811				
1812				
1813				
1814	3.540.159		2.153.210	483.4
1815	5.857.819	982.665	417.373	1.719.2
1816	9.074.439	4.018.583	860.642	1.589.3
1817	32.878.903	4.959.022	13.224.929	3.139.2
1818	5.949.719	1.415.263	206.699	2.022.3
1819	1.031.395	272.321		
1820				
1821	4.390.684	3.425.873		
1822	1.209.061	630.824		
1823	693.398			
1824	123.243	123.243		
1825	1.817.107		742.016	
1826				
1827				
1828	5.640.887	2.642.820	77.512	1.026.3
1829	9.358.348	2.668.366	516.349	1.946.1
1830	12.689.443	2.374.962	3.782.623	1.198.2
1831	19.126.404	1.057.045	4.261.981	230.4
1832	12.501.393	1.625.059	1.907.275	4.367.7
1833	3.073.007	814.770		691.6

MBER	Ttrad	Tadd	Tem	OTHER
				797.743
429.284		268.249		2.074.938
6.614.633	4.851.216	408.890		13.263.555
17.605.526	13.459.775	1.231.876		3.143.200
8.731.818	6.499.483	567.947		7.783.140
8.663.856	7.961.505	32.802		5.194.676
14.487.891	12.032.364	1.068.679	1.103.842	5.840.378
6.483.414	4.504.386	394.779		5.347.744
3.061.777	2.383.574	138.827		1.087.181
2.010.756	1.353.444	448.513		4.437.322
2.320.211	1.705.690	157.137		1.253.954
4.630.121	2.430.530	503.704		4.903.496
3.840.098	1.837.158	409.817	209.096	5.735.660
1.819.447	550.387	52.320		3.914.940
1.249.413	260.235	147.246	253.117	12.766.769
1.887.096	1.592.145	82.332		9.919.385
1.486.203	1.129.080	119.154		3.596.842
820.764	280.190	443.002		8.029.637
177.910		158.812		3.410.992

APPENDIX 3.2 *Estimates of the production of maritime transport services by shipmasters of group 2 (co*

year	GRAIN	Gtrad	Gadd	Gem
1834	1.458.614	633.713		572.3
1835	517.074	517.074		
1836	3.323.549	3.323.549		
1837	3.266.479	1.735.075		1.531.3
1838	6.165.270	2.905.967	791.028	2.247.9
1839	12.528.125	2.478.429	4.232.794	1.437.5
1840	6.763.121	4.313.740		
1841	2.008.998	778.578		809.4
1842	2.648.484	1.214.110	316.260	
1843	11.702.538	5.175.638		4.024.8
1844	7.989.216	6.512.724		482.0
1845	2.815.095	850.856		
1846	1.640.911	432.399		
1847	5.766.757	2.175.906	27.769	432.8
1848	4.588.989	2.313.939		1.023.0
1849	4.802.590	1.111.296		907.8

APPENDIX 3.3 *Estimates of the production of maritime transport services by a selection of maritime transport communities, located in* ZDZ. *Based on* STRO

year	GRAIN	Gtrad	Gadd	G
1670	16.693.748	9.599.331	1.401.218	
1671	8.414.744	6.616.338	505.920	
1672	1.524.373	799.802		
1673	5.524.703	3.529.111		
1674	9.812.351	7.553.575	844.706	
1675	2.806.703	1.958.091		
1676	5.547.102	1.276.618	59.625	
1677	9.325.264	7.419.837	837.476	
1678	7.219.301	4.737.818	502.000	
1679	15.986.887	14.418.990	214.649	
1680	20.947.315	15.433.476	1.448.890	
1681	17.301.798	13.758.630	1.114.983	

BER	Ttrad	Tadd	Tem	OTHER
2.357.188	1.622.692	302.339	237.774	3.526.648
1.851.702	1.380.707	259.234		2.224.743
3.332.546	2.298.555	346.638		5.363.383
4.427.202	2.519.348	672.555	325.417	5.546.352
2.312.984	1.105.026	548.783		6.497.615
60.948.406	59.778.905	165.183		2.179.394
1.066.538		336.210		5.704.140
917.028	150.640	54.183		4.110.379
11.245.042	787.674	9.677.652		5.455.112
2.529.455	1.799.042	54.566		4.138.780
1.385.022	468.477	150.084		1.960.270
1.230.445	865.435	38.082	203.022	921.029
2.491.224	1.547.347	470.752		810.513
3.972				1.204.199
1.368.354	610.642	453.714	237.578	2.127.710
51.724		4.607		2.909.893

TIMBER	Ttrad	Tadd	Tem	OTHER
				1.596.174
32.873				3.918.001
2.235		2.235		21.706
				597.082
2.246		2.246		448.210
				1.328.516
				2.059.824
				2.845.923
				1.014.818
3.368		3.368		3.018.431
5.126		2.246		6.344.720
				3.717.532

APPENDIX 3.3 *Estimates of the production, located in ZDZ (cont.)*

year	GRAIN	Gtrad	Gadd	Gem
1682	12.801.740	7.323.764	1.580.688	
1683	12.722.640	8.023.846	2.730.007	
1684	19.152.120	12.517.089	1.703.689	
1685	11.158.936	7.172.360	453.148	
1686	17.252.450	14.175.730	840.709	
1687	10.897.518	6.682.323	826.729	
1688	16.562.653	11.600.164	2.833.067	
1689	7.631.248	4.897.586	455.674	
1690	6.835.959	4.416.380	56.213	
1691	3.678.493	1.748.787	155.024	
1693	17.807.813	6.968.576	3.762.645	
1694	16.946.346	5.267.015	5.711.932	
1695	9.224.918	3.952.111	4.537.464	
1696				
1697	6.498.866	5.827.407		
1698	17.297.045	10.962.196	3.972.188	
1699	17.631.718	12.799.918	1.711.514	
1700	8.908.649	8.073.509	139.130	
1701	3.876.377	2.094.611	1.462.789	
1702	11.288.066	9.863.519	1.424.547	
1703	2.222.838	2.091.663	131.174	
1704	3.869.647	2.935.592		
1705	6.564.864	6.019.595	545.269	
1706				
1707				
1708				
1709				
1710	361.440	361.440		
1711	3.601.944	2.926.659		
1712	4.596.633	4.596.633		
1713	8.306.681	5.880.786		
1714	2.724.754	1.612.772	997.069	
1715	581.614			
1716				
1717	2.027.280	1.755.246		
1718	50.200	50.200		

MBER	Ttrad	Tadd	Tem	OTHER
3.368		3.368		1.559.596
				524.164
				4.835.832
				961.647
				2.737.090
793.999	786.596			3.501.537
10.689				4.127.733
2.327		2.327		2.321.930
				551.177
11.148				918.280
54.230	54.230			4.948.056
2.794		2.794		1.564.529
				3.116.045
95.052				2.945.092
				3.148.946
1.818.776	175.603	16.764		4.661.296
388.655				3.043.338
260.227	43.040	217.187		2.250.297
36.160		33.683		2.713.556
				2.836.660
14.740				1.387.971
				3.778.182
				2.760.094
23.796		22.455		1.545.917
				694.237
				1.052.549
				2.086.365
8		8		838.780
				407.835
				989.177
				616.688
				2.261.821

APPENDIX 3.3 *Estimates of the production, located in* ZDZ *(cont.)*

year	GRAIN	Gtrad	Gadd	Gem
1719	1.533.894	1.533.894		
1720	1.619.185	1.542.558		
1721	431.720	431.720		
1722	289.235	289.235		
1723	5.327.889	4.696.587		
1724	9.492.863	7.848.138	502.000	
1725	8.539.771	6.744.896	120.480	
1726	7.285.643	4.315.316		
1727	2.662.057	2.501.417	160.640	
1728	5.486.779	4.510.834	170.680	
1729	15.905.027	12.607.669	1.977.879	
1730	5.396.283	4.176.993	420.195	
1731	10.908.066	6.759.189	2.577.012	
1732	3.490.973	2.298.778	494.677	
1733	4.102.086	3.650.803	283.877	
1734	7.808.252	5.013.219	2.742.423	
1735	7.461.089	3.952.238	1.635.549	
1736	1.236.643	731.272	219.232	
1737	615.390	615.390		
1738	8.387.762	5.684.291	477.813	
1739	14.649.715	11.337.481	2.101.391	
1740	36.051.465	23.808.764	6.356.538	
1741	23.712.299	18.885.983	3.396.103	216.4
1742	8.932.198	7.925.926		
1743	10.110.296	5.416.411	630.244	
1744	13.973.719	13.739.104		
1745	2.107.989	1.682.336	224.853	
1746	3.427.408	2.878.491		
1747	8.699.892	8.699.892		
1748	7.106.814	5.579.104	1.311.850	
1749	16.332.706	13.241.485	1.741.939	700.7
1750	29.449.557	22.349.340	5.887.535	
1751	44.990.399	31.885.133	2.421.415	476.9
1752	44.130.299	25.534.227	411.640	754.6
1753	27.467.893	17.722.576	1.154.600	751.2
1754	21.539.926	16.121.346	2.030.039	

MBER	Ttrad	Tadd	Tem	OTHER
101.842		93.189		3.070.430
165.125		152.976		2.147.526
484.629	484.629			3.375.923
44.910		44.910		167.037
14.287	5.570	3.143		1.834.650
21.148		112		944.464
1.516.120	1.515.445			1.275.066
673.654		673.654		2.284.702
40.571		39.558		865.488
10.135		468		1.183.548
152.391		33.308		4.569.986
				1.910.931
116.911		114.147		3.677.250
137.275				907.315
31.616				2.148.181
23.895		2.246		3.157.757
86.823		79.248		3.983.000
69.710	19.748	49.963		1.901.284
41.542		41.542		1.289.414
1.860.911	1.588.788	233.809		14.041.597
0.949.061	140.822.265	78.401		5.049.084
189.349	97.212	92.137		5.433.861
154.086		145.157		5.258.061
135.003		128.825		24.300.389
283.551		49.976		12.827.513
385.082		158.997		17.527.635
202.445		136.205		12.727.217
0.015.869		9.550.515		28.222.132
1.148.889		823.524		21.662.418
2.490.895	1.536.532	533.854		10.879.743
333.445		158.575		13.677.994
1.801.105		120.681		17.647.296
2.358.619		50.501		15.595.953
938.903	50.635	173.485		13.603.578
1.234.983	30.381	511.096		12.194.372
6.260.865	2.025	1.185.644		32.049.651

APPENDIX 3.3 *Estimates of the production, located in* ZDZ *(cont.)*

year	GRAIN	Gtrad	Gadd	Gem
1755	11.682.895	7.713.830	2.132.904	566.4
1756	3.214.956	2.176.763	590.685	
1757	5.729.275	4.293.556		
1758	9.190.605	7.796.298	491.960	
1759	10.079.881	8.413.526	1.174.680	
1760	23.711.810	8.121.865	3.412.324	1.026.7
1761	25.842.430	9.816.804	7.832.556	2.428.7
1762	11.326.712	3.752.090	1.023.546	578.3
1763	30.930.446	21.542.994	1.559.034	
1764	31.673.738	22.547.464	2.929.289	
1765	18.505.450	14.039.883	813.878	
1766	31.932.052	23.680.578	275.445	
1767	34.386.067	22.516.351	4.864.641	1.776.3
1768	33.592.027	10.158.065	2.862.223	1.762.4
1769	46.385.382	8.667.897	4.955.429	2.193.3
1770	135.267.201	30.832.362	7.760.171	573.3
1771	35.574.219	13.581.690	10.535.654	4.321.9
1772	22.901.895	9.614.258	6.265.018	3.107.0
1773	24.641.535	12.418.848	5.627.217	2.615.3
1774	32.481.520	11.722.044	9.473.597	2.898.4
1775	35.833.840	6.183.479	14.439.365	5.467.7
1776	12.775.935	7.050.301	1.251.409	1.109.0
1777	37.791.495	13.517.383	9.057.294	6.399.6
1778	36.259.919	6.503.121	15.252.113	7.435.4
1779	27.945.266	3.685.971	4.495.200	9.420.7
1780	27.383.557	8.122.397	9.667.512	4.879.7
1781				
1782				
1783	2.989.469		1.364.187	1.045.5
1784	14.916.321		4.481.314	4.234.8
1785	27.050.551	6.106.066	968.571	5.022.9
1786	4.892.920	1.166.908	1.907.770	
1787	2.272.124	511.114	702.800	31.8
1788	4.871.843	2.225.015	471.880	
1789	16.733.669	5.091.747	2.176.328	4.746.6
1790	31.778.394	1.550.343	13.037.775	11.416.3

BER	Ttrad	Tadd	Tem	OTHER
.084.527		1.255.799		18.422.236
.007.849		1.173.923		16.350.721
.574.766		2.368.787	49.477	14.375.546
.909.768		944.496		12.845.344
.391.529		281.285		14.854.350
.206.791		777.314		11.121.463
734.079		71.460		15.685.301
700.147		255.153		16.394.747
5.895.672		2.200.357	931	21.713.799
2.064.251		1.037.205		14.090.920
3.588.153		1.575.430		17.398.128
1.652.189		117.260	2.251	15.745.195
1.724.789		151.062		13.614.570
5.383.363		488.542		22.366.466
3.954.264		461.480		16.242.033
7.613.864		6.341.722		15.271.879
.185.051		221.738		17.536.205
3.868.962		610.546		12.662.491
6.332.141		181.038		18.367.101
5.185.442		422.751		25.101.128
2.873.058		495.272		28.695.001
2.018.492				18.239.759
4.429.206		28.406		15.165.341
2.861.000		61.995		20.368.095
1.515.028		200.644		16.506.591
303.356		20.699		15.862.454
473.412		102.912		596.614
2.460.375		84.662		9.250.681
5.768.724	3.996.027	511.521		16.754.165
4.853.360	980.660	668.541		12.734.658
6.355.844		763.754		13.541.157
4.277.398		492.057		17.153.816
4.375.497	783.326	89.922		26.895.467
5.954.878		345.450		17.832.286

APPENDIX 3.3 *Estimates of the production, located in* ZDZ *(cont.)*

year	GRAIN	Gtrad	Gadd	Gem
1791	35.244.809	4.070.158	1.436.813	9.040.
1792	51.785.610	9.469.466	9.202.809	15.307.
1793	16.898.913	1.196.109	6.131.757	5.726.
1794	11.708.449	2.433.474	1.516.522	2.005.
1795				
1796	318.702			
1797				
1798				
1799				
1800				
1801				
1802				
1803	964.022	148.396		505.
1804				
1805				
1806				
1807				
1808				
1809				
1810				
1811				
1812				
1813				
1814	2.212.924		347.811	1.715.
1815	831.925		310.049	
1816	3.637.168		763.755	907.
1817	12.324.811	2.994.882	2.899.262	3.093.
1818	3.235.654	1.759.509	800.959	
1819	990.548			812.
1821	399.486			399.
1822	675.969			
1823	159.658			
1824				
1825				
1826	243.422	122.145		
1827				

MBER	Ttrad	Tadd	Tem	OTHER
5.974.517				15.869.413
4.150.633		489.881		17.176.780
2.858.499	705.512	844.753		8.326.595
2.154.219	812.278	761.761	231.023	11.897.626
				90.634
128.072				10.543
410.160	230.052			1.071.852
435.316	265.518			3.232.690
1.115.212	801.867			1.291.325
334.778		191.734		846.414
819.726	795.934			788.852
1.549.003	1.168.972	233.091		1.684.771
545.403	431.804	113.599		1.060.965
923.630	727.307	49.027		766.325
954.297	297.300	147.263		62.439
1.294.650	784.533	109.362		309
740.227	390.338	128.957		119.046
938.461	192.996	83.354		355.248

APPENDIX 3.3 *Estimates of the production, located in* ZDZ *(cont.)*

year	GRAIN	Gtrad	Gadd	Gem
1828	2.801.362		321.280	1.091.1
1829	4.153.469	444.030		2.929.0
1830	5.061.254			1.484.5
1831	3.685.112	436.740	457.937	327.9
1832	5.504.352	1.505.999		1.873.0
1833	1.174.916	425.445		
1834	464.922			
1835				
1836	833.209			
1837	1.016.961			
1838	878.163		369.674	
1839	2.223.916	231.085		725.0
1840	3.525.067	287.640		697.6
1841	768.537	569.401		
1842	459.562	282.040		
1843	842.580	293.670		
1844	1.210.871	1.042.019		
1845	154.460			
1846	82.541			
1847	447.999			
1848	1.047.729	276.890	272.860	
1849	648.652	323.834		

APPENDIX 3.4 *Estimates of the production of maritime transport services by a selection of maritime transport communities, located in* SMR*. Based on* STRO*.*

year	GRAIN	Gtrad	Gadd	Gem
1670	3.486.738	419.390		
1671	1.700.696	1.700.696		
1672	1.652.779	1.611.743		
1673	788.490	788.490		
1674	6.472.609	6.069.730		
1675				

MBER	Ttrad	Tadd	Tem	OTHER
300.969	244.969			795.566
1.178.044			501.453	1.695.347
4.218.573	138.572	167.429		2.257.870
94.145				7.546
817.036	506.546	109.665	137.744	1.571.433
37.907	15.444	22.435		862.164
416.234	226.367	1.847		857.968
60.762	60.762			179.116
221.167	219.223			89.450
174.031				260.207
273.034	272.160			186.303
172.538				76.066
68.095				332.219
191.125				73.426
350.294		135.765		127.046
444.425	288.798			840.633
31.828				147
520.227	516.479	3.749		
374.270	263.835	17.752		190.729
139.352				104.832
211.747		10.252		317.669
30.930				111.942

MBER	Ttrad	Tadd	Tem	OTHER
				43.910
673.654		673.654		281.582
				121.877
				465.586
51.670		51.647		228.269

APPENDIX 3.4 *Estimates of the production, located in* SMR *(cont.)*

year	GRAIN	Gtrad	Gadd	Gem
1676	178.874		178.874	
1677	799.947	565.427		
1678	762.024	332.620	417.373	
1679	3.505.591	1.941.179	787.047	
1680	1.178.249	1.070.300		
1681	8.304.198	2.924.251	2.815.297	
1682	2.472.387	1.526.079		
1683	6.023.371	3.173.561	864.559	1.264.0
1684	8.378.893	3.761.530	956.069	
1685	5.540.807	1.036.778	193.270	
1686	4.393.666	2.183.721		
1687	1.869.719	943.760		
1688	4.535.671	3.404.433		
1689	6.619.956	824.318		
1690	203.025	203.025		
1691	1.947.026	860.170		
1692				
1693	5.510.865	963.840	3.471.621	
1694	5.320.929	657.620	995.734	1.667.1
1695	2.101.601	1.460.634	640.966	
1696				
1697				
1698	1.112.852	94.001	200.800	
1699	7.527.216	2.829.473	3.466.584	
1700	1.792.611	1.266.249		
1701	2.055.689	1.829.789		
1702	2.133.209	2.133.209		
1703	817.088	817.088		
1704	2.341.387	2.341.387		
1705				
1706				
1707				
1708				
1709				
1710	3.407.715	3.407.715		
1711	3.614.584	3.614.584		

BER	Ttrad	Tadd	Tem	OTHER
				596.558
				1.267.702
				268.460
				201.250
				436.199
				680.969
				332.450
				512.411
				940.303
				116.357
				448.499
61.981	8.439			390.716
				1.200.092
				699.889
				140.216
31.647				149.468
				631.521
				229.100
				410.460
101.175				1.089.230
34.894				649.496
				238.840
				1.719.043
				5.729
				313.004
				21.128
815		815		194.510
				44.574
				470.709

APPENDIX 3.4 *Estimates of the production, located in* SMR *(cont.)*

year	GRAIN	Gtrad	Gadd	Gem
1712	673.980	673.980		
1713	2.910.557	1.179.577		
1714	855.135			
1715				
1716	183.100			
1717	100.400	100.400		
1718	2.235.283	1.815.893		
1719				
1720				
1721	1.346.690	342.690		
1722				
1723	1.044.235	629.085		
1724	1.171.333	1.171.333		
1725	2.449.699	1.684.689		
1726	3.376.252	470.006	287.082	
1727	607.393	607.393		
1728				
1729	9.333.416	7.390.228		
1730	1.288.040	1.288.040		
1731	6.072.470	743.530	175.700	
1732	374.197			
1733	1.240.693	361.440		
1734	914.500	686.932		
1735	2.453.375	1.612.483	84.320	
1736	569.625		233.285	125.5
1737				
1738	2.070.717		64.645	
1739	6.809.860	2.023.194	537.140	1.744.7
1740	30.541.117	7.355.096	4.952.480	4.574.8
1741	9.802.031	2.720.066	1.373.472	3.128.4
1742	1.870.802	640.436	446.896	
1743	2.595.428	802.084		376.6
1744	1.246.779			
1745				
1746				
1747	1.034.013	274.773		

MBER	Ttrad	Tadd	Tem	OTHER
				1.664.732
4.955				2.338.328
				955.753
				402.613
				1.548.773
				194.218
78.978	78.666	312		349.562
25.823		25.823		296.230
				65.666
8.503				174.641
				400.028
18.607				506.292
879.168		40.968	838.200	1.051.108
14.965		448		1.430.292
75.297		7.495		112.059
.764.982	412.601	678.727		365.884
656.178		163		873.066
				531.336
170.818	102.005	20		461.047
167.096				187.649
183.496				321.335
364.117				612.589
95.404		55.389		552.653
705.997		704.998		285.573
71.867		52.208		1.370.110
319.681	114.012		3.725	630.378
108.739				1.100.246
280.240				3.632.500
228.063		163.227		1.137.498
552.886	113.749	9.347	116	6.184.781
388.346		342.463	175	1.604.231
58.601				532.535
150.576			931	374.030
8.524				786.007
100.431				1.778.975

APPENDIX 3.4 *Estimates of the production, located in* SMR *(cont.)*

year	GRAIN	Gtrad	Gadd	Gem
1748				
1749	2.357.665	346.628		1.084.9
1750	9.636.843	4.153.289		907.1
1751	21.688.504	6.363.981		1.204.1
1752	19.727.854	3.952.482	258.581	1.989.6
1753	16.228.833	4.048.964		3.620.1
1754	4.431.756	150.600		940.9
1755	520.935			
1756				
1757				
1758	1.032.007			
1759	601.969	601.969		
1760	4.622.987	1.217.855		252.9
1761	8.064.617	3.092.360	687.672	2.384.9
1762	8.393.392		655.872	2.578.6
1763	8.433.441	1.677.605		703.5
1764	5.970.696	1.028.379		911.0
1765	3.007.105	627.738		
1766	11.263.766	2.037.396	502.000	2.092.5
1767	1.755.107			
1768	11.682.434	1.609.745	476.998	1.139.9
1769	9.896.208	339.851	387.561	2.762.1
1770	74.047.952	3.440.874	12.720	59.788.9
1771	8.678.296		596.248	5.738.5
1772	24.272.316	1.025.979	1.258.083	20.403.1
1773	9.208.552	170.680	677.741	5.236.7
1774	21.367.255	783.120	566.435	10.860.5
1775	22.865.879	573.550	2.467.842	13.822.7
1776	3.940.207		562.133	2.914.3
1777	11.477.459	1.047.291	453.152	5.728.6
1778	23.390.222	502.000	200.800	12.260.5
1779	26.605.415	524.403	799.555	7.076.5
1780	33.180.010	864.271		10.646.8
1781				
1782				
1783	9.768.989		745.310	2.287.7

MBER	Ttrad	Tadd	Tem	OTHER
869.801				3.375.729
.107.275		6.985		1.268.083
.919.635	343.863	6.928	2.537.753	2.858.564
.849.417		16.349		3.879.938
.759.386	987.055	10.198	1.516.106	3.219.813
.106.401		15.835	1.244.087	3.922.498
297.678		23.391		1.837.838
99.622			1.723	839.523
442.912				1.833.683
963.028		1.863	861.297	959.132
520.271			318.237	2.614.010
.348.112	829.391	24.791	368.110	589.022
21.729				1.262.709
595.535			110.829	4.075.252
245.144		4.678		4.554.647
.583.596		835.967	2.971.128	584.050
.714.372		873		6.146.561
.145.392		21.519	1.148.923	2.471.205
.472.840			2.745.328	2.572.602
.916.346		4.678	6.791	4.971.380
.705.245	870.080	337.056	2.956.277	16.087.623
.371.656			7.916.290	4.271.746
.634.746	1.217.269	1.397	2.295.606	5.751.331
.942.447		1.316.689	3.210.299	3.347.905
.143.521			5.722.995	4.992.561
.014.847		106.002	5.405.288	3.708.444
.149.741			6.907.087	3.153.086
.882.190		749.631	1.990.421	1.352.986
.665.995			1.649.010	3.938.775
.256.052	688.638		6.201.599	7.145.357
.141.192			4.974.230	3.051.075
.907.320			9.514.753	1.616.029
.147.898		333.650	808.118	5.990.973
.873.576		237.607	1.810.511	2.199.143

APPENDIX 3.4 *Estimates of the production, located in* SMR *(cont.)*

year	GRAIN	Gtrad	Gadd	Gem
1784	15.470.289		512.773	7.201.1
1785	10.610.748	755.510	423.336	3.098.5
1786	575.416		170.680	404.7
1787	2.105.542		150.600	525.5
1788	1.483.103	30.120		
1789	6.635.157		1.179.700	1.650.4
1790	7.776.826	301.200	210.840	5.799.1
1791	7.432.201			3.166.5
1792	13.996.888	1.232.309		8.166.3
1793	6.235.526	723.086		4.693.1
1794	9.780.964	346.628		1.007.5
1795				
1796				
1797				
1798				
1799				
1800				
1801				
1802				
1803	466.571			466.5
1804				
1805				
1806				
1807				
1808				
1809				
1810				
1811				
1812				
1813				
1814	1.787.532		770.122	702.6
1815	1.194.533			1.194.5
1816	3.415.804	399.418	183.230	1.676.2
1817	18.501.299	256.020	488.923	3.873.9
1818	3.208.027		930.146	1.304.1

BER	Ttrad	Tadd	Tem	OTHER
341.435			9.140.926	1.128.187
354.351			7.607.572	975.312
805.123	807.251		11.576.432	1.826.552
313.408			7.504.433	2.834.891
864.907			10.832.203	4.879.447
206.683		49.929	9.977.598	1.268.282
557.383		25.936	6.971.844	2.041.148
411.834			6.024.897	2.519.193
828.416	894.974	4.491	10.250.474	1.173.730
028.358			11.023.240	1.873.127
356.336			4.966.573	675.084
182.296				803.631
				1.265.706
286.785		153.754		2.032.523
166.722		85.892		2.630.499
194.968		47.504		255.498

APPENDIX 3.4 *Estimates of the production, located in* SMR *(cont.)*

year	GRAIN	Gtrad	Gadd	Gem
1819	3.644.546	125.500		2.646.[illegible]
1820				
1821	923.190	130.520		
1822				
1823	259.188			
1824				
1825				
1826	1.050.538			1.050.[illegible]
1827				
1828	4.569.096	730.410		2.330.[illegible]
1829	4.141.749		333.899	2.004.[illegible]
1830	3.083.354	658.081		550.[illegible]
1831	9.385.054	592.360	301.200	2.228.[illegible]
1832	12.293.909	233.430	89.437	5.548.[illegible]
1833	427.450			149.[illegible]
1834	1.210.065			
1835				
1836				
1837	3.012.915			2.008.[illegible]
1838	2.796.629	222.135		793.[illegible]
1839	13.620.051	251.000		10.996.[illegible]
1840	3.630.383	474.946		1.767.[illegible]
1841	422.158			413.[illegible]
1842	356.814			308.[illegible]
1843	4.872.008	303.166		2.939.[illegible]
1844	2.971.720	2.002.939		621.[illegible]
1845	467.006	467.006		
1846	1.340.987	210.865		
1847	220.291			
1848	637.255			
1849	1.943.849			938.[illegible]

MBER	Ttrad	Tadd	Tem	OTHER
631.537		105.675	308.786	693.299
822.799	302.967			10.067.747
52.012			69.103	171.427
98.048				366.162
.002.498	435.416	87.372	16.764	2.176.340
763.533			655.188	969.035
161.174				1.659.424
219.117		35.963		1.220.788
197.905				4.727.884
.140.861	261.868		296.222	2.769.462
873.727		68.563	653.962	3.004.201
356.435	355.138			1.314.276
454.052			199.511	271.150
465.841	271.115		194.726	395.521
635.440	29.493		177.355	2.718.510
804.966		29.195	576.551	171.374
.214.564			938.309	1.959.352
215.949			213.453	2.294.214
.330.375			902.449	1.538.281
668.117		35.554	494.502	824.130
486.334			477.371	690.398
64.128				1.346.512
.025.634	196.262	1.048	617.272	1.340.883
735.215			593.786	43.314
392.064		50.474		683.789
13.947				801.318
787.659			474.652	1.703.791
194.846		3.648	105.809	270.371

APPENDIX 3.5 *Estimates of the production of maritime transport services by a selection of maritime transport communities, located in* LDE. *Based on* STRO

year	GRAIN	Gtrad	Gadd	Gem
1676				
1677				
1678				
1679	209.695	209.695		
1680	241.579			
1681	596.235	596.235		
1682	1.060.189		715.497	
1683	1.968.558	506.160	1.228.270	
1684				
1685	453.678		453.678	
1686	4.567			
1687				
1688	849.653			
1689	518.735		518.735	
1690	69.326	69.326		
1691	918.648	578.787	339.861	
1692				
1693	1.168.645		1.168.645	
1694	566.435		566.435	
1695	290.174		290.174	
1696				
1697				
1698	1.560.110	325.389	620.098	
1699	2.211.005	332.620	541.394	
1700	991.424	788.959		
1701	781.850	441.760	340.091	
1702	1.131.630	296.465		
1703				
1704	90.360	90.360		
1705	398.328	267.542		
1712	181.718	181.718		
1713	460.513	460.513		
1715	25.100	25.100		
1716	357.049			
1717				

MBER	Ttrad	Tadd	Tem	OTHER
				340.174
				64.900
				30.120
				96.441
				25.169
				170.817
				117.957
				5.157
				46.287
				559.701
				5.054
				44.775
				299.988
				68.135
				79.491
				770.915
				435.933
				1.253.644
				2.092.286
				1.599
				763.228
102				2.878
				1.948
				31
				56.404

APPENDIX 3.5 *Estimates of the production, located in* LDE *(cont.)*

year	GRAIN	Gtrad	Gadd	Gem
1718				
1719				
1720	92.434	92.434		
1721				
1722				
1723	1.376.521	382.297		
1724	4.756.472	672.236		
1725	9.309.528	4.336.330		
1726	7.940.574	2.791.107		
1727	2.030.119	1.553.184		
1728	4.151.946	1.731.934		
1729	7.395.238	4.214.897	224.853	
1730	2.170.524	1.233.592		
1731	6.553.694	3.642.725		
1732	2.221.204	368.098		
1733	3.184.299	1.015.230		
1734	5.935.167	3.379.813		
1735	7.569.516	3.471.883		
1736	4.149.288	1.859.381		
1737	536.012	253.080		
1738	4.142.674	2.393.928		
1739	8.504.819	3.446.591	596.248	
1740	22.675.499	13.252.038	2.377.990	
1741	9.682.591	7.014.232	1.460.807	230.4
1742	3.681.602	2.864.450	168.640	
1743	4.133.525	1.953.498	266.060	
1744	3.429.558	3.045.138		
1745	727.900	451.800		
1746				
1747	2.304.503	1.434.278	89.941	
1748	2.788.939	1.363.017	348.523	
1749	7.317.299	2.998.838	1.180.653	
1750	5.750.912	4.713.320	983.920	
1751	10.356.738	6.184.416	1.250.866	
1752	8.471.665	4.957.952	803.957	261.0
1753	5.334.943	1.552.912	647.580	

BER	Ttrad	Tadd	Tem	OTHER
3.376				3.869
				285.084
				2.511
				98.373
				107.075
				811.580
				1.901.541
48.279		13.473		323.750
35.757		31.437		231.842
2.458				3.537.390
20.210				47.057
544.818		544.818		202.983
115.117		38.417		153.019
104.640		97.680		285.400
208.481		23.578		299.094
56.453				1.920.837
547.137		337.014		825.087
165.567		2.001.378		152.675
427.069		279.145		934.774
129.567		70.097		682.054
256.169		95.060		2.358.013
80.310		45.074		4.236.669
236.022		37.799		4.487.639
823.696		630.255		7.664.372
228.917		73.559		4.704.047
347.992		31.410		6.873.744
208.928		931		808.626
92.301		16.833		1.250.759
97.497		32.373		4.705.712
85.957		1.164		2.231.641
73.421				2.249.510
41.733				4.038.022
.063.639		1.938.665		7.447.205
369.681		97.877		1.984.983

APPENDIX 3.5 *Estimates of the production, located in* LDE *(cont.)*

year	GRAIN	Gtrad	Gadd	Gem
1754	2.645.873	1.795.205	348.523	
1755	3.656.313	2.268.645		
1756	1.081.875	737.791		
1757	14.252.435	10.517.089	685.230	
1758	6.639.478	4.476.506	486.940	
1759	6.906.347	5.772.814	230.920	
1760	21.313.372	6.655.626	2.240.286	1.989.
1761	28.175.332	7.506.720	2.494.861	1.417.
1762	25.617.544	1.665.281	683.964	357.
1763	17.276.071	8.410.246	540.101	
1764	26.517.735	13.730.062	2.285.616	1.430.
1765	17.800.457	5.655.868	399.115	337.
1766	17.282.811	8.732.259	1.622.213	716.
1767	37.371.430	22.322.744	3.144.685	730.
1768	30.984.296	12.002.212	7.912.373	685.
1769	133.477.052	10.932.267	2.763.443	612.
1770	70.519.369	45.994.937	6.229.446	3.285.
1771	22.937.672	6.965.336	6.660.628	2.807.
1772	30.710.077	16.563.900	8.053.992	3.080.
1773	14.788.481	8.361.465	1.554.298	572.
1774	23.399.732	10.223.293	5.631.727	1.591.
1775	21.484.953	3.209.837	5.281.058	2.076.
1776	8.219.174	3.309.674	2.191.229	727.
1777	28.550.203	7.012.766	9.868.930	2.022.
1778	40.353.029	5.760.685	12.985.648	2.226.
1779	21.265.833	4.214.892	9.315.001	407.
1780	52.180.408	15.372.250	24.137.260	2.383.
1781	448.314			
1782				
1783	2.742.367	115.460	1.342.303	
1784	19.252.915	4.755.351	6.191.385	2.665.7
1785	31.209.202	5.724.000	11.831.093	1.201.
1786	15.337.355		5.943.678	
1787	10.313.503	1.986.940	5.122.071	286.6
1788	15.475.360	648.480	8.524.276	
1789	31.691.006	2.156.200	14.892.449	1.824.4

MBER	Ttrad	Tadd	Tem	OTHER
.608.028		102.358		3.401.223
.116.924				4.099.156
.699.419		1.141.985		3.143.045
.823.924		31.999		6.105.412
.595.801		741.843		6.161.924
.469.396		4.678		15.209.111
.316.967		36.602		20.070.489
842.018		158.281		7.584.233
388.401		57.073		18.407.558
733.780		25.729		10.099.444
323.317		167.748		9.598.085
441.505		39.794		9.184.385
.149.791		64.839		12.122.810
.326.917		296.248		9.868.844
7.859.588		322.571		8.594.411
4.103.410		227.733		8.326.594
4.010.155				7.113.745
.381.605		79.323		8.694.133
3.273.720		150.487		7.989.059
3.496.219		107.242		14.091.701
5.744.126		152.682		9.868.440
5.537.532		118.453	231.157	10.195.852
2.191.771				1.860.341
3.825.041		42.852		5.928.429
5.332.266		53.250.918		6.394.118
5.019.553		60.124		9.667.181
1.003.665				12.054.097
77.608				
153.508				1.595.372
2.475.331		8.720		5.985.185
2.574.906		231.504		13.881.074
3.793.240		118.208		12.652.333
2.715.007		218.177		24.893.479
6.292.264		106.878		16.188.139
4.932.036		87.350		20.086.887

APPENDIX 3.5 *Estimates of the production, located in* LDE *(cont.)*

year	GRAIN	Gtrad	Gadd	Gem
1790	39.982.106	3.361.174	22.575.376	5.330.2
1791	34.960.994	6.353.119	12.840.776	6.017.2
1792	61.060.058	11.355.418	22.434.827	2.733.5
1793	16.938.643	5.966.943	3.162.468	2.571.1
1794	23.053.059	5.458.574	2.842.259	4.377.2
1795	2.079.893		2.079.893	
1796	272.086	132.049		
1797	385.061		385.061	
1798				
1799				
1800				
1801	448.153		220.880	
1802	282.004	282.004		
1803	629.625	629.625		
1804	75.300	75.300		
1805				
1806				
1807				
1808				
1809				
1810				
1811				
1812				
1813	270.493			
1814	6.913.447	306.220	1.131.449	1.512.9
1815	9.370.509	467.258	1.161.959	4.728.8
1816	29.192.722	11.533.960	4.299.646	7.941.6
1817	65.528.777	12.824.631	19.822.729	9.081.2
1818	16.325.888	2.919.463	943.535	7.196.4
1819	21.221.660	5.111.941	1.945.146	6.511.9
1820	45.180	45.180		
1821	8.022.908	2.432.236	1.632.754	1.040.5
1822	8.890.798	1.782.186	183.230	622.5
1823	7.649.822	2.437.281		
1824	3.497.936	911.584		803.8
1825	5.132.452	2.106.941	218.679	1.376.8

MBER	Ttrad	Tadd	Tem	OTHER
.951.106				9.802.304
.709.963				17.352.908
.491.700		424.591		20.694.374
.443.195		1.067.952		5.003.294
.969.676		471.261		8.032.380
				409
				348.921
33.691				1.118.382
97.946				302.143
				199.283
.030.951		158.167		5.015.097
.837.567	947.599	599.452		9.041.435
.202.355	128.588	1.518.763	208.284	8.134.372
.584.800	1.261.395	5.051		11.547.630
.975.829	120.048	76.815	594.883	5.249.553
.761.344	1.246.271	193.290		9.138.050
72.431				188.377
.100.906		740.661		28.985.135
.713.655	83.451	287.875		8.210.003
.969.309		520.071		154.813.100
.252.192	873.438	483.309		5.347.110
.939.427	280.226	249.648		14.859.085

APPENDIX 3.5 *Estimates of the production, located in* LDE *(cont.)*

year	GRAIN	Gtrad	Gadd	Gem
1826	7.456.865	4.565.936	185.504	872.7
1827				
1828	18.457.696	9.898.831		1.647.3
1829	45.750.619	20.909.661	2.256.053	6.442.7
1830	51.373.944	28.744.645	1.204.803	2.756.1
1831	45.431.114	8.501.407	6.650.331	5.442.4
1832	55.673.732	16.184.063	4.026.169	17.788.1
1833	19.359.987	9.830.699	38.161	2.852.5
1834	15.156.373	9.009.904	337.006	2.658.3
1835	3.917.279	2.643.708	266.060	
1836	13.128.660	3.881.123		387.9
1837	14.991.942	4.112.501	1.313.732	6.068.9
1838	42.272.842	16.071.224	2.788.866	12.647.5
1839	64.766.951	21.943.121	8.165.979	14.805.2
1840	37.293.397	18.541.650	123.669	2.851.1
1841	14.905.730	7.190.440	273.214	3.708.6
1842	21.990.934	8.858.682	2.335.997	4.866.5
1843	44.286.867	21.353.301	331.004	5.689.4
1844	47.762.438	28.611.583		5.945.1
1845	16.195.837	6.936.211		
1846	30.429.624	15.531.929	708.141	704.1
1847	26.936.560	9.325.407		754.6
1848	46.246.954	20.481.129		4.216.5
1849	42.884.485	9.241.767	11.243	11.841.3

Appendix to Chapter 4

	Ameland	Amsterdam	Broek	Delfzijl	Dokkum	Dordrecht
Ameland	6628	925	4	8	15	8
Amsterdam	1426	24796	41	151	342	230
Broek	6	17	541	1	6	1
Delfzijl		70	21	2357	28	1
Dokkum	22	90	5	32	2880	4

IBER	Ttrad	Tadd	Tem	OTHER
248.098	396.982	475.698	336.397	12.554.837
619.842		247.175		9.412.042
589.471		51.727		28.951.928
514.831	211.756	44.742		25.029.994
336.146				11.540.915
227.759		291.156		18.129.067
708.694		113.460	128.892	13.806.051
.424.645	1.328.937	337.129	635.657	26.907.800
.281.937	523.146	97.915		8.159.124
.009.587	151.775	344.604		29.216.846
.604.707	412.018			18.249.971
.685.796	595.807	140.401		33.968.847
.577.993		83.072		29.467.675
.167.840	367.510	220.983		17.472.121
.879.649	399.305	145.640	194.085	16.490.728
.718.608	751.812	139.390		20.389.530
.442.810	830.991	59.008		17.093.405
.641.437	984.508	8.634		11.075.759
.724.419	1.766.387	189.978	272.999	2.905.705
.650.939	557.059	327.030		5.836.578
303.178		2.158		8.594.321
.020.852	1.344.577	689.687		13.775.797
.015.257	270.953	16.654		18.848.852

	Ameland	Amsterdam	Broek	Delfzijl	Dokkum	Dordrecht
Dordrecht	25	235	6	0	5	1428
Edam	14	215	3	12	7	1
Enkhuizen	26	190	8	15	13	7
Groningen	50	218	17	153	59	10
Harlingen	90	325	3	21	44	22
Heerenveen	45	70	0	4	13	3
Hindeloopen	141	518	1	10	30	13

(*cont.*)

	Ameland	Amsterdam	Broek	Delfzijl	Dokkum	Dordrecht
Hoorn	70	380	11	6	23	14
Joure	6	64	0	2	2	0
Lemmer	9	149	2	16	16	1
Makkum	55	175	3	10	32	0
Molkwerum	14	87	0	5	0	6
Pekela	6	158	0	25	1	11
Rotterdam	45	417	8	20	27	43
Schellinkhout	8	8	3	0	0	0
Schiedam	3	73	19	3	1	5
Stavoren	77	167	16	16	62	8
Terhorne	13	40	0	0	3	3
Terschelling	146	651	10	10	50	10
Texel	27	148	31	0	7	0
Veendam	3	58	0	6	4	15
Vlieland	102	361	10	46	11	6
Warns	5	121	0	0	0	22
Workum	15	244	0	16	33	6
Woudsend	12	183	1	1	11	3
Zaandam	51	278	26	3	6	4
TOT	9140	31431	790	2949	3731	1885
% domicile	72,5%	78,9%	68,5%	79,9%	77,2%	75,8%

Domicile Ambiguity Matrix

	Edam	Enkhuizen	Groningen	Harlingen	Heerenveen	Hindeloopen
Ameland	7	11	14	74	22	52
Amsterdam	151	91	404	661	397	1175
Broek	0	1	0	1	31	2
Delfzijl	1	11	138	10	6	20
Dokkum	15	10	33	26	16	8
Dordrecht	0	2	29	17	6	13
Edam	1171	17	4	10	13	15
Enkhuizen	4	1824	21	39	13	48
Groningen	7	9	7139	58	12	17

	Edam	Enkhuizen	Groningen	Harlingen	Heerenveen	Hindeloopen
Harlingen	20	27	54	8382	51	83
Heerenveen	3	2	0	15	2179	37
Hindeloopen	18	29	12	98	15	8267
Hoorn	66	64	18	25	3	33
Joure	0	3	7	8	11	13
Lemmer	0	6	14	45	94	47
Makkum	2	21	20	62	7	25
Molkwerum	3	3	12	22	0	61
Pekela	4	6	149	99	2	7
Rotterdam	20	18	47	59	31	22
Schellinkhout	2	9	0	21	19	3
Schiedam	3	12	16	32	22	2
Stavoren	17	27	5	35	21	137
Terhorne	2	0	3	15	97	23
Terschelling	23	64	8	125	4	59
Texel	4	7	12	4	0	16
Veendam	1	2	76	22	2	3
Vlieland	23	49	60	130	13	49
Warns	0	40	0	26	23	68
Workum	5	8	12	42	12	77
Woudsend	1	16	47	25	18	83
Zaandam	31	22	2	19	7	25
TOT	1604	2411	8356	10207	3147	10490
% domicile	73,0%	75,7%	85,4%	82,1%	69,2%	78,8%

Domicile Ambiguity Matrix

	Hoorn	Joure	Lemmer	Makkum	Molkwerum	Pekela
Ameland	51	6	45	20	33	17
Amsterdam	610	262	578	383	217	270
Broek	8	14	0	21	10	1
Delfzijl	3	6	1	6	3	45
Dokkum	5	3	25	17	29	17
Dordrecht	23	3	5	10	5	73
Edam	76	0	1	19	1	12
Enkhuizen	42	11	8	5	18	52

(*cont.*)

	Hoorn	Joure	Lemmer	Makkum	Molkwerum	Pekela
Groningen	11	13	7	41	0	306
Harlingen	27	14	26	52	43	126
Heerenveen	5	30	60	15	6	18
Hindeloopen	17	39	16	31	127	9
Hoorn	4856	1	9	12	17	22
Joure	13	1280	52	2	7	1
Lemmer	3	63	3557	13	6	11
Makkum	3	6	10	2476	34	3
Molkwerum	6	13	0	19	2339	5
Pekela	5	1	7	6	0	8006
Rotterdam	40	15	32	23	38	66
Schellinkhout	55	0	2	1	30	2
Schiedam	11	5	8	7	3	32
Stavoren	12	8	5	13	46	7
Terhorne	0	5	9	3	11	4
Terschelling	69	44	22	65	49	28
Texel	9	2	2	30	0	0
Veendam	2	6	11	2	1	143
Vlieland	37	13	16	33	25	46
Warns	6	1	3	13	46	0
Workum	12	39	35	109	27	7
Woudsend	13	26	63	24	6	12
Zaandam	32	10	23	16	29	15
TOT	6062	1939	4638	3487	3206	9356
% domicile	80,1%	66,0%	76,7%	71,0%	73,0%	85,6%

Domicile Ambiguity Matrix

	Rotterdam	Schellinkhout	Schiedam	Stavoren	Terhorne	Terschelling
Ameland	22	0	3	34	7	234
Amsterdam	365	10	44	413	264	1358
Broek	2	6	0	19	0	8
Delfzijl	2	2	14	6	0	24
Dokkum	7	19	1	11	1	62

	Rotterdam	Schellinkhout	Schiedam	Stavoren	Terhorne	Terschelling
Dordrecht	31	0	2	11	4	36
Edam	9	21	1	11	2	47
Enkhuizen	5	4	9	55	5	87
Groningen	11	7	33	25	40	51
Harlingen	20	3	5	61	23	124
Heerenveen	8	0	2	38	28	29
Hindeloopen	14	10	0	146	8	59
Hoorn	31	48	13	27	15	128
Joure	3	0	2	10	4	8
Lemmer	9	0	6	10	21	47
Makkum	13	2	9	32	3	43
Molkwerum	2	3	0	53	1	15
Pekela	28	3	29	22	1	6
Rotterdam	2634	8	32	42	5	48
Schellinkhout	1	334	1	14	0	39
Schiedam	37	10	742	17	1	34
Stavoren	7	21	3	4224	22	70
Terhorne	1	1	0	4	1159	5
Terschelling	20	63	19	75	14	10516
Texel	10	0	1	9	1	25
Veendam	14	0	9	1	0	7
Vlieland	34	11	17	52	9	272
Warns	4	2	1	28	1	22
Workum	19	5	6	53	7	74
Woudsend	13	0	0	64	36	24
Zaandam	11	18	12	20	0	68
TOT	3387	611	1016	5587	1682	13570
% domicile	77,8%	54,7%	73,0%	75,6%	68,9%	77,5%

Domicile Ambiguity Matrix

	Texel	Veendam	Vlieland	Warns	Workum	Woudsend
Ameland	7	0	81	6	13	44
Amsterdam	86	101	632	135	621	560
Broek	0	0	25	0	3	0
Delfzijl	0	7	17	0	18	7

(*cont.*)

	Texel	Veendam	Vlieland	Warns	Workum	Woudsend
Dokkum	1	4	3	3	23	11
Dordrecht	0	35	18	15	14	6
Edam	2	2	44	3	12	1
Enkhuizen	6	0	84	15	11	3
Groningen	0	132	69	1	46	23
Harlingen	4	32	104	6	42	74
Heerenveen	0	0	11	0	34	29
Hindeloopen	3	2	62	123	61	27
Hoorn	25	1	111	5	22	6
Joure	0	0	7	1	15	26
Lemmer	2	3	11	7	11	100
Makkum	1	2	19	14	72	10
Molkwerum	0	0	34	38	17	5
Pekela	4	65	31	0	5	4
Rotterdam	9	69	79	5	49	33
Schellinkhout	1	0	40	0	0	0
Schiedam	1	7	30	1	7	4
Stavoren	7	1	73	36	40	83
Terhorne	0	0	3	0	11	25
Terschelling	24	1	205	20	50	15
Texel	691	0	46	0	0	0
Veendam	2	7630	1	0	2	2
Vlieland	24	2	9272	10	33	25
Warns	0	0	12	1506	10	11
Workum	0	13	32	36	4039	115
Woudsend	0	2	28	10	69	3385
Zaandam	3	0	76	20	8	21
TOT	903	8111	11260	2016	5358	4655
% domicile	76,5%	94,1%	82,3%	74,7%	75,4%	72,7%

Domicile Ambiguity Matrix

	Zaandam	TOT alt	
Ameland	15	1778	17,5%
Amsterdam	187	12165	24,8%
Broek	9	192	20,8%
Delfzijl	0	467	14,2%
Dokkum	5	508	13,0%
Dordrecht	5	634	23,5%
Edam	11	586	25,0%
Enkhuizen	6	810	23,5%
Groningen	6	1432	14,3%
Harlingen	25	1551	13,5%
Heerenveen	9	514	16,0%
Hindeloopen	28	1667	14,4%
Hoorn	39	1245	16,9%
Joure	1	268	14,8%
Lemmer	3	725	14,5%
Makkum	6	694	18,0%
Molkwerum	4	428	13,4%
Pekela	16	701	7,5%
Rotterdam	19	1369	25,5%
Schellinkhout	17	276	31,2%
Schiedam	23	429	26,8%
Stavoren	38	1080	16,9%
Terhorne	0	281	16,3%
Terschelling	26	1969	13,6%
Texel	16	407	27,0%
Veendam	4	399	4,7%
Vlieland	51	1570	12,6%
Warns	0	465	19,1%
Workum	5	1064	17,3%
Woudsend	10	801	16,1%
Zaandam	1021	856	31,3%
TOT	1605		16,9%
% domicile	63,6%		6,2% (St. Dev.)

Domicile Ambiguity Matrix

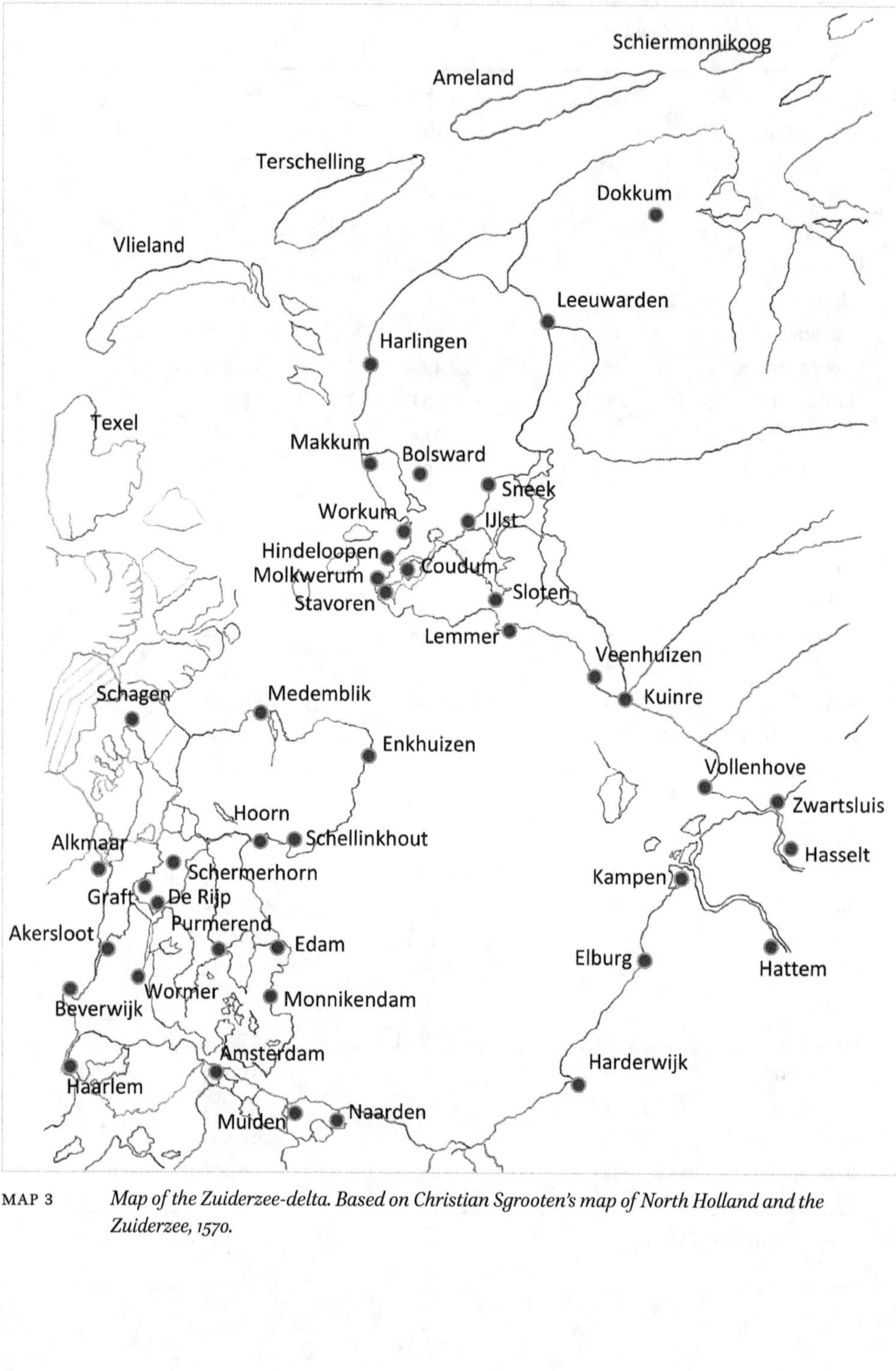

MAP 3 *Map of the Zuiderzee-delta. Based on Christian Sgrooten's map of North Holland and the Zuiderzee, 1570.*

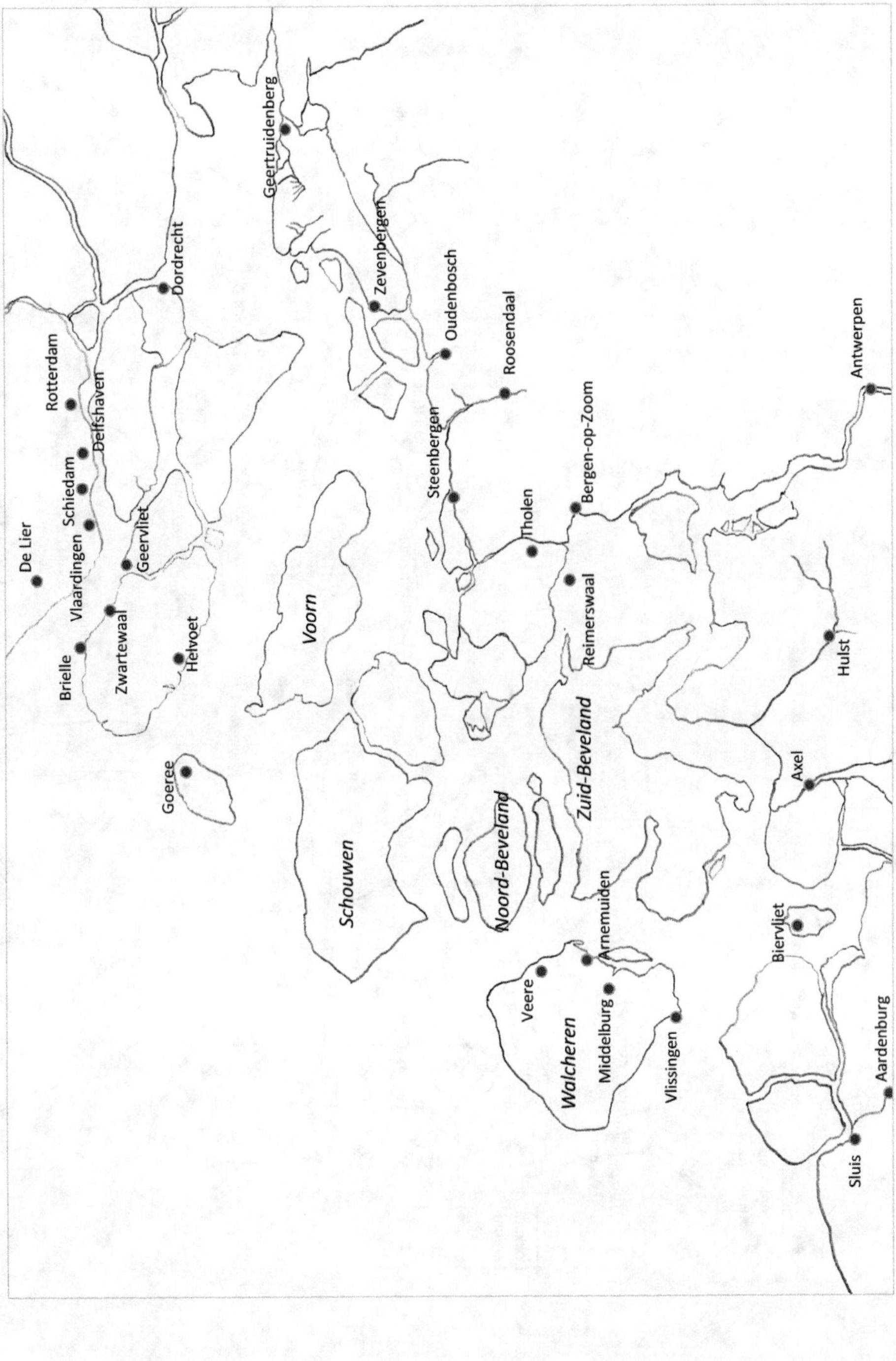

MAP 4 *Map of the Scheldt-Meuse-Rhine delta. Based on Jacob van Deventer's map of Zelandia, 1560.*

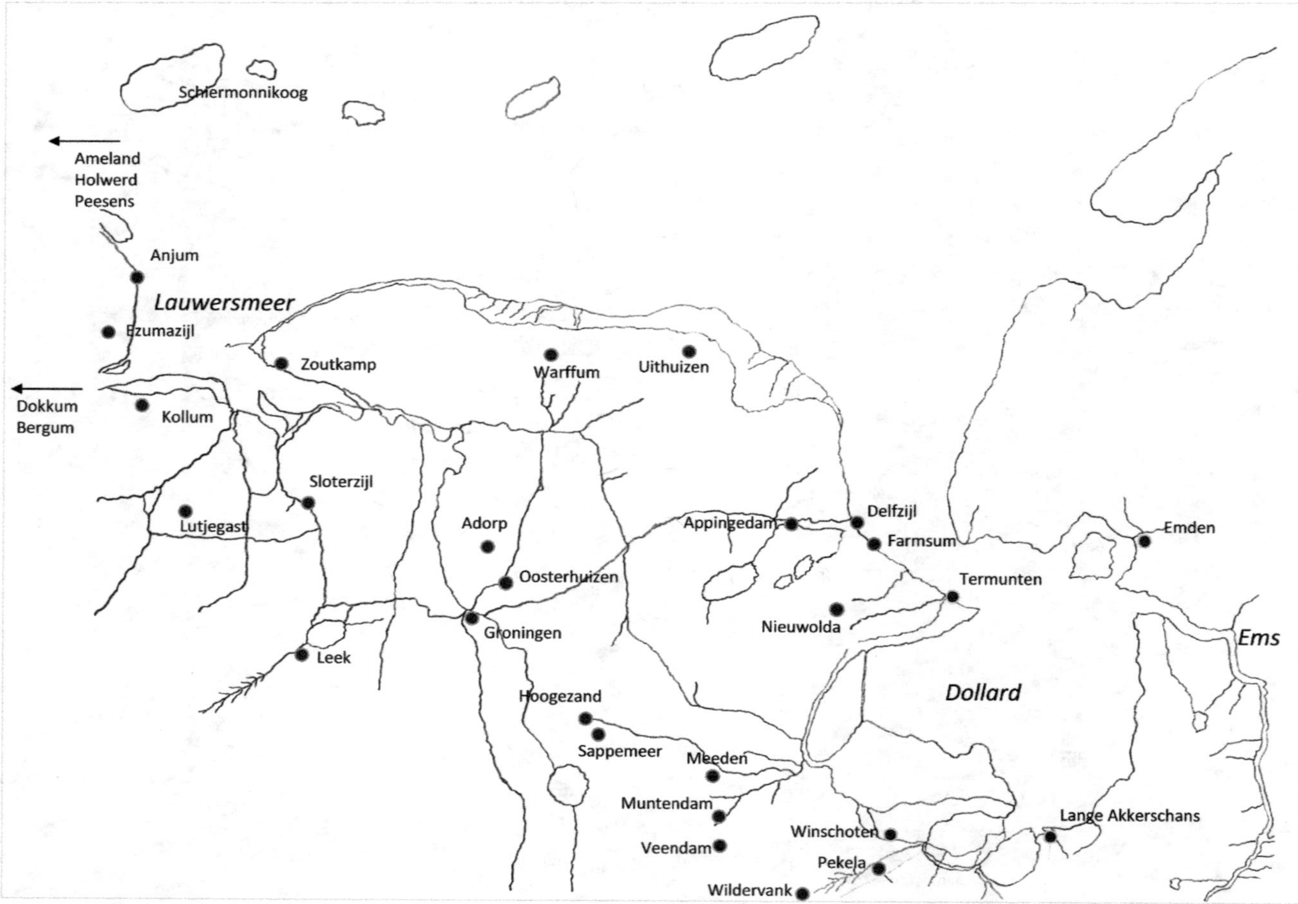

MAP 5 *Map of the Lauwers-Dollard-Emsdelta. Based on Bortholdo Wicheringe's map of Groninga Dominium, 1635.*

Bibliography

Archives

Gemeentearchief Rotterdam (GAR), Handschriftenverzameling 33.01, nr. 3307: Ordonnantien en vrachtlijsten, met aantekeningen op handelsgebied.

Groninger Archieven, 1325 Gilden, nrs. 115–117.

Groninger Archieven, 1032 Verzameling afschriften en reproducties, nr. 1094: Leekster schippersgildeboek.

Groninger Archieven, Plakkaten en Ordonnanties, nr. 540: wijziging van de willekeur (=gilderol) voor de schippers van Veendam, Widervank en in de Oldambten.

Groninger Archieven, 1534 Volle Gerecht van de stad Groningen, nr. 4435: Willekeur voor de schippers van Oude en Nieuwe Pekela 1792; nr. 4435: Reglement van het oude schippersgilde te Pekela 1794.

Groninger Archieven, 740: Verzameling publicaties van de overheid 1595–1814, nr. 401: Nieuwe ordonnantie op het Veer van Termunter-zil na Embden, waer nae een ijder, soo Schippers, als Passagiers, sich sullen hebben te reguleren.

Groninger Archieven, bibliotheek, 7026. Placcaat-boek Groningen 3356 (1677).

Stadsarchief Amsterdam, 366: Gilden en het Brouwerscollege, nrs. 390–398.

Stadsarchief Amsterdam: Zeebrieven.

Tresoar, 13–45 Nedergerecht Wymbritseradeel, Hypotheekboeken. Daarin: Contract van onderlinge verzekering tegen ongelukken ter zee, tussen schippers te Woudsend behorende 29 mei 1724.

Tresoar, 344 Verzameling Fries Genootschap, nr. 355: Stukken betreffende de beurtvaart en het schippersgilde te Joure.

Tresoar, Zeebrieven Stavoren / Zeebrieven Workum.

Wymbritseradiel, 01. Overheidsarchieven Wijmbritseradeel, 01.01.10 Gritenijbestuur tot 1812, Handel, scheepvaart en veeteelt. Daar: Assurantie-overeenkomsten tussen de zeevarende schippers van Heeg, Woudsend en Hommerts … nr. 228: 1705–1720 (Woudsend, Heeg, Hommerts), nr. 229: 1735 (Heeg).

Zeeuws Archief, 1001 Stad Middelburg. Ordonnanties: nr. 4/113.

Nationaal Archief (The Hague), 1.01.47.09: Admiraliteitscolleges XIII Hinxt.

Staatsarchiv Osnabrück, Nr. 7294 Hzgl. Arenbergisches Archiv zu Meppen, Dep 62 b: Liste der 1807 in der Großen Schiffersgilde zu Papenburg verzeichneten Schiffe.

Primary Sources

Allen, Robert C. and Tommy Murphy. Before the Metre, the Gram, the Litre: Building a Rosetta Stone of Weights and Measures in the Early Modern World. URL: http://www.nuffield.ox.ac.uk/users/murphy/measures/before_metre.htm

Asaert, Gustaaf. *Documenten voor de geschiedenis van de Antwerpse scheepvaart voornamelijk de Engelandvaart (1404–1485)*. Brussel: Wetenschappelijk Comité voor Maritieme Geschiedenis, Koninklijke Academie voor Wetenschappen, Letteren en Schone Kunsten van België, 1985.

Corpus of First Names in the Netherlands, edited by Meertens Instituut. URL: http://www.meertens.knaw.nl/nvb/english/

Database Dutch Craft Guilds, Study Global ID: hdl:10411/10101.

Dataset of the *Convooigeldregisters* (1580–1584): Clé Lesger, 2004/2007. Amsterdam Convooigeld 1580 and 1584. Date created: unknown. [Available at] Data Archiving and Networked Services—DANS. Persistent identifier: urn:nbn:nl:ui:13-xuk-8xl.

Distances between ports. Pub. 151. Prepared and published by the National Imagery and Mapping Agency. Eleventh Edition, Bethesda, Maryland, 2001.

Heeringa, K. and J.G. Nanninga (eds.). *Bronnen tot de Geschiedenis van den Levantschen Handel (1590–1826)*, 's-Gravenhage: Nijhoff, 1910–1966.

Jenks, Stuart. *Das Danziger Pfundzollbuch von 1409 und 1411*. Köln – Weimar – Wien: Böhlau, 2012. (Quellen und Darstellungen zur Hansischen Geschichte, Band LXIII).

Lindblad, Jan Thomas. *Dutch Entries in the pound-toll registers of Elbing*. Den Haag: Instituut voor Nederlandse Geschiedenis, 1995.

Namen en naamkunde in Nederland en elders, edited by Meertens Instituut. URL: http://www.meertens.knaw.nl/nnn/

Piet de Buck (†), Sebastiaan Kerkvliet en Milja van Tielhof. *Amsterdamse notariële akten over de Archangelvaart 1594–1724* (http://resources.huygens.knaw.nl/archangel)

Poelman, Huibert Antonie. *Bronnen tot de geschiedenis van den Oostzeehandel, 1ste deel: 1122–1499*. 's-Gravenhage: Martinus Nijhoff, 1917.

Smit, Homme Jacob. *Bronnen tot de geschiedenis van den handel met Engeland, Schotland en Ierland, 1150–1585*. 's-Gravenhage: Nijhoff, 1928–1950.

Smit, Johannes Gradus. *Bronnen voor de economische geschiedenis van het Beneden-Maasgebied. Tweede deel: Rekeningen van de Hollandse tollen, 1422–1534*. Den Haag: Instituut voor Nederlandse Geschiedenis, 1997.

Sneller Zeger Willem And Willem Sybrand Unger. *Bronnen tot de geschiedenis van den handel met Frankrijk, 753–1585*. 's-Gravenhage: Nijhoff, 1930–1942.

Sound Toll Registers Online: http://www.soundtoll.nl

Studer, Paul. *The port books of Southampton, or (anglo-french) accounts of Robert Florys, water-bailiff and receiver of petty customs, 1427–1430*. Southampton: Cox and Sharland, 1913.

Tangelder, Franciscus Bernardinus Maria. *Nederlandse Rekeningen in de Pondtolregisters van Elbing 1585–1602*. Den Haag: Martinus Nijhoff, 1972.

Thesaurus of Frisian names, created by Tresoar, the Frisian Historical and Literary Centre. Version used in this monograph is version 7, dated July 2010. URL: http://www.tresoar.nl

Unger, Willem Sybrand. *De tol van Iersekeroord: documenten en rekeningen 1321–1572*. 's-Gravenhage, Martinus Nijhoff, 1939.

van der Aa, A.J. *Aardrijkskundig woordenboek der Nederlanden*, Gorinchem. Jacobus Noorduyn, 1839–1851.

Welling, George Maria. *Database of the Paalgeldregisters, 1771–1787*. http://www.let.rug.nl/welling/paalgeld/appendix.html

———. *Version history of the Reconstruction of the Johansen Soundtolregisters Database 1784–1795, Version 3.02 of April 2010*. http://www.let.rug.nl/welling/sont/johansen.htm

Winkelman, P.H. *Bronnen voor de geschiedenis van de Nederlandse Oostzeehandel in de zeventiende eeuw*. Den Haag: Instituut voor Nederlandse Geschiedenis, 1977–1983.

Secondary Literature

't Hart, W.G. and Paul C. Van Royen. "Het smakschip "De Neufville van der Hoop". Een onderzoek naar de rendabiliteit van de Nederlandse vrachtvaart in de achttiende eeuw." *Economisch- en sociaal-historisch jaarboek* 48 (1985): 150–168.

't Hart, Marjolein. "De Nederlandse Republiek en de betekenis van steden." In *Een wonder weerspiegeld: de Nederlandse Republiek in Europees perspectief*, edited by Karel Davids and Jan Lucassen (Amsterdam: Aksant, 2005), 53–89.

Alanen, Aulis J. *Der Außenhandel und die Schifffahrt Finnlands im 18. Jahrhundert unter besonderer Berücksichtigung der Umbruchsperiode der Handelsfreiheit im Bottnischen Meerbusen und der großen Seekriege*. Helsinki, 1957.

Asaert, Gustaaf. *De Antwerpse Scheepvaart in de XVe eeuw (1394–1480): bijdrage tot de ekonomische geschiedenis van de stad Antwerpen*. Brussel: Paleis der Akademiën, 1973.

Baasch, Ernst. "Hamburgs Seeschifffahrt und Waarenhandel vom Ende des 16. bis zur Mitte des 17. Jahrhunderts." *Zeitschrift des Vereins für hamburgische Geschichte*, 1893: 295–420.

———. *Die Börtfahrt zwischen Hamburg, Bremen und Holland*. 1898.

Bakker, G. "Scheepsverzekering te Woudsend Ao. 1705." *Jaarboek Fries Scheepvaart museum en Oudheidkamer*. Sneek: Vereniging Fries Scheepvaart Museum, 1967–1968.

Barrett, Ward. "World bullion flows, 1450–1800." In *The Rise of Merchant Empires: Long-distance trade in the early modern world, 1350–1750*, edited by James D. Tracy, 224–254. Cambridge: University Press, 1990.

Berkenvelder, F.C. "Frieslands handel in de late Middeleeuwen." *Economisch-Historisch Jaarboek* 29 (1963): 136–187.

Bird, J. *Centrality and Cities*. London and Boston, 1977.

Blondé, Bruno. "At the cradle of the transport revolution?: Paved roads, traffic flows and economic development in eighteenth-century Brabant" *Journal of Transport History* 31–1 (2010): 89–111.

Bloothooft, Gerrit. "Assessment of systems for nominal retrieval and historical record linkage." *Computers and the Humanities* 32–1 (1998): 39–56.

Boekel, Pieter Nicolaas. *De zuivelexport van Nederland tot 1813*, Utrecht: Drukkerij Fa. Schotanus & Jens, 1929.

Boelmans Kranenburg, H.A.H. De *Zeevisscherij van Holland in den tijd der Republiek*. Amsterdam, 1946.

Boon, P. "West Friesland and the Sound (1681–1720): Sound Toll Registers, Sound Toll Tables and the Facts for West Friesland." In *From Dunkirk to Danzig: Shipping and Trade in the North Sea and the Baltic 1350–1850. Essays in honour of J.A. Faber on the occasion of his retirement as professor of social and economic history at the University of Amsterdam*, edited by W.G. Heeres, 171–189. Hilversum: Verloren, 1988.

Boon, Piet. "Schippers en hun schepen: de koopvaardij van Wijdenes en Oosterleek omstreeks 1560." *Tijdschrift voor Zeegeschiedenis* 2–1 (1983): 3–16.

———. *Bouwers van de zee: Zeevarenden van het Westfriese platteland, c.1680–1720*. Den Haag: Stichting Hollandse Historische Reeks, 1996. (Hollandse Historische Reeks, vol. 26).

Boschma-Aarnoudse, Corrie. *Tot verbeteringe van de neeringe deser stede: Edam en de Zeevang in de late Middeleeuwen en de 16e eeuw*. Hilversum: Verloren, 2003.

Bosse, Ewald. *Norwegens Volkswirtschaft vom Ausgang der Hansaperiode bis zur Gegenwart mit besonderer Berücksichtigung der internationalen Handelsbeziehungen. Erster Teil*. Jena: Gustav Fischer, 1916.

Brakel, S. van. "Schiffsheimat und Schifferheimat in den Sundzollregistern." *Hansische Geschichtsblätter* 21 (1915): 211–228.

Broeze, Frank J.A. "Rederij", In *Maritieme geschiedenis der Nederlanden, vol. 3: achttiende eeuw en eerste helft negentiende eeuw, van ca. 1680 tot 1850–1870*, edited by Frank J.A. Broeze, Jaap R. Bruijn & Femme S. Gaastra, 106–116. Bussum: De Boer Maritiem, 1977.

Broeze, Frank. "The ports and port system of the Asian Seas: An overview with historical perspective from c.1750," *The Great Circle* 18–2 (1996): 73–96.

Bruijn, Jaap R. "De vaart in Europa." In *Maritieme Geschiedenis der Nederlanden, Vol. 2: Zeventiende eeuw, van 1585 tot ca. 1680*, edited by Leo Akveld, Simon Hart & W.J. van Hoboken. Bussum: De Boer Maritiem, 1978.

———. "Productivity, profitability, and costs of private and corporate Dutch ship owning in the seventeenth and eighteenth centuries." In *The Rise of Merchant Empires. Long-distance trade in the early modern world 1350–1750*, edited by James D. Tracy, 174–194. Cambridge: University Press, 1990.

Brulez, Wilfrid. "De scheepvaartwinst in nieuwe tijden." *Tijdschrift voor geschiedenis* 92–1 (1979): 1–19.

Brünner, E.C.G. "De waarde der Skibsfarttabellen van Nina Ellinger Bang voor de kennis der handelsgeschiedenis van Holland in de 16e eeuw." *Bijdragen voor Vaderlandsche Geschiedenis en Oudheidkunde*, vijfde reeks 9 (1922): 269–280.

Buck, Henning. *Die Arenberger in Emsland.* Sögel: Emsländischer Heimatbund, 2003.

Burghardt, A.F. "A hypothesis about gateway cities." *Annals of the Association of American Geographers* 61 (1971): 269–285.

Christensen, A.E. "Der handelsgeschichtliche Wert der Sundzollregister. Ein Beitrag zur seiner Beurteilung." *Hansische Geschichtsblätter* 59 (1934): 28–142.

Christensen, Aksel E. *Dutch Trade to the Baltic about 1600: studies in the Sound Toll Register and Dutch shipping records.* Copenhagen: Einar Munksgaard / The Hague: Martinus Nijhoff, 1941.

Cornelisse, Charles. *Energiemarkten en energiehandel in Holland in de late Middeleeuwen.* Hilversum: Verloren, 2008.

Corten, A. *Herring and Climate: changes in the distribution of North Sea herring due to climate fluctuations.* PhD Dissertation Rijksuniversiteit Groningen, 2001.

Cox, Joost C.M. *Hebbende privilege van stede.' De verlening van stadsrechtprivileges in Holland en Zeeland (13de–15de eeuw). PhD Dissertation* Universiteit Leiden, 2011.

Davids, Karel A. *Zeewezen en wetenschap: De wetenschap en de ontwikkeling van de navigatietechniek in Nederland tussen 1585 en 1815*, Amsterdam/Dieren: De Bataafsche Leeuw, [1985].

Davids, Karel. "Maritime labour in the Netherlands, 1570–1870." In *"Those emblems of hell?" European sailors and the maritime labour market, 1570–1870*, edited by P. Van Royen, Jaap Bruijn & Jan Lucassen, 41–71. St. John's, 1997.

De Buck P. and J. Th. Lindblad. "Navigatie en negotie. De Galjootsgeldregisters als bron bij het onderzoek naar de geschiedenis van de Oostzeehandel in de achttiende eeuw" *Tijdschrift voor Zeegeschiedenis*, 1 (1990): 27–48.

De Graaff, C. "De Woudsenders en hun zeeverzekeringen, 1750–1803." *Jaarboek Fries Scheepvaartmuseum en Oudheidkamer 1998.* Sneek: Vereniging Fries Scheepvaart Museum, 1999.

De Lange, H.J. *Eb en Vloed in Groningen: De stad als zeehaven.* s.l.: Uitgeverij Noordboek, 2004.

De Moor, Tine. "The Silent Revolution: A New Perspective on the Emergence of Commons, Guild, and Other Forms of Corporate Collective Action in Western Europe." *International Review of Social History* 53 (supplement) (2008): 179–212.

De Peuter, Roger. *Brussel in de achttiende eeuw: Sociaal-economische structuren en ontwikkelingen in een regionale hoofdstad.* Brussel: VUB Press, 1999.

De Vries, Jan and Ad Van der Woude. *Nederland 1500–1815. De eerste ronde van moderne economische groei*. Amsterdam: Balans, 1995 [English version: *The First Modern Economy: Success, Failure and Perseverance of the Dutch Economy, 1500–1815*. Cambridge: Cambridge University Press, 1997.]

De Vries, Jan. *Barges and Capitalism: passenger transportation in the Dutch economy 1632–1839*. Wageningen, 1978. (AAG Bijdragen, 21).

———. *The Dutch Rural Economy in the Golden Age, 1500–1700*. New Haven and London: Yale University Press, 1974.

De Vries, Johannes. *De Economische Achteruitgang der Republiek in de achttiende eeuw*, Amsterdam: Drukkerijen Elleman Harms, 1959.

De Wit, Annette. *Leven, werken en geloven in zeevarende gemeenschappen: Schiedam, Maassluis en Ter Heijde in de zeventiende eeuw*. Amsterdam: Aksant, 2008.

Dedieu, Jean-Pierre et al. "Navigocorpus: A database for shipping information—a methodological and technical introduction." *International Journal of Maritime History* XXIII-2 (2011): 331–359.

Den Heijer, Henk J. *De geoctrooieerde compagnie: De VOC en de WIC als voorlopers van de naamloze vennootschap*. Deventer: Kluwer, 2005.

Desfeuilles, Paul. "La Navigation à travers le Sund en 1784: d'après les registres de la Douane danoise." *Annales: Histoire, Sciences Sociales* XIV-3 (1959): 492–520.

———. "La Navigation à travers le Sund et la guerre russo-suédoise de Finlande (1788–1790)." *Vierteljahrschrift für Sozial- und Wirtschaftsgeschichte* XLVIII-1 (1961): 56–74.

Diepeveen, W.J. *De Vervening in Delfland en Schieland tot het einde der zestiende eeuw*. Leiden: Eduard Ijdo, 1950.

Doursther, Horace. *Dictionnaire universel des poids et mesures anciens et modernes, contenant des tables des monnaies de tous les pays*, Amsterdam: Meridian Publishing, 1965 [reprint: Bruxelles, 1840].

Draper, Maarten and Jerem Van Duijl. "Shipmasters from the West Frisian Islands in the Sound Toll Registers, 1737–1800." URL: http://www.soundtoll.nl/images/files/Draper%20and%20van%20Duijl.pdf

Duijvendak, Maarten and Hendrik Boels. *Geschiedenis van Groningen, II. Nieuwe Tijd*. Zwolle: Waanders, 2008.

Ellinger Bang, Nina and Knud Korst. *Tabeller over skibsfart og Varetransport gennem Oresund 1497–1660*, I: Nina Ellinger Bang (red.), *Tabeller over Skibsfarten*. Copenhagen, Gyldendalske Boghandel—Nordisk Forlag; Leipzig, Otto Harrassowitz, 1922; II. Ibidem (red.), *Tabeller over Varetransporten A*. Copenhagen, Gyldendalske Boghandel—Nordisk Forlag; Leipzig, Otto Harrassowitz, 1922; II. Nina Ellinger Bang, Knud Korst (red.), *Tabeller over Varetransporten B*. Copenhagen, Gyldendalske Boghandel—Nordisk Forlag; Leipzig, Otto Harrassowitz, 1933. *Tabeller over skibsfart og Varetransport gennem Oresund 1661–1783 og gennem Storebalt 1701–1748*, I: Nina Ellinger Bang, Knud Korst (red.), *Tabeller over Skibsfarten*. Copenhagen: Gyldendalske Boghandel—Nordisk Forlag; Leipzig: Otto Harrassowitz, 1930.

Epstein, S.R. Craft guilds, apprenticeship, and technological change in preindustrial Europe. *Journal of Economic History* 58–3 (1998): 684–713.

Faber, Johannes A. "Shipping to the Netherlands during a turbulent period 1784–1810." In *From Dunkirk to Danzig: Shipping and Trade in the North Sea and the Baltic, 1350–1850*, edited by W.G. Heeres et al., 95–106. Hilversum: Verloren Publishers, 1988.

———. "Structural changes in the European economy during the eighteenth century as reflected in the Baltic trade." In *From Dunkirk to Danzig: Shipping and Trade in the North Sea and the Baltic, 1350–1850*, edited by W.G. Heeres et al., 8–94. Hilversum: Verloren Publishers, 1988.

———. *Drie Eeuwen Friesland: Economische en sociale ontwikkelingen van 1500 tot 1800*. Leeuwarden: DeTille, 1973.

Floud, Roderick. *An Introduction to Quantitative Methods for Historians. Second Edition*. London, 1979.

Fuchs, J.M. "De Amsterdamse beurtvaart op de Duitse Rijnhavens." *Jaarboek van het Genootschap Amstelodamum* 50 (1958): 30–45.

Fuchs, Johannes Marius. *Beurt- en wagenveren*. Den Haag: L.J.C. Boucher, 1946.

Gaastra, Femme. "Zeilende kustvaart van circa 1750 tot circa 1920. Het fenomeen van de Groninger scheepvaart in de negentiende eeuw." *Tijdschrift voor Zeegeschiedenis* 26–2 (2007): 154–165.

Geerdink van der Worp, E.C. *De Sociteit Compact of het Schippersgilde te Leek, Deel I van 1764 tot 1811*. Leek: [s.n.], 1994.

Gerding, Michiel A.W. *Vier Eeuwen Turfwinning: De verveningen in Groningen, Friesland, Drenthe en Overijssel tussen 1550 en 1950*, Utrecht, HES, 1995.

Go, Sabine. *Marine Insurance in the Netherlands 1600–1870: a comparative institutional approach*. PhD Dissertation Vrije Universiteit Amsterdam, 2009.

Gorski, Richard. *Maritime labour. Contributions to the history of work at sea, 1500–2000*. Amsterdam: Aksant, 2007.

Gras, Norman Scott Brien. *The evolution of the English corn market from the twelfth to the eighteenth century*. Cambridge, 1915.

Greif A., P. Milgrom and B.R. Weingast. "Coordination, commitment, and enforcement: the case of the merchant guild." *Journal of Political Economy*. 102–4 (1994): 745–776.

Hart, Simon, "Rederij", *in* L.M. Akveld, S. Hart & W.J. van Hoboken (eds.). *Maritieme geschiedenis der Nederlanden, vol. 2: zeventiende eeuw, van 1585 tot 1680*, Bussum, De Boer Maritiem, 1977, pp. 106–125.

Heckscher, E.F. "Öresundsräkenskaperna och deras behandlings," *Historisk Tidskrift*, 1942, 170–186.

Houtman-De Smedt, H. "De Zuidelijke Nederlanden en de Oostzee in de 18e eeuw (Oostenrijkse Periode)." In *The interactions of Amsterdam and Antwerp with the Baltic region, 1400–1800 = De Nederlanden en het Oostzeegebied: papers presented at the third international conference of the "Association Internationale d'Histoire des Mers Nordiques de l'Europe", Utrecht, August 30th–September 3rd 1982*, edited by Wieringa, Wiert Jan, 105–119. Leiden: Nijhoff, 1983.

Hutchison, Ragnhild. *In the doorway to development: an enquiry into market oriented structural changes in Norway ca. 1750–1830*. Ph D Dissertation European University Institute, 2010.

Hybel, Nils Introduction to *Cogs, Cargoes and Commerce: maritime bulk trade in Northern Europe 1150–1400*, edited by Lars Berggren, Nils Hybel & Annette Landen. Toronto: Pontifical Institute of Mediaeval Studies, 2002.

Ibelings, Bart. "De route 'binnendunen'. De scheepvaartroute langs de goudse sluis en tol, de Wassenaerse Gouwesluis en de Spaarndammertol (13e tot 16e eeuw)." In *Vander Rekeninghe. Bijdragen aan het symposium over onderzoek en editieproblematiek van middeleeuws rekeningmateriaal, gehouden in Utrecht 27 en 28 februari 1997*, edited by Dick De Boer et al., 221–254 and 323–339. Den Haag, 1998.

Jackson, G. "Early Modern Seaport Studies: Highlights and Guidelines." In *European Seaport Systems in the Early Modern Age—A Comparative Approach. International Workshop. Proceedings*. Porto, 2007.

Jackson, Gordon. "The Significance of Unimportant Ports." *International Journal of Maritime History*, XIII-2 (2001): 1–17.

Jeannin, P. "Les comptes du Sund comme source pour la construction de l'activité économique en Europe (XVIe–XVIIIe siècle)." *Revue historique* CCXXXI (1964): 55–102; 307–340.

Johansen, Hans Chr. "Scandinavian Shipping in the late eighteenth century in a European perspective." *The Economic History Review*, New Series, 45–3 (1992): 479–493.

———. *Shipping and Trade between the Baltic and Western Europe, 1784–95*. Odense, 1983.

Kappelhoff, Bernd. "Grundzüge der Wirtschaftsgeschichte Papenburgs von den Anfangen bis 1945." In *Geschichte der Stadt Papenburg*, edited by Wolf-Dieter Mohrmann, 319–475. Papenburg: Verlag der Stadt Papenburg, 1986.

Kerkmeijer, W. "Groningers onder neutrale vlag (1795–1810)." *Tijdschrift voor zeegeschiedenis*, 13–3 (1994): 149.

Kerkmeijer, Wicher. "Groningen en de scheepvaart." *Tijdschrift voor zeegeschiedenis* 26–2 (2007): 138–139.

Kernkamp, J.H. *De handel op den vijand, 1572–1609*. Utrecht, 1931–1934.

Ketner, Frans. *Handel en scheepvaart van Amsterdam in de vijftiende eeuw*. Leiden, 1946.

Kiedel, Klaus-Peter. "'Baut, schifft getrost, verlieret nie den Mut!': Papenburger Schiffbau aund Schiffahrt in vier Jahrhunderten." In *Geschichte der Stadt Papenburg*, edited by Wolf-Dieter Mohrmann, 265–317. Papenburg: Verlag der Stadt Papenburg, 1986.

Kirby, David Gordon. *The Baltic World 1772–1993: Europe's Northern Periphery in an age of change*. London: Longman, 1995.

Knoppers, Jake Th. *Dutch trade with Russia from the time of Peter I to Alexander I: a quantitative study in eighteenth century shipping*. Montréal: ICES, 1976.

Knoppers, Jake V.Th. and Frits Snapper. "De Nederlandse scheepvaart op de Oostzee vanaf het einde van de 17de tot het begin van de 19de eeuw." *Economisch- en sociaal-historisch jaarboek* 41 (1978): 115–153.

Knotter, Ad. "De Amsterdamse scheepvaart en het Noordhollandse platteland in de 16de en 17de eeuw. Het probleem van de arbeidsmarkt." *Holland* XVI (1984): 281–290.

Knottnerus, Otto Samuel. "Räume und Raumbeziehungen im Ems-Dollart Gebiet." In *Rondom Eems en Dollard. Rund um Ems und Dollart. Historische verkenningen in het grensgebied van Noordoost-Nederland en Noordwest-Duitsland. Historische Erkundungen im Grenzgebiet der Nordostniederlande und Nordwestdeutschlands*, edited by Otto Samuel Knottnerus et al., 11–42. Groningen / Leer: Van Dijk & Foorthuis REGIO-Projekt Groningen, Schuster Verlag Leer, 1992.

Kotilaine, J.T. "Competing Claims: Russian foreign trade via Arlhangel'sk and the eastern Baltic ports in the 17th century." *Kritika: Explorations in Russian and Eurasian History* 4–2 (2003): 279–311.

Kotilaine, Jarmo. *Russia's foreign trade and economic expansion in the seventeenth century: windows on the world*. Leiden: Brill, 2005.

Krömer, Eckart. *Kleine Wirtschaftsgeschichte Ostfrieslands und Papenburgs*. Norden: Soltau-Kurier-Norden, 1991.

Le Bouëdec, G. "Economie atlantique et profil des villes-ports en Europe (XVIIe–XIXe siècle)." In *Villes atlantiques dans l'Europe occidentale du Moyen-Age au XXe siècle*, edited by Guy Saupin, 1–12. Rennes, 2006.

———. "Intra-european coastal shipping from 1400 to 1900. A long forgotten sector of development." In *A Deus ex Machina revisited, Atlantic colonial trade and European development*, edited by P. Emmer, O. Petré-Grenouilleau and J. Roitman, 89–107. Leiden, 2006.

———. "Small ports from the 16th to the early 19th century and the local economy of the French Atlantic coast." *International Journal of Maritime History* 21 (2009): 103–126.

Le cambiste universel, ou traité complet des changes, monnaies, poids et mesures, de toutes les nations commercantes et de leurs colonies; avec un exposé de leurs banques, fonds-public et papiers-monnaies. Traduit et calculé aux unités francaises sur la seconde édition. Augmenté de tableaux des monnaies d'or et d'argent, d'un apercu sur la lettre de change et les opérations de la Bourse de Paris, Aillard, Bessange, 1823.

Lesger, Clé. "Intraregional trade and the port system in Holland, 1400–1700." In *The Dutch Economy in the Golden Age: Nine Studies*, edited by Karel Davids and Leo Noordegraaf, 189–199. Amsterdam: Nederlands Economisch-Historisch Archief, 1993.

———. *Handel in Amsterdam ten tijde van de Opstand. Kooplieden, commerciële expansie en verandering in de ruimtelijke economie van de Nederlanden ca. 1550–ca. 1630*. Hilversum: Verloren, 2001.

———. *Hoorn als stedelijk knooppunt: Stedensystemen tijdens de late middeleeuwen en vroegmoderne tijd*. Hilversum: Verloren, 1990.

Lilja, Sven. "Small Towns in the periphery: population and economy of small towns in Sweden and Finland during the early-modern period." In *Small Towns in Early-Modern Europe*, edited by Peter Clark, 50–76. Cambridge: University Press, 1995.

Lindblad, Thomas. "Dutch Trade in Narva in the Eighteenth Century." In *Around Peter the Great: Three Centuries of Russian-Dutch Relations*, edited by Carel Horstmeier et al., 103–114. Groningen: INOS, 1997.

———. "Nederland en de Oostzee, 1600–1850." In *Goud uit Graan: Nederland en het Oostzeegebied, 1600–1850*, edited by Remmelt Daalder et al., 8–27. Zwolle: Waanders Uitgevers, 1998.

Ljungman Axel Vilhelm. "The Great Bohuslän herring-fisheries." *Report of the Commissioner of Fish and Fisheries*, IX (1878): 220–239.

Lootsma, Sipke. "De Zeevaart van Hindeloopen in de zeventiende en achttiende eeuw." *Economisch-Historisch Jaarboek*, 21 (1940) 218–296.

Lucassen, Jan. "A Multinational and its Labor Force: The Dutch East India Company, 1595–1795." *International labor and working-class history* 66 (2004): 12–39.

———. "Zeevarenden.", *Maritieme Geschiedenis der Nederlanden*, Vol. 2, 126–158.

———. *Naar de kusten van de Noordzee: trekarbeid in Europees perspectief, 1600–1900*. Gouda: [s.n.], 1984.

Martin, Ron. "Roepke Lecture in Economic Geography—Rethinking regional path dependence: beyond lock-in to evolution." *Economic Geography* 86–1 (2010): 1–27.

Meilink, P.A. "Rekening van het Lastgeld in Amsterdam, Waterland en het Noorderkwartier van Holland in 1507." *Bijdragen en Mededeelingen van het historisch genootschap* 44 (1923): 187–230.

Meyer, J.W. and B. Rowan. "Institutionalized organizations: Formal Structure as Myth and Ceremony." *American Journal of Sociology* 83–2 (1977): 340–363.

Middelhoven, P.J. "De Amsterdamse veilingen van Noord-Europees naaldhout, 1717–1808: een bijdrage tot de Nederlandse prijsgeschiedenis." *Economisch en Sociaal-historisch Jaarboek* 41 (1978): 86–114.

Mokyr, Joel. "The Industrial Revolution and The Netherlands: Why did it not happen?" *De Economist* 148–4 (2000): 503–520.

Molema, Jan et al. *400 jaar Pekela*. Bedum: Stichting 400 jaar Veenkoloniën, 2006.

Morineau, Michel. "Le commerce de la Baltique dans ses rapports avec le commerce hors de la Baltique (du milieu du XVIe siècle à la fin du XVIIIe)." In *The interactions of Amsterdam and Antwerp with the Baltic region, 1400–1800 = De Nederlanden en het Oostzeegebied: papers presented at the third international conference of the "Association Internationale d'Histoire des Mers Nordiques de l'Europe", Utrecht, August 30th–September 3rd 1982*, edited by Wieringa, Wiert Jan, 31–42. Leiden: Nijhoff, 1983.

Morozova, T. Ju. "Organizacija i obrabotka original'noj informacii v russkich gazetach" In *Rossija v XVIII Stoletii*, edited by Ryčalovskij, E.E., 24–37. Moskva, 2002.

Nelkenbrecher, Johann Christian. *Taschenbuch eines Banquiers und Kaufmans, enthält die Beschaffenheit und Vergleichung der Münzen, Maasse, Gewichte, Wechsel-Course und anderer zum Handel gehöriger Dinge der vornehmsten Handels-Orte, 6. Auflage, umgearb., vermehrt und verbessert durch G. Wever*. 1786.

Nieuwe Nederlandsche Jaerboeken, vervolg der Merkwaerdigste Geschiedenissen.... 5 (1770): 898–914.

North, Douglass C. *Institutions, institutional change and economic performance*. Cambridge: University Press, 1990.

Ogilvie, Sheilagh. "Guilds, Efficiency, and Social Capital: Evidence from German Proto-Industry." *The Economic History Review*. New Series. 57–2 (2004): 286–333.

Oldewelt, W.F.H. "De scheepvaartstatistiek van Amsterdam in de 17e en 18e eeuw." *Jaarboek Amstelodamum* 45 (1953): 114–151.

Ordonnantie der Heeren borgemeesteren ende raadt in Groningen, gestatueert op de beurt-schepen varende op Amsteldam, Hamborch ende Bremen: midts-gaders den taxt van de vrachten, Groningen: by Frans Bronchorst, 1663.

Ormrod, David. *The Rise of Commercial Empires: England and the Netherlands in the Age of Mercantilism, 1650–1770*. Cambridge: University Press, 2003.

Paesie, Ruud. *Lorrendrayen op Africa: de illegale goederen- en slavenhandel op West-Afrika tijdens het achttiende-eeuwse handelsmonopolie van de West-Indische Compagnie, 1700–1734*. Amsterdam: De Bataafsche Leeuw, 2008.

Parmentier, Jan. *De maritieme handel en visserij in Oostende tijdens de achttiende eeuw: een prosopografische analyse van de internationale Oostendse handelswereld, 1700–1794*. PhD dissertation Universiteit Gent, 2001.

Pierenkemper, Toni. *Umstrittene Revolutionen: Industrialisierung im 19. Jahrhundert*. Frankfurt-am-Main: Fischer, 1996.

Polónia, Amélia. "Les petits ports dans le système portuaire européen à l'Age Moderne (XVIe–XVIIIe siècles)." *Revista da Faculdade de Letras* HISTÓRIA *Porto III Série*, 1 (2008): 27–51.

Postan, M.M. *Medieval Trade and Finance*, Cambridge: Cambridge University Press, 1973.

Posthumus, Nicolaas Wilhelmus. *De Oosterse handel van Amsterdam: het oudst bewaarde koopmansboek van een Amsterdamse vennootschap betreffende de handel op de Oostzee 1485–1490*. Leiden: Brill, 1953.

Rachel, Hugo. *Die Handels-, Zoll- und Akzisepolitik Preußens 1740–1786*, Dritter Band, Erste Hälfte. Berlin, 1928.

Reinders Folmer-van Prooijen, Corrie. " 'Met de rapen in de pot': Het hoekerschip Maria Elisabeth in de Europese vaart, 1720–1732." In *In het kielzog. Maritiem-historische studies aangeboden aan Jaap R. Bruijn bij zijn vertrek als hoogleraar zeegeschiedenis aan de Universiteit Leiden*, edited by Leo Akveld et al., 222–232. Amsterdam, 2003.

Rodrigue Jean-Paul, Claude Comtois and Brian Slack. *The Geography of Transport Systems (second edition)*. London and New York: Routledge, 2010.

Röhlk, Frauke. *Schifffahrt und Handel zwischen Hamburg und den Niederlanden in der zweiten Hälfte des 18. und zu Beginn des 19. Jahrhunderts*. Wiesbaden: Steiner Verlag, 1973.

Roorda, Reinder Sierks. [*Genealogische collectie Roorda, bestaande uit brieven, aantekeningen, artikelen in kranten en tijdschriften, stambomen en kwartierstaten, van ca. 1930 tot ca. 1965*]. [Leeuwarden: Rijksarchief, 1973].

Rutte, Reinout. "Stadslandschappen: Een overzicht van de stadswording in Nederland van de elfde tot de vijftiende eeuw." In *Stadswording in de Nederlanden: Op zoek naar overzicht*, edited by Reinout Rutte, Hildo van Engen, 143–169. Hilversum: Verloren, 2005.

Scammell, Geoffrey. *The World Encompassed: The first European Maritime Empires, c. 800–1650*. London: Methuen, 1981.

Scheltjens, Werner. "Maße und Gewichte: Konvertierungsmöglichkeiten auf Basis der Sound Toll Registers Online." *Jahrbuch des österreichischen Arbeitskreises für Stadtsgeschichtsforschung, in print.*

———. "Het ontstaan van een geintegreerde maritieme transportruimte in de Lage Landen, ca. 1300–1800." *Revue belge de la Philologie et d'Histoire* vol. 92-2 (2014): 293–363.

———. "The Changing geography of demand for Dutch maritime transport in the eighteenth century" *Histoire & Mesure* vol. XXVII-2 (2012): 3–47.

———. "The influence of spatial change on operational strategies in early-modern Dutch maritime shipping: a case-study on Dutch maritime shipping in the Gulf of Finland and on Archangel, 1703–1740" *International Journal of Maritime History* XXIII-1 (2011): 115–149.

———. "The operational structure of historical maritime shipping: a general review of knowledge discovery tools." *Self-organizing Networks and GIS Tools: Cases of Use for the Study of Trading Cooperation (1400–1800)*, Special Issue of *The Journal of Knowledge Management, Economics and Information Technology*, edited by Ana Crespo Solana and David Alonso García. II-3 (2012): 357–389.

———. "The Volume of Dutch Baltic Shipping at the End of the eighteenth Century: a new estimation based on the Danish Sound Toll Registers." *Scripta Mercaturae* XLIII-1/2 (2009): 83–109.

———. *De invloed van ruimtelijke verandering op operationele strategieën in de Nederlandse scheepvaart: een case-study over de Nederlandse scheepvaart in de Finse Golf en op Archangel, 1703–1740*. Groningen: Barkhuis Publishing, 2009.

Scheltjens, Werner and Hans Van Koningsbrugge. *Van onze reporter ter plaatse: Russisch-Nederlandse betrekkingen in de Sint-Petersburgse Tijdingen 1728–1775*. Groningen: NRAC, 2003.

Scheltjens, Werner and Jan Willem Veluwenkamp. "Sound Toll Registers Online: introduction and first research examples." *International Journal of Maritime History* XXIV-1 (2012): 301–330.

Scheltjens, Werner and Kurt Dopfer. "Unified rule approach and the semantic enrichment of economic movement data." In *Towards a Universal Ontology of Geographic Space*, edited by T. Podobnikar and M. Ceh, 229–247. IGI Global, 2012.

Scherer, Hermann. *Der Sundzoll. Seine Geschichte, sein jetziger Bestand und seine staatsrechtlich-politische Lösung. Nebst einem Anhang über die Sundzollfreiheit der Pommerschen und Preussischen Städte*. Berlin, 1845.

Schöffer, I. "De vonnissen in averij grosse van de Kamer van Assurantie en Averij te Amsterdam in de 18e eeuw. Onderzoek naar hun economisch-historische waarde voor de geschiedenis van handel en scheepvaart van Amsterdam op de Oostzee 1700–1770" *Economisch-Historisch Jaarboek* 26 (1956): 73–132.

Schroor, Meindert. *Geschiedenis van de stad Dokkum: hart van Noordelijk Oostergo*. [Dokkum], Stichting Historia Doccumensis, 2004.

Schut, Engbert. *Geschiedenis van de Joodse Gemeenschap in de Pekela's 1683–1942*. Assen/Maastricht: Van Gorcum, 1991.

Semenov, Aleksej. *Izuchenie istoricheskikh svedenii o rossiiskoy vneshney torgovle i promyshlennosti s poloviny XVII-go stoletiya po 1858 god* (3 parts bound in 2 vols). Newtonville: Oriental Research Partners, 1997 [reprint St. Petersburg 1859].

Sluyterman, Keetie. *Drie eeuwen De Kuyper 1695–1995: een geschiedenis van jenever en likeuren*. Schiedam: De Kuyper, 1995.

Smit, Homme Jacob. "De registers van den biertol te Amsterdam." *Bijdragen en Mededeelingen van het historisch genootschap* 38 (1914): 1–97.

———. "Het Kamper Pondtolregister van 1439–1441." *Economisch-historisch Jaarboek: bijdragen tot de economische geschiedenis van Nederland* 5 (1919): 209–296.

Snapper, Frits. "De zeescheepvaart in het Maasgebied, 1756–1774", in *Tijdschrift voor Zeegeschiedenis*, vol. 3, 1984, pp. 15–30.

———. "Een voorlopige reconstructie van de scheepvaartstatistiek van Holland, 1741–1794" *Economisch- en sociaal-historisch jaarboek* 48 (1985): 118–129.

———. *Oorlogsinvloeden op de overzeese handel van Holland, 1551–1719*. s.l.: s.n., 1959.

Sneller, Zeger Willem. "De Hollandsche korenhandel in het Sommegebied in de 15e eeuw." *Bijdragen voor Vaderlandsche Geschiedenis en Oudheidkunde* 6 (1925): 161–178.

———. "Wijnvaart en Wijnhandel tusschen Frankrijk en de Noordelijke Nederlanden." *Bijdragen voor vaderlandsche geschiedenis en oudheidkunde* 1 (1924): 198.

Stabel, Peter. *Dwarfs among Giants: the Flemish urban network in the Late Middle Ages*. Leuven / Apeldoorn: Garant, 1997.

Steensgaard, Niels. "The growth and composition of the long-distance trade of England and the Dutch Republic before 1750" In *The Rise of Merchant Empires: Long-distance trade in the early modern world, 1350–1750*, edited by James D. Tracy, 102–152. Cambridge: University Press, 1990.

Steinberg, Philip. *The Social Construction of the Ocean*, Cambridge: University Press, 2001. (Cambridge Studies in International relations, vol. 78).

Straubel, Rolf. "Stettin als Handelsplatz und wirtschaftlicher Vorort Pommerns im spätabsolutistischen Preußen" *Jahrbuch für die Geschichte Mittel- und Ost-Deutschlands* 50 (2004): 131–190.

Tractaat noopende de tollen in den Orisont en Noorwegen, tussen sijn Koninklijke Majesteit van Denemarken en Noorwegen, ter eenre, en de Hoog Mog. Heeren Staaten Generaal der vereenigde Nederlanden ter andere zijde, geslooten tot Koppenhage den vijftienden Juny 1701. Mitsgaders de Tolrolle van de jare 1645, in gevolge van het tweede Articul van dit jegenwoordig Tractaat. In 's Gravenhage: Bij Paulus Scheltus, ordinaris drukker van de Hoog Mog Heeren Staten Generaal der vereenighde Nederlanden, 1701.

Tracy, James D. Introduction to *The Rise of Merchant Empires: Long-distance trade in the early modern world 1350–1750*, edited by James D. Tracy. Cambridge: Cambridge University Press, 1990.

Unger, Richard W. "Beer: A New Bulk Good of International Trade." *Cogs, Cargoes and Commerce: maritime bulk trade in Northern Europe 1150–1400*, edited by Lars Berggren, Nils Hybel & Annette Landen, 113–127. Toronto: Pontifical Institute of Mediaeval Studies, 2002.

Unger, Richard. "The Tonnage of Europe's merchant fleets, 1300–1800" *The American Neptune: a quarterly journal of maritime history* LII (1992): 247–261.

———. *Dutch Shipbuilding before 1800*. Assen/Amsterdam, 1978.

———. *Ships and Shipping in the North Sea and Atlantic, 1400–1800*. Ashgate: Variorum, 1997.

Unger, Willem Sybrand. "De publikatie der Sonttabellen voltooid." *Tijdschrift voor Geschiedenis*, 71 (1958): 147–205.

———. "Middelburg als handelsstad (XIIIe tot XVIe eeuw)." *Archief: vroegere en latere mededeelingen voornamelijk in betrekking tot Zeeland* 1935: 1–177.

van der Schoor, Arie. *Stad in Aanwas: Geschiedenis van Rotterdam tot 1813*. Zwolle: Waanders Uitgevers, 1999.

Van der Wee, Herman. "Structural changes in European long-distance trade, and particularly in the re-export trade from south to north, 1350–1750." In *The Rise of Merchant Empires: Long-distance trade in the early modern world, 1350–1750*, edited by James D. Tracy, 14–33. Cambridge: University Press, 1990.

Van Der Woude, Adrianus Maria. *Het Noorderkwartier: Een regionaal historisch onderzoek in de demografische en economische geschiedenis van westelijk Nederland van de late middeleeuwen tot het begin van de negentiende eeuw*. Wageningen: Veenman en Zonen, 1972.

Van Eeghen, I.H. *Archief van de gilden en het Brouwerscollege. Versie 80.1*. Amsterdam, 2012 [1951]. URL: http://stadsarchief.amsterdam.nl/archieven/archiefbank/printversie/366.nl.pdf

Van Eyck van Heslinga, E.S. "De vlag dekt de lading. De Nederlandse koopvaardij in de Vierde Engelse Oorlog" *Tijdschrift voor Zeegeschiedenis*, 2 (1982): 102–113.

Van Herwaarden, Jan et al. *Geschiedenis van Dordrecht tot 1572*. Hilversum: Verloren, 1996.

Van Lottum, Jelle and Jan Lucassen. "Six cross-sections of the Dutch maritime labour market: A preliminary reconstruction and its implications (1607–1850)." In *Maritime Labour: Contributions to the History of Work at Sea (1500–2000)*, edited by Richard Gorski, 13–42. Amsterdam: Aksant, 2007.

Van Lottum, Jelle. *Across the North Sea: The impact of the Dutch Republic on international labour migration c. 1550–1850*. Amsterdam: Aksant, 2007.

Van Riemsdijk, J. *Het brandersbedrijf te Schiedam in de 17de en 18de eeuw*. Schiedam: Roelants, 1916.

Van Royen, Paul C. "The "National" Maritime Labour Market: Looking for common characteristics", In *"Those emblems of hell?" European sailors and the maritime labour market, 1570–1870*, edited by Paul Van Royen, Jaap Bruijn and Jan Lucassen, 1–9. St. John's: s.n., 1997.

———. *Zeevarenden op de koopvaardijvloot omstreeks 1700*. Amsterdam: De Bataafsche Leeuw, 1987.

———. "Varend volk" In *Goud uit Graan: Nederland en het Oostzeegebied, 1600–1850*, edited by Daalder, Remmelt et al., 82–97. Zwolle: Waanders Uitgevers, 1998.

Van Tielhof, Milja. *De Hollandse Graanhandel 1470–1570: koren op de Amsterdamse molen*. Den Haag: Stichting Hollandse Historische Reeks, 1995.

———. *The 'Mother of all Trades': The Baltic Grain Trade in Amsterdam from the Late 16th to the early 19th Century*. Leiden – Boston – Köln: Brill, 2002.

Van Zanden, Jan Luiten and Milja van Tielhof. "Roots of Growth and Productivity in Dutch Shipping, 1500–1800" *Explorations in Economic History* 46 (2009): 389–403.

Van Zanden, Jan Luiten. *Arbeid tijdens het handelskapitalisme: Opkomst en neergang van de Hollandse economie 1350–1850*. Bergen: Octavo, 1991.

Vance, James E. Jr. *The Merchant's World: The Geography of Wholesaling*. Englewood Cliffs: Prentice-Hall, Inc., 1970.

Vanelslander, Thierry et al. *Ruimtelijk-economische en logistieke analyse: de Vlaams-Nederlandse delta in 2040*, Antwerpen/Rotterdam, 2011. URL: http://www.vndelta.eu/documenten/analyse-delta-2040-van-ua-en-erasmus-universiteit (23 Mei 2013).

Veluwenkamp, Jan Willem and P.W. Klein. "The role of the entrepreneur in the economic expansion of the Dutch Republic." In *The Dutch Economy in the Golden Age: Nine Studies*, edited by Karel Davids and Leo Noordegraaf, 27–54. Amsterdam: Nederlands Economisch-Historisch Archief, 1993.

Veluwenkamp, Jan Willem. "Het Nederlandse handelsstelsel in de vroegmoderne tijd. Oude en nieuwe visies." *Leidschrift* 23 (2008): 63–76.

———. "International business communication patterns in the Dutch commercial system, 1500–1800." In *Your humble servant. Agents in Early Modern Europe*, edited by H. Cools, M. Keblusek and B. Noldus, 121–134. Hilversum: Verloren, 2006.

———. "Merchant Colonies in the Dutch Trade System (1550–1750)." In *Kapitaal, ondernemerschap en beleid: Studies over economie en politiek in Nederland, Europa en Azië van 1500 tot heden*, edited by C.A. Davids, W. Fritschy and L.A. Van der Valk, 141–164. Amsterdam: NEHA, 1996.

———. *Archangel: Nederlandse ondernemers in Rusland, 1550–1785*. s.l.: Uitgeverij Balans, 2000.

Verhoeven, Gerrit. "Een divertissant somertogje: Transport innovations and the rise of short-term pleasure trips in the Low Countries, 1600–1750." *Journal of Transport History* 30 (2009): 78–97.

Vogel, Walther. "Zur Größe der europäischen Handelsflotten im 15., 16. und 17. Jahrhundert. Ein historisch-statistischer Versuch." In *Forschungen und Versuche zur Geschichte des Mittelalters und der Neuzeit. Festschrift Dietrich Schäfer zum siebzigsten Geburtstag dargebracht von seinem Schülern*, 268–333. Jena: Fischer, 1915.

Wagenaar, Jan. *Amsterdam, in zyne opkomst, aanwas, geschiedenissen, voorregten, koophandel, gebouwen, kerkenstaat, schoolen, schutterye, gilden en regeeringe*. Aphen a/d Rijn: Repro-Holland; Amsterdam: Buijten & Schipperheijn, 1760–1802 [reprint 1971–1972].

Ward, Robin. *The World of the Medieval Shipmaster: law, business and the sea, c.1350–c.1450*. Woodbridge: The Boydell Press, 2009.

Wätjen, Hermann. *Die Niederländer im Mittelmeergebiet zur Zeit ihrer höchsten Machtsstellung*. Berlin: Curius, 1909.

Wegener Sleeswyk, André. *De Gouden Eeuw van het Fluitschip*. Franeker: Van Wijnen, 2003.

Wegener Sleeswyk, Rienk. "Omvang en financiering van de Friese vrachtvaart in de 18de eeuw." In *Van Fries Wijdtschip tot Workumer Aak: historie en interviews*, edited by Gerben Groenhof, 52–72. Sneek: Stichting 300 jaar scheepstimmerwerf De Hoop, 1996.

Welling, George Maria. "Price-supply relations on the Amsterdam staple market 1778." In *From Dunkirk to Danzig: Shipping and Trade in the North Sea and the Baltic, 1350–1850*, edited by W.G. Heeres, et al., 457–469. Hilversum: Verloren, 1988.

———. *The Prize of Neutrality: trade relations between Amsterdam and North America 1771–1817: a study in computational history*. [s.l.]: [s.n.], 1998.

Willemsen, Renatus Theodorus Henricus. *Enkhuizen tijdens de Republiek: een economisch-historisch onderzoek naar stad en samenleving van de 16e tot de 19e eeuw*. Hilversum: Verloren, 1988.

Wrigley, Edward Anthony (ed.). *Identifying People in the Past*. London: Edward Arnold, 1973.

Ypma, Ijsbrand Nicolaas. *Geschiedenis van de Zuiderzeevisserij*. Amsterdam, s.n., 1962.

Index

www.ingramcontent.com/pod-product-compliance
Lightning Source LLC
LaVergne TN
LVHW010543300926
849162LV00002B/218